SYSTEM FAULT DIAGNOSTICS, RELIABILITY AND RELATED KNOWLEDGE-BASED APPROACHES

VOLUME 1

FAULT DIAGNOSTICS AND RELIABILITY

SYSTEM FAULT DIAGNOSTICS, RELIABILITY AND RELATED KNOWLEDGE-BASED APPROACHES

VOLUME 1

FAULT DIAGNOSTICS AND RELIABILITY

Proceedings of the First European Workshop on Fault Diagnostics, Reliability and Related Knowledge-Based Approaches, Island of Rhodes, Greece, August 31–September 3, 1986

Edited by

SPYROS TZAFESTAS
National Technical University of Athens, Greece

MADAN SINGH
UMIST, Manchester, England

and

GÜNTHER SCHMIDT
Munich Technical University, F.R.G.

D. Reidel Publishing Company

A MEMBER OF THE KLUWER ACADEMIC PUBLISHERS GROUP

Dordrecht / Boston / Lancaster / Tokyo

Library of Congress Cataloging in Publication Data

CIP

European Workshop on Fault Diagnostics, Reliability, and Related Knowledge-Based Approaches (1st: 1986: Rhodes, Greece)
System fault diagnostics, reliability and related knowledge-based approaches.

"Proceedings of the 'First European Workshop on Fault Diagnostics, Reliability, and Related Knowledge-Based Approaches' (Island of Rhodes, Greece, August 31–September 3, 1986."
Includes index.
Contents: v. 1. Fault diagnostics and reliability — v. 2. Knowledge-based and fault-tolerant techniques. 1. Fault location (Engineering)—Congresses. 2. Reliability (Engineering)—Congresses. 3. System analysis—Congresses. I. Tzafestas, S. G., 1939– II. Singh, Madan G. III. Schmidt, Günther, 1935– . IV. Title.
TA169.6.E95 1986 620'.0044 87-13014

ISBN-13: 978-94-010-8243-3 e-ISBN-13: 978-94-009-3929-5
DOI: 10.1007/978-94-009-3929-5

Published by D. Reidel Publishing Company,
P.O. Box 17, 3300 AA Dordrecht, Holland.

Sold and distributed in the U.S.A. and Canada
by Kluwer Academic Publishers,
101 Philip Drive, Assinippi Park, Norwell, MA 02061, U.S.A.

In all other countries, sold and distributed
by Kluwer Academic Publishers Group,
P.O. Box 322, 3300 AH Dordrecht, Holland.

P R E F A C E

This 2-volume work represents the proceedings of the First European Workshop on Fault Diagnostics, Reliability and Related Knowledge-Based Approaches held in the Island of Rhodes, Greece (August 31-September 3, 1986). This Workshop was organized in the framework of a joint research project sponsored by the Commission of the European Communities under the Stimulation Action Programme.

The principal aim of the Workshop was to bring together people working on the numeric and symbolic (knowledge-based) treatment of reliability and fault diagnosis problems, in order to promote the interaction and exhange of ideas, experiences and results in this area.

The workshop was a real success, with 55 papers presented and 70 participants. A second Workshop of the same nature has been decided to be held in Manchester (UMIST), England, in April 1987.

The two volumes contain sufficient amount of information which reflects very well the state-of-the-art of the field, and shows the current tendency towards knowledge-based (expert systems) and fault-tolerant approaches.

Volume 1 contains the contributions on fault diagnostics and reliability issues (numeric treatment), and Volume 2 the contributions on knowledge-based and fault-tolerant techniques.

We are grateful to the Commission of the European Communities for having sponsored the Workshop, and to all authors for their high quality contributions and presentations.

December 1986

Spyros Tzafestas
Madan Singh
Günther Schmidt

C O N T E N T S

CONTENTS OF VOLUME 2

PART 1: KNOWLEDGE-BASED FAULT DIAGNOSIS, SIMULATION, SUPERVISION AND CONTROL

PART 2: FAULT-TOLERANT, SELF-CHECKING AND OTHER DIGITAL TECHNIQUES

CONTRIBUTORS

VOLUME 1: SYSTEM FAULT DIAGNOSTICS AND RELIABILITY

Anagnostou, M.A.
Athans, M.

Borne, P.
Botsaris, C.
Bourey, J.P.
Burge, S.E.

Cao, Q.
Cecchin, T.
Corbeel, D.
Craye, E.

Dauphin-Tanguy, G.
Dialynas, E.N.
Dimitrov, B.N.

Frank, P.
Fripp, R.N.

Gentina, J.C.

Hamad, M.

Isermann, R.

Jones, A.H.

Labath, N.A.

Machias, A.V.
Mudge, S.K.
Müller, P.C.

Nivolianitou, Z.

Papastavrou, J.
Patton, R.J.
Petrov, P.G.
Philippou. A.
Porter, B.
Pouliezos, A.
Poulsen, L.

Ragot, J.

Sauter, D.
Schneeweiss, W.G.
Singh, I.
Souflis, J.L.
Staroswiecki, M.
Stavrakakis, G.

Tanaka, S.
Tzafestas, S.G.

Venetsanopoulos, A.N.
Ventzas, D.E.

Willcox, S.W.
Wünnenberg, J.

VOLUME 2: KNOWLEDGE-BASED AND FAULT-TOLERANT TECHNIQUES

Alexandridis, N.A.
Andritsakis, A.P.
Angelopoulos, S.

Batzias, F.A.
Beke, L.
Bieker, B.
Bourbakis, N.G.
Burge, S.E.

Cosmidis, E.T.

David, J.M.
Delorme, C.
Deplanche, A.M.

Efremidis, S.
Elloy, J.P.

Farris, C.D.

Gasdik, I.
Giambiasi, N.
Glaros, N.A.

Halatsis, C.
Halse, R.G.
Hassapis, G.D.

Jiang, Q.
Jones, A.H.

Kayafas, E.A.
King, R.E.
Kostarakis, P.
Koubias, S.
Krivine, J.P.

Lappus, G.
Layzell, P.J.
Lbath, R.
Loucopoulos, P.
Lucas, R.J.

Maier, R.
Moraitis, C.

Nikolos, D.

Oussalah, C.

Panagiotopoulos, T.
Papadopoulos, G.
Papakonstantinou, G.
Paschalis, A.M.
Philokyprou, G.
Piana, G.
Preece, C.

Roux, P.

Santucci, J.F.
Saumont, S.
Schmidt, G.K.
Sideri, M.
Skordalakis, E.
Stamatopoulos, G.

Trahanias, P.
Tsanakas, P.
Tzafestas, S.G.

Vanwassenhove, C.
Ventzas, D.E.

PART 1: INVITED PAPERS

EXPERIENCES WITH PROCESS FAULT DETECTION METHODS VIA PARAMETER ESTIMATION

Rolf Isermann
Institut für Regelungstechnik
Technical University of Darmstadt
Schlossgraben 1
D-6100 Darmstadt
F. R. Germany

ABSTRACT. Fault detection and fault diagnosis of technical processes become more important in the course of progressive automation. Computer based fault supervision methods are developed which allow the early detection and localization of process faults during normal operation. The use of process models enables to estimate process state variables and parameters which are influenced by faults. The contribution concentrates on fault diagnosis based on process parameters. The general procedure is outlined, comprising parameter estimation, feature extraction, fault decision and classification. Experimental results are given for the detection of several faults in an electrical drive/centrifugal pump set and an steam heated heat exchanger.

KEYWORDS. Fault detection; supervision; reliability; safety; process models; parameter estimation; fault decision; d.c.motor; centrifugal pump; heat exchanger.

1. INTRODUCTION

The main interest in the automation of technical processes during the last decades could be observed in measurement, feedforward and feedback control, alarm monitoring and protection, documentation, manual operation and optimization. Good progress could be made in the equipment and the performance for measurement and control. The improvements in process control are on one side based on the development of the components for sensors, transducers, control systems and actuators and on the other side on the understanding and modelling of process dynamics and on applied control theory.

Also the components for *process supervision* including alarm monitoring and protection were developed. However, the implemented methods are

S. Tzafestas et al. (eds.), System Fault Diagnostics, Reliability and Related Knowledge-Based Approaches, Vol. 1, 3–33.

still rather simple and exist mostly in the limit value checking of some easily available single signals. In contrast to the field of control, methods based on modern dynamic systems theory are hardly applied. But there is now an increasing interest in the field of process supervision, which includes failure detection and diagnosis, see for example Willsky (1976), Himmelblau (1978), Pau (1981), Isermann (1981), (1984a), Gertler and Singer (1985). The reasons are:

. Increasing demands on reliability and safety of technical processes in general,
. high investment capital for processes,
. increasing interlinkage of processes,
. increasing complexity of processes,
. unmanned automation,
. errors due to manual operations.

The goal of supervision is to detect process changes and faults during normal operation and to take actions to avoid a damage of the process or an accident with human beings. For this a supervision loop is implemented which contains the tasks:

- fault detection and diagnosis
- fault evaluation
- decision on operating state (stop or change)
- fault elimination.

An essential prerequisite for improving the supervision is an early *process fault detection* and *localization* such that there is enough time for the following actions:

- fault elimination
- prevention of further fault expansion
- operation in other mode
- scheduling of next maintenance and repair.

Fig. 1 shows a process which can be described by a mathematical process model

$$\underline{Y} = f[\underline{U},\underline{N},\underline{X},\underline{\Theta}] \tag{1}$$

where $\underline{U}(t)$ and $\underline{Y}(t)$ are measurable input and output variables, $\underline{\Theta}$ constant or slow time varying process parameters, $\underline{X}(t)$ process state variables and $\underline{N}(t)$ disturbance variables. A process fault generally directly causes changes $\Delta\underline{\Theta}(t)$ of process parameters or changes $\Delta\underline{X}(t)$ of process state variables, depending on the type of fault. These changes then lead to changes $\Delta\underline{Y}(t)$ of the output signals, according to the process dynamics and statics. Fault detection methods may be classified with respect to the use of following quantities, cf. Isermann (1981, 1984a):

(1) Measurable signals $\underline{U},\underline{Y}$
(2) State variables $\underline{X}(t)$, mostly nonmeasurable
(3) Process parameters $\underline{\Theta}(t)$, mostly nonmeasurable
(4) Characteristic quantities $\eta = f[\underline{U},\underline{Y},\underline{\Theta}]$.

The usual way is to check the limit of the measurable signals $\underline{Y}(t)$. A normal state is if

$$\underline{Y}_{min} < \underline{Y}(t) < \underline{Y}_{max}. \tag{2}$$

This is called *absolute value limit check*. It can also be applied to the trend $\dot{\underline{Y}}(t)$ of the signals of the signals.

Another way is to *analyse the signals* $\underline{Y}(t)$, which often consist of lower frequency components $\underline{Y}_{LF}(t)$ and higher frequency components $\underline{Y}_{HF}(t)$. Especially the analysis of higher frequency components (vibrations) by autocorrelation or spectral analysis can give additional information on the inner state of a process.

If both, output and input variables are available *characteristic quantities* η like efficiencies, consumption per time, wear per time, etc. can be calculated. Changes $\Delta\eta$ may give overall information on internal changes. However, a detailed fault detection and localization mostly is not possible.

If a *process model* is known, one can try to estimate the usually unmeasurable process *state variables* $\underline{X}(t)$ or *process parameters* $\underline{\Theta}(t)$ based on the measurable inputs $\underline{U}(t)$ and outputs $\underline{Y}(t)$ by using state variable or process parameter estimation methods and to detect changes $\Delta\underline{X}(t)$ or $\Delta\underline{\Theta}(t)$. This has following advantages:

- In terms of signal flow the state variables $\underline{X}(t)$ or process parameters $\underline{\Theta}(t)$ are in many cases closer to the process faults. Hence the process faults may be detected earlier and localized more precisely than by limit value checking of $\underline{Y}(t)$.

- A process fault usually causes changes of several output variables $\Delta Y_i(t)$ with different signs and dynamics. The model based fault detection now takes into account all these entailed changes, provides a data reduction and determines (theoretically) the state variable or the process parameter which has been changed directly by the fault. Hence, it can be expected that a significant change $\Delta X_j(t)$ or $\Delta\Theta_j(t)$ can be extracted and that the fault detection selectivity is improved.

- Closed loops generally compensate for changes $\Delta\underline{Y}(t)$ of the outputs by changing the inputs $\underline{U}(t)$. Therefore deviations by faults cannot be recognized by limit value checking. Model based fault detection methods automatically consider the relations between inputs and outputs and therefore are also applicable in closed loops.

- If one wants to detect more process faults one can try to implement more sensors which as directly as possible measure the required information. However, in general one needs many more sensors and also non-robust sensors. This may become expensive and more sensitive to sensor based faults. Model based fault detection methods need only some few and possibly robust sensors.

The price one has to pay for these advantages is the higher effort in modelling the process, the problem solving method itself (software-package) and the computer capacity.

The application of these model based fault detection methods is not restricted to continuous process supervision in normal operation, but can also be used on request to enable:

. preventive maintenance
. maintenance on request (instead on fixed schedules)
. remote diagnosis of processes
. automatic inspection of products without disassembling in manufacturing.

This contribution concentrates on model based fault detection methods based on parameter estimation methods. As a survey of various process fault detection methods was given in Isermann (1984a) including a detailed reference list, the field is not further reviewed here. In section 2 the principle of the method is described. This is followed by methods for the fault diagnosis. The third section shows results which were obtained for an electrically driven centrifugal pump system. Then in section 4 experiments with a heat exchanger are discussed.

2. FAULT DIAGNOSIS BASED ON PROCESS PARAMETERS

2.1 General procedure

A genralized scheme of technical fault diagnosis is shown in Fig. 2. Three phases can be distinguished

(1) *Data processing*
The measured signals are processed by methods of filtering and estimation such that a reduced information results which is suitable for fault detection and diagnosis. The reduced information for example exists in filtered signal components, correlation functions, or in parameter or state-variable estimates, if process models are applied.

(2) *Fault detection*
Based on the reduced process information features are extracted which allow the detection of faults in the process. Then changes of these features are determined with reference to the normal process. These changes are then used to recognize the event of a fault and the time of its event. For this task statistical decision methods may be used.

(3) *Fault diagnosis*
After a fault event has been detected, the features and their changes are submitted to a classification procedure with the goal to determine the fault type, fault location, fault size and the cause of the fault. Therefore classification respectively pattern recognition methods can be used. (Sometimes fault decision and fault classification are merged together, so that a clear separation is not possible).

This general procedure holds as well for fault diagnosis methods based on *signal models* and *process models* in nonparametric or parametric form. In the case of parametric process models the reduced information may consist in *process model parameters* or *process model state variables*.

In the following only *fault diagnosis methods with process model parameters* are considered. A more specific scheme for this class of fault diagnosis is shown in Fig. 3. It basically agrees with Fig. 2. The fundamental idea is that many process faults appear as changes of process coefficients $\underline{p}$ like resistances, capacities, inductivities, mass, stiffness, etc. These process coefficients are contained in the parameters $\underline{\Theta}$ of a process model. Process model parameters are understood as constants or time-dependent coefficients in the process which appear in the mathematical description of the relationship between the input and output signals, the process model. A distinction can be made between *static process models*, e. g. in form of a polynomial equation

$$Y(U) = \beta_o + \beta_1 U + \beta_2 U^2 + \dots \qquad (3)$$

and *dynamic process models* which, for processes with lumped parameters, are usually differential equations

$$a_o y(t) + a_1 \dot{y}(t) + a_2 \ddot{y}(t) + \dots + y^{(n)}(t)$$
$$= b_o u(t) + b_1 \dot{u}(t) + b_2 \ddot{u}(t) + \dots + b_m u^{(m)}(t) \qquad (4)$$

in the simplest case linearized about one operating point (of course also nonlinear differential equations can be used). The process model parameters

$$\underline{\Theta}^T = [\beta_o \ \beta_1 \ \beta_2 \ \vdots \ldots]$$

$$\underline{\Theta}^T = [a_o \ldots a_{n-1} \vdots b_o \ldots b_m] \qquad (5)$$

are now more or less intricate relationships of mostly several physical process coefficients $\underline{p}$. To determine these process coefficients and their changes the following procedure was developed, compare the single blocks in Fig. 3:

a) Establishment of the process model for the measurable input and output signals

$$Y(t) = f[U(t), \underline{\Theta}] \qquad (6)$$

by theoretical modelling.

b) Determination of the relationship between the model parameters Θ_i and the process coefficients p_j: $\underline{\Theta} = f(\underline{p})$.

c) Estimation of the model parameters Θ_i based on the measured signals Y(t) and U(t).

d) Calculation of the process coefficients

$$\underline{p} = f^{-1}(\underline{\Theta}) \qquad (7)$$

e) Determination of changes Δp_j of the process coefficients with reference to normal values.

f) Fault decision based on the process coefficient changes and fault signatures of a model for the faulty process in which the relationships between process faults and changes of the process coefficients Δp_j is established.

g) Fault classification to determine the fault type, localization and size.

Therefore this fault detection is based on:
. Theoretical process modelling,
. parameter estimation for continuous-time models,
. statistical decision and classification.

Prerequisites for the application of this method are:
. The process models have to describe the process behaviour precisely,
. powerful parameter estimation methods are available which yield precise parameter estimates,

. sufficient excitation of process dynamics or statics by appropriate input signals is possible,
. unique determination of the determination of the required process coefficients (exceptions possible, see Isermann (1984a, and section 4)

2.2 Parameter estimation

The basic process model is developed by theoretical process modelling. Therefore the physical laws which govern the process behaviour must be known. By stating the balance equations, the physical-chemical state equations and the phenomenological laws continuous-time models for the dynamics result in ordinary or partial differential equations. Hence, the structure of the models is known and the parameters Θ_i have to be estimated in order to detect changes during normal operation. A survey of parameter estimation methods for linear continuous-time models based on sampled signals was given by Young (1981), see also the summary in Isermann (1984a). At that time mainly the least squares and instrumental variables method were developed. The least squares method is still the most robust method and recommended if the superimposed noise is small. The numerical properties can be improved by using a square root filtering approach. The differential equation Eq. (4) of a stable linear process with lumped parameters and transfer function

$$G_p(s) = \frac{y(s)}{u(s)} = \frac{B(s)}{A(s)} = \frac{b_o + b_1 s + \ldots + b_m s^m}{a_o + a_1 s + \ldots + s^n} \tag{8}$$

can be written in the form

$$y^n(t) = \underline{\psi}^T(t)\underline{\Theta} + e(t) \tag{9}$$

with the vector of the signals and their derivatives

$$\underline{\psi}^T(t) = [-y^{(n-1)}(t) \; -y^{(n-2)}(t) \; \ldots \; -y^{(1)}(t) \; - $$
$$- \; y(t) \; u^{(m)}(t) \; u^{(m-1)}(t) \; \ldots \; u^{(1)}(t) \; u(t) \;] \tag{10}$$

and the parameter vector

$$\underline{\Theta}^T = [a_{n-1} \; a_{n-2} \; \ldots \; a_1 \; a_o \vdots b_m \; b_{m-1} \; \ldots \; b_1 \; b_o] \tag{11}$$

Measurements of the input and output signals are made at discrete times $t = kT_o$, $k = 0,1,2,\ldots,N$ with T_o the sampling time. The deviations

$$y(t) = Y(t) - Y_{oo};\ u(t) = U(t) - U_{oo} \tag{12}$$

are determined and also the derivatives. Then N + 1 equations are obtained

$$y^{n}(k) = \underline{\Psi}^{T}(k)\ \hat{\underline{\Theta}} + e(k) \tag{13}$$

with e(k) as the equation error, resulting in a vector equation

$$\underline{y}^{n} = \underline{\Psi}\ \hat{\underline{\Theta}} + \underline{e}. \tag{14}$$

Minimization of the loss function

$$V = \sum_{k=0}^{N} e^{2}(k) = \underline{e}^{T}\underline{e} \tag{15}$$

yields the well-known nonrecursive LS-estimation equation

$$\hat{\underline{\Theta}} = [\underline{\Psi}^{T}\underline{\Psi}]^{-1}\ \underline{\Psi}^{T}\ \underline{y}^{n}. \tag{16}$$

A certain problem is the determination of the signal derivatives. State variable filters with transfer function

$$F(s) = \frac{1}{1 + f_{1}s + \ldots + f_{n}s^{n}} \tag{17}$$

have shown to be well suitable, Geiger (1984). They simultaneously provide the time derivation without differentiation and filter the noise, Young (1981). For the filter parameters

$$f_{i} \approx \hat{a}_{i} \tag{18}$$

is recommended. The LS-method results only then in consistent estimates, if the noise is zero. Otherwise biased estimates are obtained. The numerical properties of the basic LS-method can be improved by bringing it into a square root filter form. Various nonrecursive and recursive algorithms are possible, see Goedecke (1985), Geiger (1985).

2.3 Determination of process coefficients

It was already shown in Isermann (1984a) that a unique determination of process coefficients

$$\underline{p} = \underline{f}^{-1}\ (\underline{\Theta})$$

is not always possible and that the use of dynamic models instead of static models allows to monitor more process coefficients. To get some more insight two simple examples of lumped parameter processes are considered:

Example 1: First order elctrical circuit.

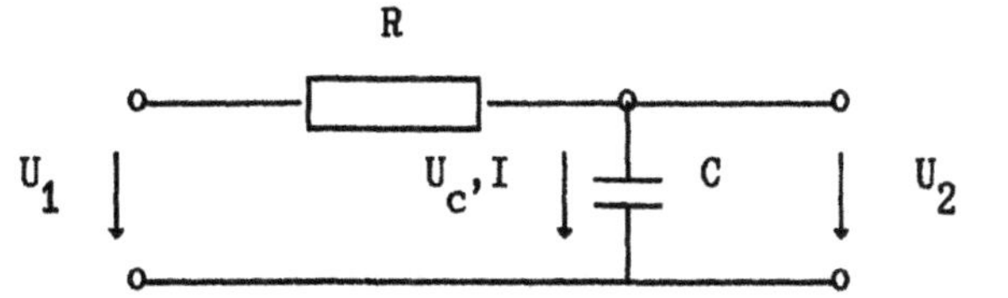

$$U_1(t) - U_R(t) - U_c(t) = 0; \; U_R = RI; \; I = C\, \dot{U}_c(t)$$

a) input: voltage U_1; output: voltage U_2

$$a_1\, \dot{U}_2(t) + U_2(t) = b_o\, U_1(t); \; a_1 = RC; \; b_o = 1.$$

The dymanic model has only one parameter a_1 which contains 2 process coefficients. Hence the process coefficients cannot be determined uniquely. One of them, either R or C has to be assumed as known. The static model (d/dt=0) does not give any information on R and C (however, if $b_o \neq 1$ a fault of the condensator (leakage current) is indicated).

b) input: voltage U_1; output: current I

$$a_1\, \dot{I}(t) + I(t) = b_1\, \dot{U}_1(t); \; a_1 = RC; \; b_1 = C.$$

In this case the dynamic model has 2 parameters a_1 and b_1 from which the process coefficients R and C can be determined uniquely.

Example 2: Second order mechanical system.

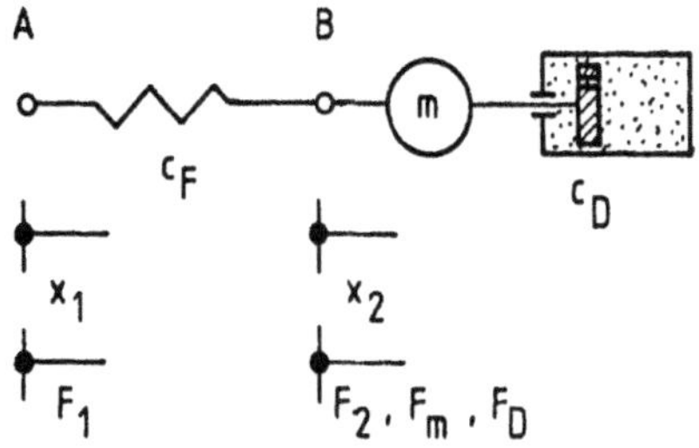

Forces at point B: $F_1 + F_2 + F_m + F_D = 0$

spring: $F_1 = c_F x_1$; $F_2 = -c_F x_2$

mass: $F_m = -m\,\ddot{x}_2$

damper: $F_D = -c_D\,\dot{x}_2$

a) input: displacement x_1; output: displacement x_2

$$a_2\ddot{x}_2(t) + a_1\dot{x}_2(t) + x_2(t) = b_o x_1(t)$$
$$a_2 = m/c_F;\ a_1 = c_D/c_F;\ b_o = 1.$$

From 2 parameters a_1, a_2 the 3 process coefficients m, c_F, c_D and c_F cannot be determined uniquely. (However, a fault of the spring or the damper is indicated if $b_o \neq 1$).

b) input: force F_1; output: displacement x_2

$$a_2\ddot{x}_2(t) + a_1\dot{x}_2(t) + x_2(t) = b_o F_1(t)$$
$$a_2 = m/c_F;\ a_1 = c_D/c_F;\ b_o = 1/c_F.$$

The 3 parameters a_1, a_2, b_o allow to determine the 3 process coefficients m, c_F, c_D uniquely. From the static model only $c_F = \hat{b}_o$ can be estimated.

These examples show:

a) It depends on the selected input and output variables, whether the process coefficients can be determined uniquely from the process model parameters. There exists an *identifiability condition* for the process coefficients, which is for linear single-input, single-output passive lumped processes:
In order to determine the process coefficients uniquely the measured variables have to be selected as follows with regard to the signal flow diagram:
   - either as input varaible which influences the state variable between the process coefficients
   - or as directly measured state variable between the process coefficients.

b) if not all process coefficients can be determined some of them have to be assumed as known.

c) Dynamic models allow to determine more process coefficients than static models.

For higher order an more complex processes it depends on the individual model structure and the measurable signals which process coefficients can be determined.

2.4 Fault decision and classification

After the process coefficients $\underline{p}$ are determined, significant changes Δp_j have to be detected with reference to normal values as a basis for the fault decision and the fault classification. As the process coefficients usually appear as stochastic variables a *Bayes decision* can be applied. The observed coefficients at time j are described by the vector $\hat{\underline{p}}(j)$ and for time j = 1,2,...,N by the process coefficient matrix.

$$\hat{\underline{p}} = [\hat{\underline{p}}(1)\ \hat{\underline{p}}(2)\ \dots\ \hat{\underline{p}}(N)]. \tag{19}$$

Now it is assumed that the $\hat{\underline{p}}(j)$ have a normal distribution and are statistically independent. The Bayes decision then leads to following algorithms, see Geiger (1985):

Phase I: Determination of the normal state (training)

$$\hat{\mu}_{pio} = \frac{1}{N'} \sum_{j=1}^{N'} \hat{p}_i(j) \qquad 1 \le i \le 1 \tag{20}$$

$$\hat{\sigma}^2_{pio} = \frac{1}{N'} \sum_{j=1}^{N'} [\hat{p}_i(j) - \hat{\mu}_{pio}]^2 \tag{21}$$

Phase II: Detection of changes

$$\hat{\mu}_{pi}(k) = \frac{1}{N} \sum_{j=1}^{N} p_i(k-j) \tag{22}$$

$$\hat{\sigma}^2_{piI}(k) = \frac{1}{N} \sum_{j=1}^{N} [p_i(k-j) - \hat{\mu}_{pio}]^2 \tag{23}$$

$$\sigma^2_{piII}(k) = \frac{1}{N} \sum_{j=1}^{N} [p_i(k-j) - \hat{\mu}_{pi}(k)]^2 \tag{24}$$

$$d_i(k) = \frac{\sigma^2_{piI}(k)}{\sigma^2_{pio}} - \ln \frac{\sigma^2_{piII}(k)}{\sigma^2_{pio}} - 1 \tag{25}$$

$$d_2(k) \le 2 \ln \frac{NP_o}{1-P_o} \quad \text{no change} \tag{26}$$

$$d_i(k) > \ln \frac{NP_o}{1-P_o} \quad \text{change of } p_i. \tag{27}$$

Hereby d_i is called test quantity and P_o is the apriori probability of the faultless case ($0 \leq P_o < 1$). As a result, significant changes of single process coefficients p_i or combinations of several process coefficients p_i are indicated. By a comparison with the fault signatures the process coefficients changes can be classified and the fault diagnosis can be performed, see section 2.1.

3. FAULT DIAGNOSIS OF A D.C. MOTOR-CENTRIFUGAL PUMP

The described fault diagnosis based on parameter estimation was investigated for a centrifugal pump with a water circulation system, driven by a speed controlled direct current motor, see Fig. 4. A detailed description of this engine is given in Isermann (1984a) and Geiger (1985). Therefore only a brief outline is given here. The emphasis of this section lies in new results on the detection of implemented faults.

3.1. Mathematical model of the process dynamics

Following variables are measured:

U_2 armature voltage

I_2 armature current

ω angular velocity

H pump total head

$\dot{V}$ volume flow rate

The basic equations after some simplifying assumptions are, see Geiger (1985):

a) armature circuit

$$L_2 \frac{dI_2(t)}{dt} = -R_2 I_2(t) - \Psi\omega(t) + U_2(t) \tag{28}$$

b) mechanics of motor and pump

$$I\frac{d\omega}{dt}(t) = \Psi I_2(t) - c_{RO} - \rho g h_{TH1} \omega(t)\, \dot{V}(t) \tag{29}$$

c) hydraulics of the pump

$$H(t) = h_{NN}\omega^2(t) \tag{30}$$

d) hydraulics of the pipe system

$$a_B \frac{d\dot{V}(t)}{dt} = -h_{RR}\dot{V}^2(t) + H(t). \tag{31}$$

The model is nonlinear and contains 9 process coefficients

$$\underline{p}^T = [L_2, R_2, \Psi, I, c_{RO}, h_{TH1}, h_{NN}, a_B, h_{RR}]. \tag{32}$$

For the parameter estimation the equations are brought into the form

$$y_j(t) = \underline{\Psi}_j^T(t)\, \hat{\underline{\Theta}}_j, \quad j = 1,2,3,4 \tag{33}$$

where

$$\left.\begin{aligned} y_1(t) &= dI_2(t)/dt & y_2(t) &= d\omega(t)/dt \\ y_3(t) &= H(t) & y_4(t) &= dV(t)/dt \end{aligned}\right\} \tag{34}$$

$$\left.\begin{aligned} \underline{\Psi}_1^T(t) &= [I_2(t), \omega(t), U_2(t)] \\ \underline{\Psi}_2^T(t) &= [I_2(t), 1, \omega(t)V(t)] \\ \underline{\Psi}_3^T(t) &= [\omega^2(t)] \\ \underline{\Psi}_4^T(t) &= [H(t), V^2(t)] \end{aligned}\right\} \tag{35}$$

$$\left.\begin{aligned} \underline{\Theta}_1^T &= [a_{11}\; a_{12}\; b_1] \\ \underline{\Theta}_2^T &= [a_{21}\; a_{22}\; a_{23}] \\ \underline{\Theta}_3^T &= [a_{31}] \\ \underline{\Theta}_4^T &= [a_{41}\; a_{42}] \end{aligned}\right\} \tag{36}$$

Based on the parameter estimates

$$\hat{\underline{\Theta}} = [\hat{\underline{\Theta}}_1^T\; \hat{\underline{\Theta}}_2^T\; \hat{\underline{\Theta}}_3^T\; \hat{\underline{\Theta}}_4^T]^T \tag{37}$$

all process coefficients $\underline{p}$ can be determined uniquely

$$\left.\begin{aligned}
&\hat{L}_2 = \frac{1}{\hat{b}_1}; \ \hat{R}_2 = -\hat{a}_{11}\hat{L}_2 \\
&\hat{\Psi} = -\hat{a}_{12}\hat{L}_2; \ \hat{I} = \hat{\Psi}/\hat{a}_{21} \\
&\hat{c}_{RO} = -\hat{a}_{22}\hat{I}; \ \hat{h}_{TH1} = -\hat{a}_{23}\hat{I}/\rho g \\
&\hat{h}_{NN} = \hat{a}_{31}; \ \hat{a}_B = 1/\hat{a}_{41} \\
&\hat{h}_{RR} = -\hat{a}_{42}\hat{a}_B
\end{aligned}\right\} \quad (38)$$

3.2 Experimental Results

The d.c. motor is controlled by an a.c./d.c. converter with cascade control of the speed and the armature current as auxiliary control variable. The manipulated variable is the armature current U_2. A microcomputer DEC-LSI 11/23 was connected on-line to the process. For the experiments the reference value W(t) of the speed control has been changed stepwise with a magnitude of 750 rpm every 2 min. The operating point was n = 1000 rpm, H = 5.4 m and $\dot{V}$ = 6.48 m^3/h. The signals were sampled with sampling time T_0 = 5 msec and 20 msec over a period of 2.5 and 10 sec, so that 500 samplings were obtained. These measurements were stored in the core memory. Then the parameters were estimated. The available computation time was about 100 sec. Hence one set of parameters and process coefficients was obtained every 120 sec. For the training phase 50 coefficient sets were used.

Table 1 gives an overview of significant changes of process coefficients for 19 different artificially generated faults.
A selection of experiments will now be considered in more detail.

A. Faults in the d.c. motor

Experiment A3: *Change of carbon-brushes*, Fig. 5
A reduction of the brush contact surface of 50 % increases mainly the armature resistance R_2. With increasing operating time R_2 decreases due to the grinding effect.

Experiment A5: *Disturbance of the air cooling*, Fig. 6.
A stepwise reduction (25 %, 50 %, 75 %, 100 %) of the air flow (e.g. due to plugging by dirt) leads to a temperature change of the whole motor and therefore to an increase of the resistance in the armature circuit and excitation circuit. Therefore R_2 increases and the magnetic flux linkage Ψ decreases. This is an example where the coefficients move in opposite directions.

Table 1: Detected significant changes of process coefficients after appearance of a fault. $P_0 = 0.5$.

In brackets the sequence of appearance is shown.

EXPERIMENT	L_2	R_2	$\bar{\Psi}$	I	C_{RO}	H_{TH1}	H_{KN}	A_B	H_{RR}
A1: EXC. RESISTANCE INCREAS.	-	-	-(1)	-	-	-	-	-	-
A2: ARM. RESISTANCE INCREAS.	-	+(1)	-(2)	-	-	-	-	-	-
A3: AFFECTED BRUSHES	-	+(1)	-(2)	-	-	-	-	-	-
A4: NEW BRUSHES	-	-	-	-	-	-	-	-	-
A5: INSUFFICIENT COOLING	-	+(2)	-(1)	-	-	-	-	-	-
A6: COLD DRIVE	-	-	+(1)	-	-	-	-	-	-
K1: SHAFT DISPLACEMENT	-	-	-	-(1)	-	-	-	-	-
P1A: BEAR. WITHOUT GREASE	-	-	-	-	-(1)	-	-	-	-
P1B: BEAR. WITH DIRT	-	-	-	+(3)	+(2)	+(1)	-	-	-
P2: SIDE THRUST COMPEN. DEF.	-	-	-	-	-	-	-	-	-
P3: SPLITRING CLEARANCE INCR.	-	-	-	-	-	+(1)	-(2)	-	-
P4: AFFECTED IMPELLER	-	-	-	-(3)	-	+(2)	-(1)	-	-
P5A: PUMP CASING DEFECTIVE I	-	-	-	-	+(1)	-	-	-	-
P5B: PUMP CASING DEFECTIVE II	-	-	-	-	-(3)	+(2)	-	+(1)	-
F1A: LITTLE CAVITATION	-	-	-	-	-	-	-	+(1)	-
F1B: MEDIUM CAVITATION	-	-	-	-	-	+(3)	-(2)	-(1)	-(4)
F2: INSUFFICIENT VENTING	-	-	-	-	-	+(4)	-(2)	-(1)	-(3)
F3: FLUID TEMPERATURE INCR.	-	-	-	-	-	-	-(1)	-	-(2)
F4: VALVE POSIT. INCREAS.	-	-	-	-	-	+(2)	-	-	+(1)

B. Faults in the centrifugal pump

Experiment P1a: *Ball bearing without lubrication*, Fig. 7.
If the grease of the ball bearing is washed out, the adhesive friction coefficient c_{RO} first decreases by about 30 %.

Experiment P1b: *Ball bearing with dirt*, Fig. 8.
By adding fine iron cuttings to the ball bearing the coefficient c_{RO} increases its mean as well as its variance.

Experiment P3: *Increase of slot clearance*, Fig. 9.
An increase of the clearance of the slot between the pump wheel and the pump case increases the internal losses. Therefore h_{TH1} increases and h_{NN} decreases.

C. Faults in the fluid part

Experiment F1a: *Small cavitation*, Fig. 10.
A small cavitation and gas bubble generation in the pump by lowering the entrance pressure is indicated by an increasing coefficient a_B, which is the time constant of the pipesystem. There was no noise.

Experiment F1b: *Medium cavitation*, Fig. 11.
The effects of F1a are now much stronger.

Experiment F4: *Increased resistance in the pipesystem*, Fig. 12.
The resistance in the pipe could be increased by closing the valve from 100 % to 80 %. as a result the hydraulic friction coefficient h_{RR} increases and also the internal loss coefficient h_{TH1} of the pump (due to the new operating point). For a more detailed description and discussion of all experiments see Geiger (1985).

These experiments have demonstrated that in all cases where a significant change of process coefficients could be expected the fault could be detected. In these cases a fault diagnosis could also be performed, based on the patterns given in Table 1.

3.3 Mathematical model of the process statics

It is now considered which process coefficients can be determined if only the static behaviour is taken into account and the same signals are measured as in the dynamic case. With d/dt = 0 it follows from Eq. (28) to (31):

$$R_2 I_2 = \Psi\omega + U_2 \quad (39)$$

$$\Psi I_2 = c_{Ro} + \rho g h_{TH1} \omega \dot{V} \quad (40)$$

$$H = h_{NN}\omega^2 \quad (41)$$

$$H = h_{RR}\dot{V}^2 \quad (42)$$

Hence the coefficients L_2, I and a_B cannot be determined anymore. Eq. (41) and (42) allow directly to calculate h_{NN} and h_{RR}. By *changing the speed* ω signals for different operating points can be measured. Theoretically Eq. (39) then yields R_2 and Ψ. However, as $R_2 I_2 \ll \Psi\omega$, R_2 can hardly be obtained with sufficient accuracy. Ψ can be determined approximately. Based on Eq. (40) then c_{RO} and h_{TH1} can be estimated.

In comparison to the dynamic model the static model allows to determine 5 process coefficients instead of 9. Therefore following faults of Table 1 cannot de detected: A2, A3, A5, K1, F1a and b, F2. A second type of experiments can be perforemd by varying the pipeline resistance h_{RR} for the case of ω = const., e.g. by changing the *valve position*. Then R_2 can be estimated and also the coefficients Ψ, c_{Ro}, h_{TH1}, h_{NN}, h_{RR}, i.e. 6 process coefficients. Following faults cannot be detected: K1, F1a and b, F2. However, the accuracy of these estimates has to be examined by experiments.

These considerations show that the use of dynamic models allows to get more information than the use of static models and that therefore more faults can be detected.

Some other examples where fault detection based on process parameters was investigated are:
- Hydraulic drive of a machine tool, Hohmann (1977)
- Jet engines, Baskiotis, Raymond and Rault (1979)
- El. motor for machine tools, Filbert and Metzger (1982).

4. FAULT DIAGNOSIS OF A TUBULAR HEAT EXCHANGER

The fault diagnosis via parameter estimation was then investigated for a process which cannot be modelled with high accuracy. An industrial size steam heated heat exchanger was implemented in a pilot plant, consisting of an electric powered steam generator, a steam-condensate circulation, a water circulation and a cross-flow heat exchanger to transport the heat from water to air, see Fig. 13.

4.1. Mathematical model of the process dynamics

Following variables are measured, Fig. 14:

$\dot{M}_S$ mass flow of the steam

$\dot{M}_F$ mass flow of the fluid (water)

ϑ_{Fi} inlet temperature of the fluid

ϑ_{Fo} outlet temperature of the fluid

The fluid outlet temperature ϑ_{Fo} is considered as output variable, the other three measured variables as input variables. To model the dynamic behaviour, the heat exchanger is subdivided into the tubular section, the water head, a transport delay and the temperature sensor. The basic balance equations for a heated tube are obtained by considering an infinitesimal small element of length dz and assuming distributed parameters for the fluid and lumped parameters for the tubewall, see Isermann (1969) and Goedecke (1986):

$$\frac{\partial h_F(z,t)}{\partial t} + w_F(t)\,\frac{\partial h_F(z,t)}{\partial z} = \frac{1}{A_F\rho_F}\,\frac{\partial \dot{Q}(z,t)}{\partial z} \tag{43}$$

$$\dot{Q}_{SW}(z,t) - \dot{Q}_{WF}(z,t) = \frac{d}{dt}\,\dot{Q}_W(z,t) \tag{44}$$

and after introducing the phenomenological equations for the heat transfer and $h_F = c_F\vartheta_F$

$$\frac{\partial \vartheta_F(z,t)}{\partial t} + w_F(t)\,\frac{\partial \vartheta_F(z,t)}{\partial z} = \frac{U_F\alpha_{WF}}{A_F\rho_F c_F}\,[\vartheta_W(z,t) - \vartheta_F(z,t)] \tag{45}$$

$$\alpha_{SW}U_W[\vartheta_S(z,t) - \vartheta_W(z,t)] - \alpha_{WF}U_F[\vartheta_W(z,t) - \vartheta_F(z,t)] = A_W\rho_W c_W\,\frac{d\vartheta_W(z,t)}{dt} \tag{46}$$

The various symbols are:

A	cross-sectional area	w	velocity
c	specific heat capacity	z	location
h	enthalpy	α	heat transfer coefficient
l	length of tubes	ϑ	temperature
M, $\dot{M}$	mass, mass flow	ρ	density
Q, $\dot{Q}$	heat, heat flow		
r	evaporation heat		
U	periphery		

and the subscripts:

F	Fluid (water)	W	Wall
K	Condensate	i	inlet
S	Steam	o	outlet

In addition balance equations are stated for the steam space and the shell tube. Then the equation system is linearized around the operation point, several simplifying neglections are made and by applying the Laplace transform with respect to time t and location z one obtains three transfer functions, e.g.

$$G_{\vartheta F}(s) = \frac{\Delta\vartheta_{Fo}(s)}{\Delta\vartheta_{Fi}(s)} = e^{-T_t s} \cdot e^{-\kappa_F \frac{T_{WF}s}{1+T_{WF}s}} \tag{47}$$

with $$T_t = \frac{1}{w_F} \; ; \; K_F = \frac{T_t}{T_F} = \frac{\alpha_{WF} l \, U_F}{w_F A_F \rho_F c_F} \; ; \; T_{WF} = \frac{A_W \rho_W c_W}{\alpha_{WF} U_F} \tag{48}$$

This transcendental transfer function can be approximated by a lumped parameter model with dead time

$$G_{\vartheta\vartheta}(s) \approx \tilde{G}_{\vartheta F}(s) = K_\vartheta \frac{1 + T_{D\vartheta}s}{1 + T_{1\vartheta}s} e^{-T_t s} \tag{49}$$

$$T_{1\vartheta} = \frac{\kappa_F}{1-e^{-\kappa_F}} T_{WF} \; ; \; T_{D\vartheta} = T_{1\vartheta} e^{-\kappa_F} \; ; \; K_\vartheta \approx 1 \tag{50}$$

Similarly one obtains the approximate transfer functions

$$\tilde{G}_{S\vartheta}(s) = \frac{\Delta\vartheta_{Fo}(s)}{\Delta\dot{M}_S(s)} = \frac{K_D}{(1+T_{1D}s)(1+T_{2D}s)} e^{-T_{tD}s} \tag{51}$$

with

$$\left. \begin{aligned} K_D &= \frac{r}{\dot{M}_S c_F} \; ; \; T_{1D} = \frac{1}{w_F} \left[1 + \frac{A_W \rho_W c_W}{A_F \rho_F c_F}\right] \\ T_{2D} &= \frac{A_W \rho_W c_W}{\alpha_{WF} U_F} \cdot \frac{1}{\left[1 + \frac{A_W \rho_W c_W}{A_F \rho_F c_F}\right]} \end{aligned} \right\} \tag{52}$$

and

$$\tilde{G}_{F\vartheta}(S) = \frac{\Delta\vartheta_{Fo}(s)}{\Delta\dot{M}_F(s)} = \frac{K_F}{[1+T_{1F}s][1+T_{2F}s]} e^{-T_{tF}s} \tag{53}$$

with

$$\left.\begin{aligned} K_F &= \frac{\vartheta_W-\vartheta_F}{\dot{M}_F}\ \frac{1}{w_F}\ \frac{\alpha_{WF}U_F}{A_F\rho_F c_F} \\ T_{1F} &= T_{1D}\ ;\ T_{2F} = T_{2D} \end{aligned}\right\} \tag{54}$$

These equations show that it is not possible to determine all process coefficients uniquely. In the case of $\tilde{G}_{\vartheta\vartheta}$ for example 2 parameter estimates compare to 9 process coefficients, and in the case of $\tilde{G}_{S\vartheta}$ three parameter estimates compare to 10 process coefficients. By assuming some of the process coefficients to be known e.g. following process coefficients can be determined:

$$\left.\begin{aligned} \tilde{G}_{\vartheta\vartheta}\ :\ \alpha_{WF} &= \frac{\dot{M}_F c_F}{I\ U_F}\left[-\ \ln\frac{T_{D\vartheta}}{T_{1\vartheta}}\right] \\ A_W\rho_W c_W &= \frac{\dot{M}_F c_F}{I}\ [T_{1\vartheta} - T_{D\vartheta}] \end{aligned}\right\} \tag{55}$$

$$\left.\begin{aligned} \tilde{G}_{S\vartheta}\ :\ \alpha_{WF} &= \frac{A_F\rho_F c_F}{T_{2D}\ U_F}\left[1 - \frac{1}{T_{1D}w_F}\right] \\ A_W\rho_W c_W &= \frac{T_{1D}M_F c_F}{I} - A_F\rho_F c_F \\ r &= K_D M_F c_F \end{aligned}\right\} \tag{56}$$

$$\left.\begin{aligned} \tilde{G}_{F\vartheta}:\ \ \alpha_{WF} &= \frac{A_F\rho_F c_F}{T_{2F}U_F}\left[1 - \frac{1}{T_{1F}w_F}\right] \\ A_W\rho_W C_W &= \frac{T_{1F}M_F c_F}{I} - A_F\rho_F c_F \\ \vartheta_W - \vartheta_F &= K_F T_{2F} M_F - \frac{w_F}{I}\ \frac{T_{1F}w_F}{T_{1F}w_F - I} \end{aligned}\right\} \tag{57}$$

4.2 Experimental results

The parameter estimation experiments are based on transient functions of the fluid outlet temperature ϑ_{Fo} after steps in the input variables ϑ_{Fi}, $\dot{M}_S$ and $\dot{M}_F$ in the direction of decreasing temperature ϑ_{Fo}. The operating point was

$$\dot{M}_F = 3000 \text{ kg/h}; \ \dot{M}_S = 50 \text{ kg/h}; \ \vartheta_{Fi} = 60 \ °C; \ \vartheta_{Fo} \approx 70 \ °C$$

The sampling time was $T_0 = 50$ msec, the time period of one experiment 360 sec, so that 720 samplings were used. For the parameter estimation the method of total least squares in a recursive form by using a digital state variable filter for the determination of the derivation was applied. The determination of the normal state (training phase) was based on 60 transients for each transfer function. Then 30 transients were carried out for each transfer function and for each of 4 artificially generated faults:

F1: air (inertgas) in the steam space
F2: opened condensate valve
F3: closed condensate valve
F4: plugged tube

(Altogether 540 experiments were carried out, lasting about 150 operating hours)

Fig. 15 shows a) one measured transient function and b) the corresponding time history of the parameter estimates. A good convergence of the parameter estimates was obtained in all cases. A verification of the measured and the calculated transient function shows very good agreement Fig. 15c. In Table 2 the parameter estimates are given for $G_{S\vartheta}$. Table 3 indicates that for each of the 4 faults different changes of the parameter estimates are obtained. Corresponding results were obtained for changes of the fluid inlet temperature, $\tilde{G}_{\vartheta\vartheta}$ and changes of fluid flow, $\tilde{G}_{F\vartheta}$. Based on the parameter estimates $\hat{K}_D$, $\hat{T}_{1D}$, $\hat{T}_{2D}$ the obtainable process coefficients due to eq. (56) are shown in Table 4.

The differences between the coefficients used for theoretical modelling and the estimated ones for the normal state can be at least partially explained by the usual tolerances of the tabulated values and by the assumption and neglection in the modelling. The differences between the faulty state and the normal state are quite significant. However, they can only partially be explained by the physical effects of the faults. One reason is also that the investigated faults are not directly mapped by the obtainable process coefficients. In addition the calculated process coefficients are obviously rather sensitive to changes of the process parameter estimates and to the values of the coefficients which have to be assumed as known. This is a known fact for processes with heat transfer. Therefore it is recommended that the fault detection for this type of process, which is a higher order distributed parameter system and is approximated by lower order lumped

Table 2 Parameter estimates for steam flow changes $\tilde{G}_{S\vartheta}$.

Fault	Mean Standard Derivation	$\hat{K}_D$ [Kh/kg]	$\hat{T}_{1D}$ [sec]	$\hat{T}_{2D}$ [sec]
None	μ	0.1708	12.38	7.21
	σ	0.0032	1.63	1.07
F1	μ	0.1896	7.26	7.26
	σ	0.00716	0.73	0.73
F2	μ	0.1268	7.62	7.62
	σ	0.00369	0.35	0.35
F3	μ	0.1899	13.89	3.81
	σ	0.0042	0.82	0.44
F4	μ	0.1689	13.65	6.01
	σ	0.00322	1.50	0.81

Table 3 Changes of the parameter estimates of table 2

Fault	$\hat{K}_D$	$\hat{T}_{1D}$	$\hat{T}_{2D}$
F1	$>$	$\ll$	$=$
F2	$\ll$	$\ll$	$>$
F3	$>$	$>$	$\ll$
F4	$=$	$>$	$<$

Table 4 Some process coefficients for steam flow changes

Process coefficients	Theoretical modelling	Normal state	Faulty state F1	F2	F3	F4
r[kJ/kg]	2309	2145	2382	1593	2385	2122
$A_W\rho_W c_W$[kJ/Km]	3.806	7.343	0.200	0.703	9.455	3.715
α_{WF}[kJ/Km^{2}S	2.692	1.267	0.058	9.186	2.752	1.712

models, is based on the model parameter estimates like gain and time constants. This means that a detailed therotical modelling is not necessary. The case study has shown that the considered four faults could be detected by using patterns of changes according to Table 3. This holds for all three transfer functions. Hence, it is sufficient to use only one transfer function. The most significant changes were obtained for steam flow changes, $\tilde{G}_{S\vartheta}$. The use of static models allows only to recognize changes from the normal state. Hence a fault detection and diagnosis in this case is only possible by applying dynamic models.

5. CONCLUSION

The use of mathematical models and measurable input and output signals allows to estimate usually unmeasurable internal quantities like parameters which can then be used for process fault detection. These quantities directly express changes in the process generated by process faults. The model based fault detection takes into account all entailed changes of input and output signals, provides a data reduction and tries to determine the process parameter which has been affected by a fault. Therefore process faults can be detected earlier and localized more precisely.

A generalized procedure for process fault diagnosis based on process parameters was described. It was shown how the process coefficients can be determined from the estimated process model parameters and how this can be followed by a fault decision and a fault diagnosis.

The methods were first investigated and tested experimentally for a d.c. motor-centrifugal pump system which can be modelled as a nonlinear lumped system. Based on 5 measured signals 9 process coefficients could be monitored. Then 19 process faults were generated artificially. In all cases where a significant change of process coefficients could be expected the fault could be detected and localized. As a further application an industrial size heat exchanger was used as a representative of a process with distributed parameters which can only be modelled approximately by lower order lumped models. It was shown that 2 measured signals are sufficient to monitor 3 process model parameters and to detect 4 artificially generated faults.

The case studies have demonstrated that the calculated process coefficients $\underline{p} = f^{-1}(\hat{\underline{\theta}})$ allow a detailed fault detection and diagnosis, if the process can be modelled precisely with lumped parameters (so called well-defined processes) and if enough measured signals are available so that the identifiability condition is satisfied. If the process can only be described by approximate models (so called not well-defined processes) and if only few signals are measurable the process parameter estimates $\hat{\underline{\Theta}}$ contain still valuable information for fault detection. However, the fault diagnosis can then not be as detailed as with reliable process coefficients estimates. Both examples have shown that the dynamic process behaviour yields considerably more information on process faults than the static behaviour.

REFERENCES

Baskiotis, C., J. Raymond and A. Rault (1979). *Parameter identification and discriminant analysis for jet engine mechanical state diagnosis.* IEEE Conference on Decision and Control, Fort Lauderdale.

Filbert, D. and K. Metzger (1982). *Quality test of systems by parameter estimation,* 9th IMEKO-Congress, Berlin.

Frank, P.M., Janßen, K. (1986). *Entdeckung von Komponentenfehlern in dynamischen Systemen durch hierarchische Zustandsschätzung.* Automatisierungtechnik **44**, 23-31.

Geiger, G. (1984). *Fault identification of a motor-pump system using parameter estimation and pattern classification.* Proceedings of the 9th IFAC Congress, Budapest, Pergamon Press, Oxford.

Geiger, G. (1985). *Technische Fehlerdiagnose mittels Parameterschätzung und Fehlerklassifikation am Beispiel einer elektrisch angetriebenen Kreiselpumpe.* Dissertation Technische Hochschule Darmstadt. VDI-Fortschrittsbericht. Reihe 6, VDI-Verlag, 1985.

Goedecke, W. (1985). *Fault detection in a tubular heat exchanger based on modelling and parameter estimation.* 7th IFAC-Symposium on Identification and System Parameter Estimation, York, 3-7 July 1985.

Goedecke, W. (1986). *Fehlererkennung an einem thermischen Prozeß mit Methoden der Parameterschätzung.* Internal report, Institut für Regelungstechnik, Technische Hochschule Darmstadt.

Gertler, J. and D. Singer (1985). *Augmented models for statistical fault isolation in complex dynamic systems.* American Control Conference, Boston.

Himmelblau, D.M. (1978). *Fault Detection and Diagnosis in Chemical and Petrochemical Processes.* Elsevier, Amsterdam.

Hohmann, H. (1977). *Automatic monitoring and failure diagnosis for machine tools* (in German). Dissertation, Technische Hochschule Darmstadt.

Isermann, R. (1969). *Einfache mathematische Modelle für das dynamische Verhalten beheizter Rohre.* Wärme, 75, 89-93.

Isermann, R. (1981). *Fault detection methods for the supervision of technical processes.* Process Automation, 1, pp. 36-44. Oldenbourg, Munich.

Isermann, R. (1984a). *Process fault detection based on modelling and estimation.* A survey. Automatica 20, 387-404.

Isermann, R. (1984b). *Fehlerdiagnose mit Prozeßmodellen.* Technisches Messen 51, 345-355.

N.N. (1984). *Verfahren und Systeme zur technischen Fehlerdiagnose.* VDI/VDE-GMR-Bericht 1, Workshop in Langen, F.R. Germany. VDI-Verlag Düsseldorf.

Pau, L.F. (1981). *Failure Diagnosis and Performance Monitoring.* Marcel Dekker, New York.

Willsky, A.S. (1976). *A survey of design methods for failure detection systems.* Automatica, 12, 601.

Young, P.C. (1981). *Parameter estimation for continuous-time models - a survey.* Automatica, 17, 23.

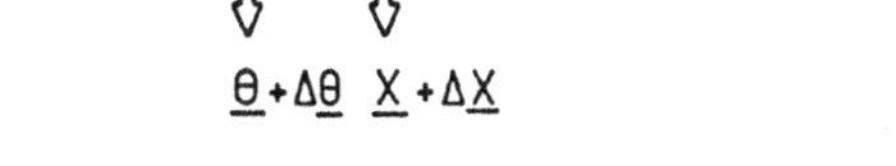

Fig. 1: Representation of a process which is influenced by a fault

$\underline{U}$, $\underline{Y}$: measurable input and output variables
$\underline{\Theta}$: process parameters
$\underline{X}$: process state variables
$\underline{N}$: disturbance variables (noise)

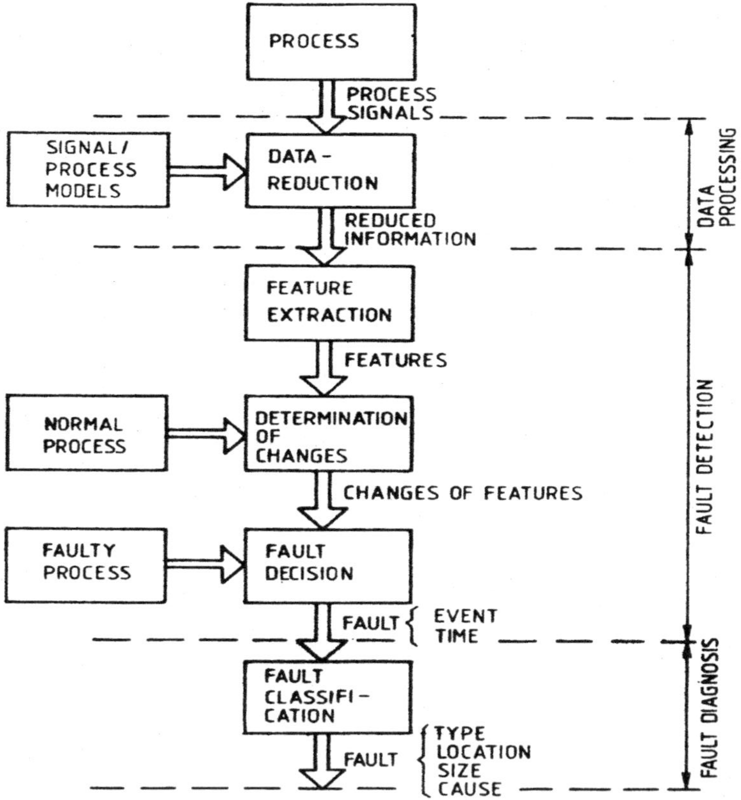

Fig. 2: Generalized scheme of technical fault diagnosis.

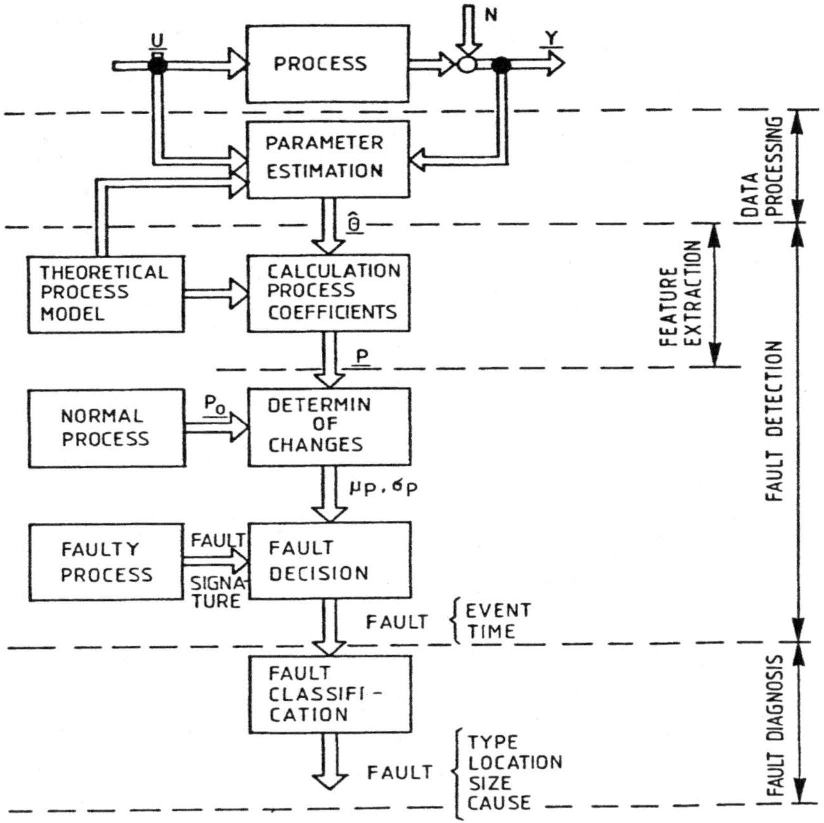

Fig. 3: Fault diagnosis based on parameter estimation and theoretical modelling.

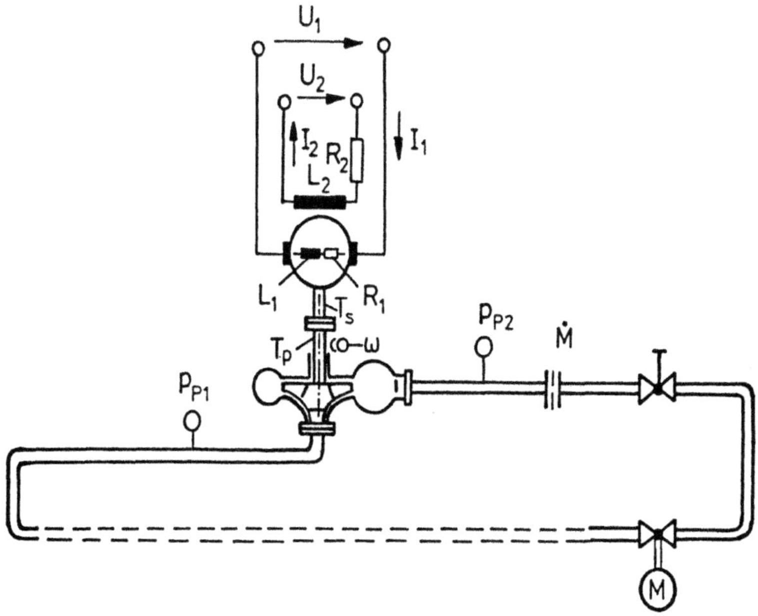

Fig. 4: Scheme of the speed controlled d.c.motor and centrifugal pump
Motor: $P_{max} = 4kW$; $n_{max} = 3000$ rpm
Pump: $H = 39m$; $\dot{V}_{max} = 160\ m^3/h$; $n_{max} = 2600$ rpm

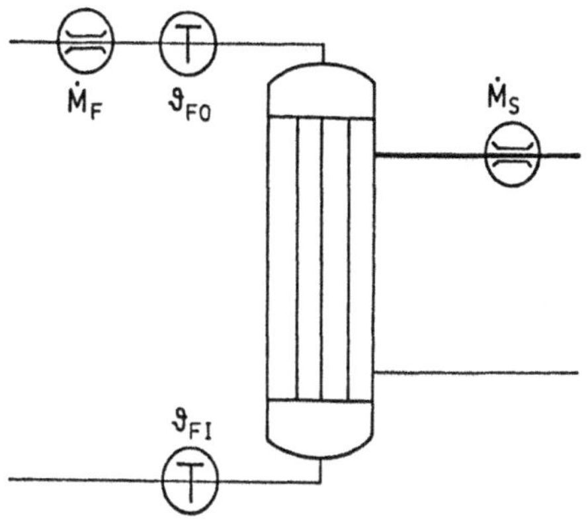

Fig. 14: Tubular heat exchanger and measured variables

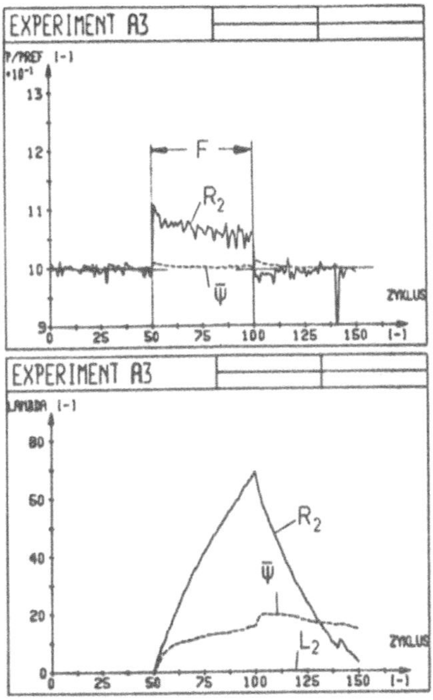

Fig. 5: Change of process for a reduction of the brush contact surface of about 50 %.

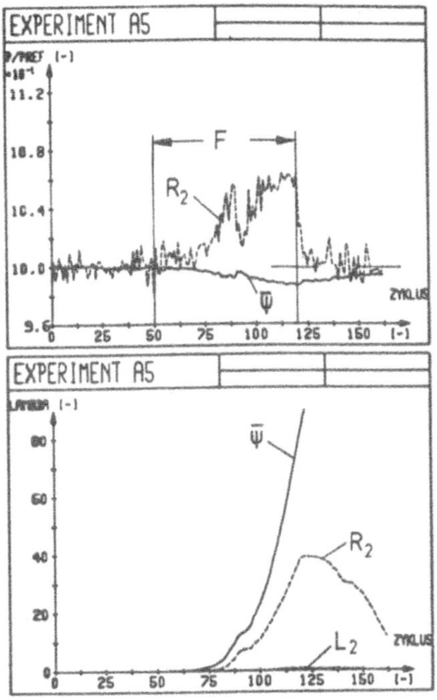

Fig. 6: Change of process coefficients for a reduction of the cooling air flow.

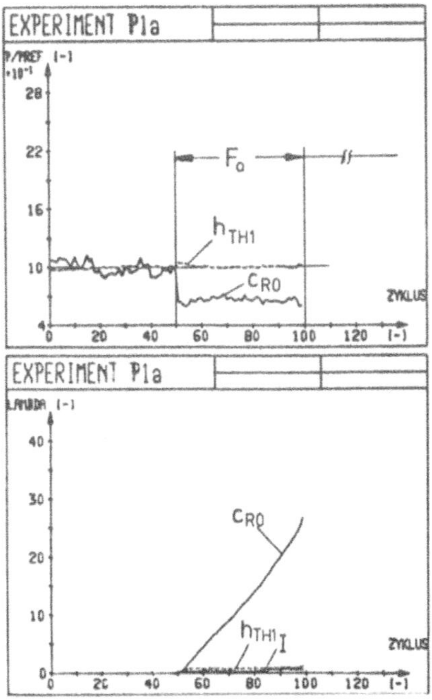

Fig. 7: Change of process coefficients after washing out the lubrication from the ball bearings.

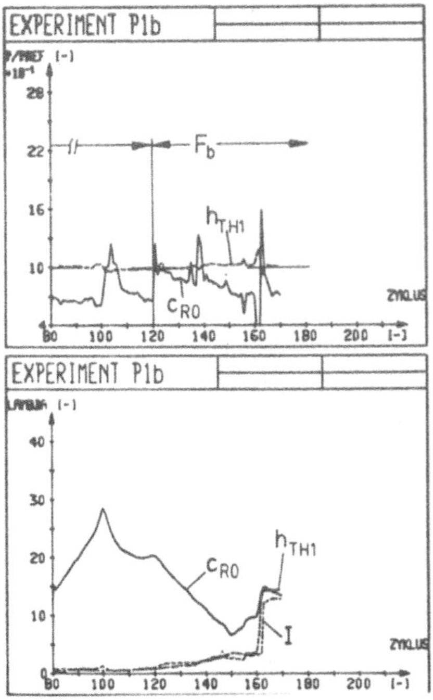

Fig. 8: Change of process coefficients after the ball bearings get dirty by iron cuttings.

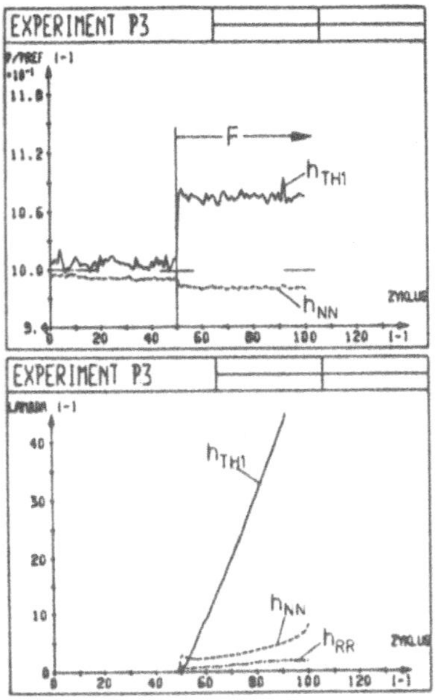

Fig. 9: Change of process coefficients after increasing the slot clearance of the pump.

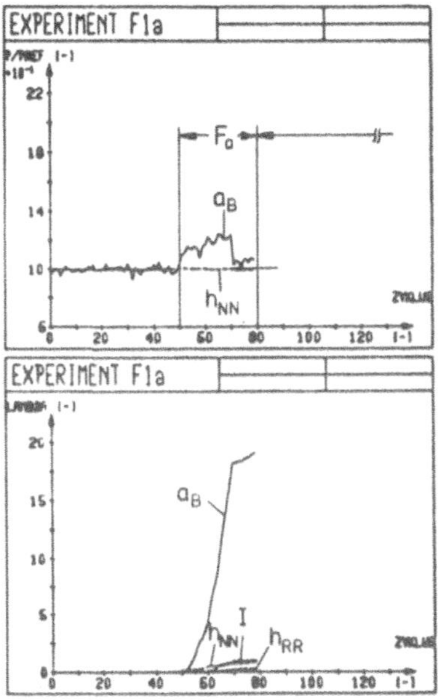

Fig. 10: Change of process coefficients in the case of small cavitation and gas bubble generation in the pump.

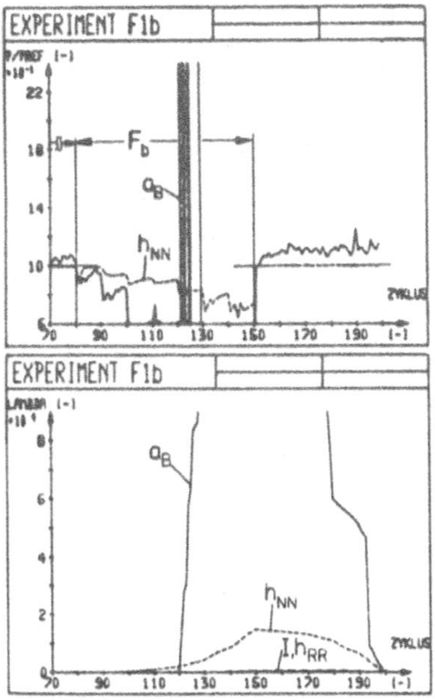

Fig. 11: Change of process coefficients in the case of medium cavitation and gas bubble generation in the pump.

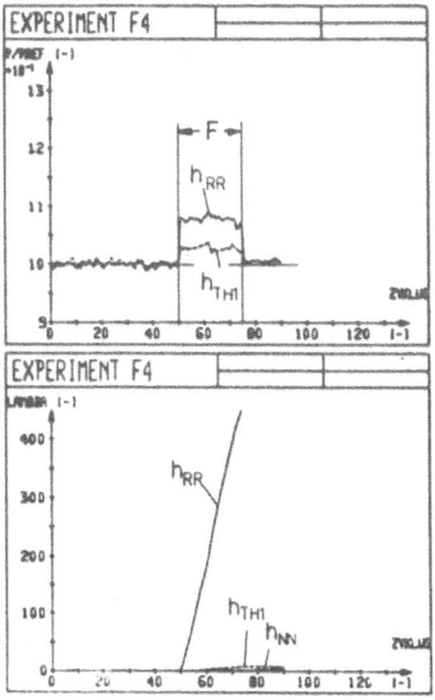

Fig. 12: Change of process coefficients after closing the valve position in the pipe from 100 % to 80 %.

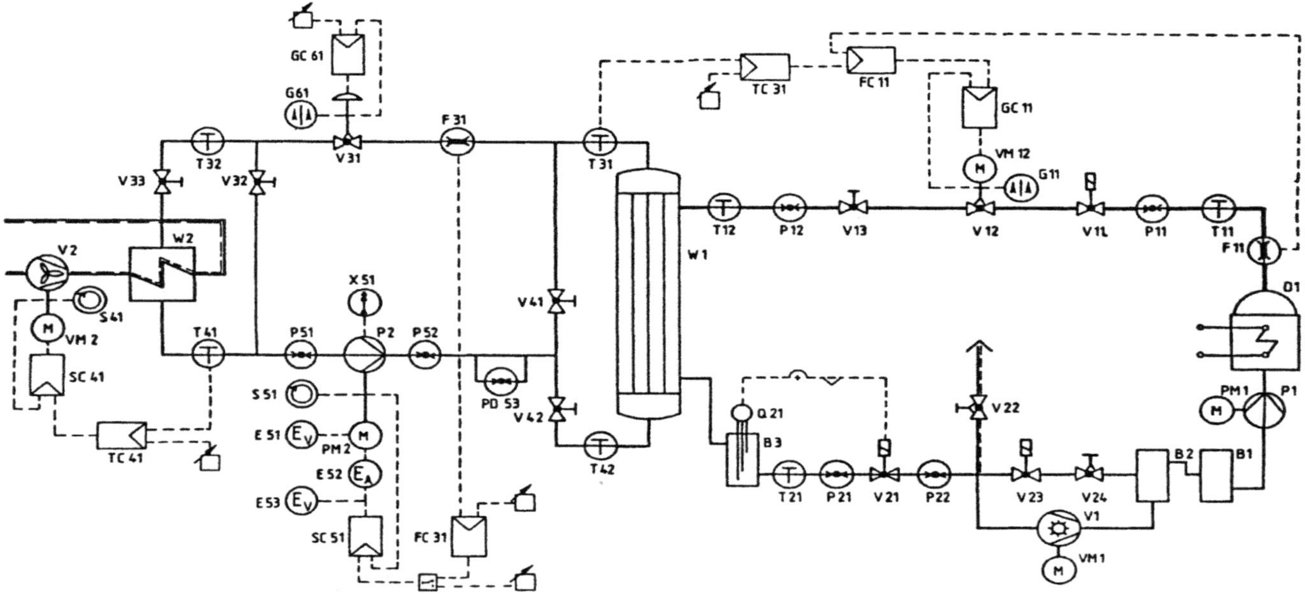

Fig. 13: Scheme of the thermal pilot plant with the steam heated heat exhanger

Fig. 14: see 3 pages before

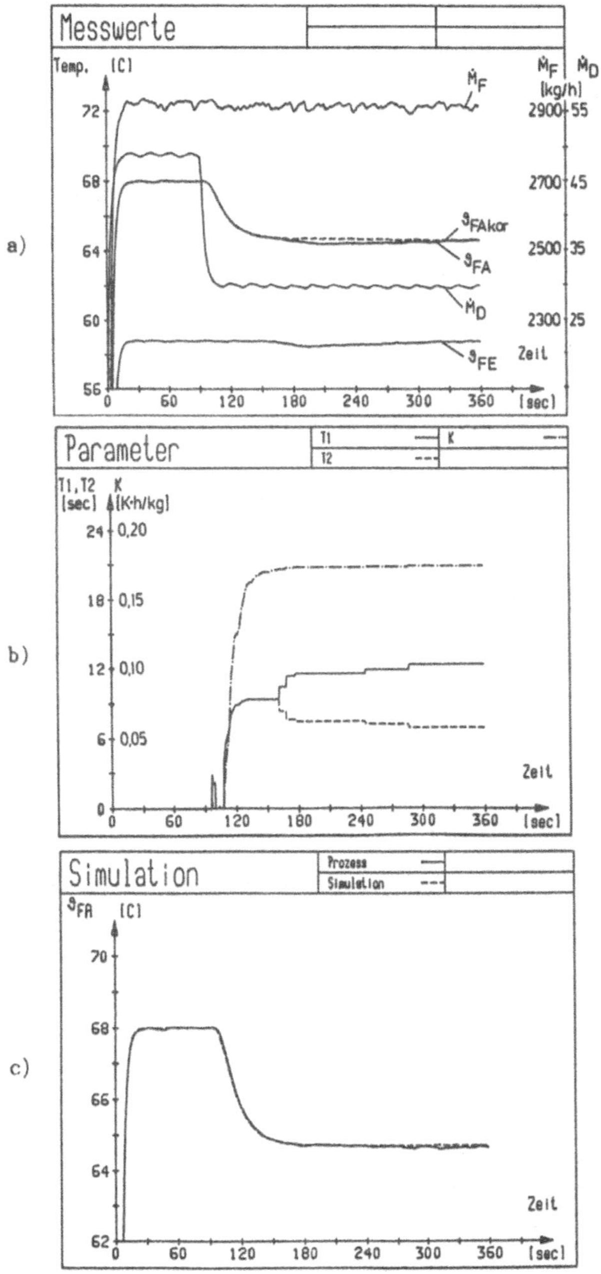

Fig. 15: Results for a change of the steam flow $\Delta\dot{M}_S$.

Heat exchanger with no faults:

a) measured transient function

b) parameter estimates

c) measured and simulated transient function

FAULT DIAGNOSIS IN DYNAMIC SYSTEMS VIA STATE ESTIMATION - A SURVEY

P. M. Frank
University of Duisburg
Department of Electrical Engineering
Measurement and Control Group
Bismarckstraße 81
4100 Duisburg 1
Federal Republic of Germany

ABSTRACT. This paper gives a survey on methods for the detection and localization of sensor and component faults of uncertain dynamic systems. In contrast to the commonly used techniques of hardware redundancy these methods make use of analytical redundancy and, thereby, allow to detect and localize faults with the aid of a digital computer. They comprise single, multiple or hierarchical state estimation using Luenberger observers or Kalman filters. An issue of particular relevance is the consideration of parameter uncertainties or parameter variations of the process. Several proposals are discussed to reduce the effects of parameter variations. Moreover, results from computer simulations and a practical technical application to the control of an inverted pendulum are reported.

KEYWORDS. Fault diagnosis; instrument fault detection; component fault detection; analytical redundancy; state estimation; robustness

1. INTRODUCTION

The permanent expansion of automation, its application to larger and more complex systems and the resulting increased demand for safety have enhanced the interest in advanced techniques of failure detection, isolation and accommodation (FDIA) in dynamic systems. Logically, failure tolerance is one of the prerequisites for safety and reliability and constitutes one of the fundamental design features in complex technological systems. In a broad sense, a failure is understood as any kind of anomaly or malfunction that leads to an undesired performance of the system under consideration. Such failures can, in principle, occur either in the sensors or in the functional components (such as actuators) of the system. We will therefore distinguish more precisely between instrument failure detection (IFD) and component

S. Tzafestas et al. (eds.), System Fault Diagnostics, Reliability and Related Knowledge-Based Approaches, Vol. 1, 35–98.

failure detection (CFD).

All failure detection schemes are in one way or another based on the use of *redundancy*, i.e. on static or dynamic relationships among measured variables. Traditional failure detection schemes use *hardware* redundancy generated by installing multiple like sensors for the same measured variable. Naturally, such backing-up of sensors leads to an according enlargement of equipment causing additional volume, weight, complexity and costs. In the past decade a great deal of work has therefore been devoted to the alternative approach of *analytical* (or *functional*) redundancy that is a priori contained in the system in form of the *inherent* (possibly dynamic) relationships among the measured variables. For surveys and further references see, e.g., Willsky (1976), Walker (1983), Frank and Keller (1984), Isermann (1984) and Merrill (1985). This development was undoubtedly favoured by the achievements of modern control theory and, maybe to an even larger extend, by the spectacular progress of computer technology which has become capable and reliable enough to be used for the sophisticated task of supervision.

It is by the nature of the analytical redundancy approach that it requires some sort of *mathematical model* of the system under consideration. Based on the model one generates (and evaluates) signals that are accentuated by the occurance of a fault. The entire FDIA process consists then of the following three steps:

(i) the *generation* of so-called *residuals*, i.e. of functions that carry information about faults

(ii) the *decision* process that evaluates the residuals and monitors if and where a fault has occured

(iii) the *accommodation* process by which the normal system operation is restored.

Looking to the literature on the failure detection and isolation (FDI) problem based on analytical redundancy one can see that the wide variety of published methods can roughly be devided into two major groups

(i) parameter estimation methods
(ii) state (or output) estimation methods.

In this paper we discuss FDI methods based on state (or output) estimation. The basic scheme of FDI using state estimation is shown in Fig. 1.

While the *parameter estimation* approach employs on-line identification of the mathematical model in order to determine the physical coefficients of the process, the *state estimation* approach restricts to the on-line reconstruction of sets or subsets of states or measured variables taking the mathematical model to be given. This is done with the aid of observers or Kalman filters whose estimates or innovations are then used for residual generation.

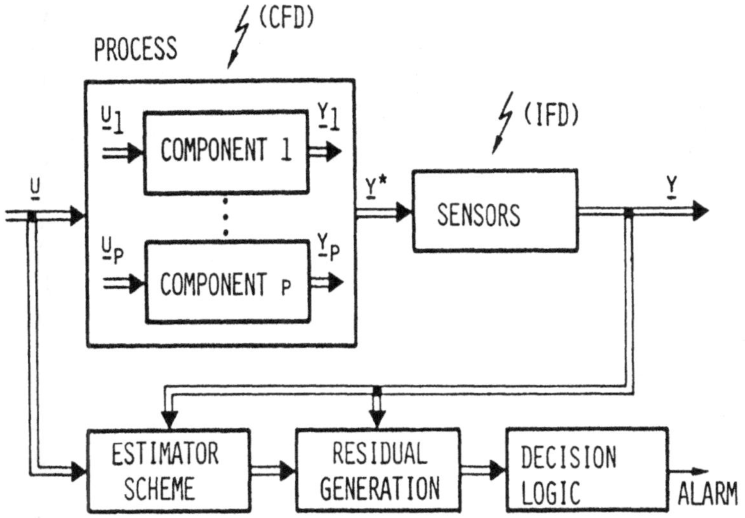

Fig.1: Basic scheme of FDI using estimation

Comparing the parameter and state estimation approaches it might seem at first glance that for both, IFD and CFD, the parameter estimation approach is superior because it provides a complete parametric picture of the system at any time thus allowing to discover all kinds of changes in the system. However, going one step further it turns out that there are some weak points in this procedure. First, what is really identified are not the *physical* coefficients of the process but the parameters of its *mathematical* model. These actually depend on the choice of the model and there is in general no unique relationship backward from the mathematical parameters to the physical coefficients. Second, though well apt for simple cases, the standard parameter estimation methods fail when the process contains *essential* nonlinearities. Third, in many practical situations there is obviously no reason why the mathematical model should permanently be re-identified rather than just once (off-line at the beginning) and the corresponding outputs are directly used for fault indication. The latter saves considerable computational expenditure and time. Most of the FDI methods proposed in the literature have adopted this approach.

In this paper we discuss FDI methods based on state (or output) estimation. Besides noise, an issue of fundamental relevance in any observer scheme for FDI is the sensitivity with respect to uncertainties (or unknown variations) of the process parameters. There is always a difference between the *actual* process parameters and those of the *nominal* model used in the observers or Kalmann filters. The effect of parameter variations obscures the effect of failure detection in that it acts like noise as a sourse of false alarms. The question of *parameter sensitivity* is thus a decisive point in the application of FDI schemes based on state estimation and has therefore long been of concern in the literature (see, e.g., Clark (1977)). In recent years enlarged attempts were made to *design* observer schemes with increased robustness with respect to parameter variations in the process (see, e.g., Frank and Keller (1980 and 1984), Watanabe e.a.(1981), Watanabe and Himmelblau (1982), Elmadbouly e.a. (1983), Weiß e.a. (1985), Lou e.a. (1986), Belkoura (1983), Chow and Willsky (1984), Wünnenberg e.a. (1986)). The sensitivity and robustness problem will be a major concern in this paper.

The paper is organized as follows. In section 2 the literature on FDI using state estimation methods is reviewed. Sections 3 and 4 outline basic observer schemes for IFD and CFD and decision logic. The sensitivity with respect to parameter variations is addressed in section 5. This forms the starting point of the design of robust observer schemes that are described in section 6. In order to demonstrate the power and the limits of the described methods some results from simulation studies and application to a laboratory setup of a real system are reported and discussed in section 7.

2. HISTORICAL REVIEW

To the author's knowledge, the idea to replace hardware redundancy by analytical redundancy was origianted by Beard (1971) and Meier e.a. (1971). Beard proposed methods of self-reorganization to maintain closed loop stability. The problem of identifying failures and changes in the system sensors was solved by comparison of the outputs of observers. Meier e.a. investigated the usefulness of functional redundancy to detect aircraft control data instrument failures. The functional redundancy was achieved from the functional relations that exists among different measurements. These were checked for consistency with the aid of two Kalman filters and several algebraic relations.

Mehra and Peschon (1971) introduced a general procedure for fault detection and diagnosis in dynamic systems with the aid of a *single* Kalman filter (similar as later did Willsky/Jones 1976, Cardy/Rozna 1980 and Belkoura 1983). An innovation sequence was generated and subjected to statistical tests of whiteness, mean and covariance. Knowing the time histories of the output variables under normal conditions, the deviations under faulty conditions are detected by statistical decision theory.

More software expenditure is needed for failure accommodation using Bayesian decision theory as proposed by Montgomery and Caglayan (1976), see also Willsky e.a. (1974 and 1975). They provide a bank of m parallel Kalman filters designed for the set of m - 1 possible failure modes and for normal operation. The erroneous instrument is detected with the aid of m-array hypothesis testing. A moving window of the innovation of each Kalman filter drives a detector that calculates the likelihood ratio for each hypothesis corresponding to a possible failure mode.

Deckert e.a. (1977) describe a functional redundancy scheme combined with dual sensor redundancy of the process. Failure identification is accomplished on the basis of the functional relationships among the outputs of dissimilar instruments by performing sequential probability ratio tests of the differences among the outputs. A similar concept was discussed by Onken and Stuckenberg (1979). Using doublicated sensors and state observers for the generation of analytical redundancy, they achieved the quality of a triplex system. Montgomery and Tabak (1979) replaced Kalman filters by a bank of so-called UD factorized filters where the filter gains were computed by a factorization of the covariance matrix.

Cunningham and Poyneer (1977) designed a dual sensor dual computer system for fail safe for single sensor failures in oder to obtain fail-operative flight control for mission essential sensors, or fail safe for both sensor types after one failure. Three concepts for the design of the diagnostic filters were proposed: the first uses five diagnostic filters which model physical relationships using sensor outputs. Concept II uses Kalman filters instead of observers. Concept III employs band pass Kalman filters.

Several contributions to IFD with state estimation methods using either observers or Kalman filters were made by Clark (1975, 1977, 1978, 1980, 1982). In 1977 Clark published the dedicated observer scheme (DOS) for IFD using a bank of Luenberger observers, each driven by one sensor output. If none of the sensors fails, all the reconstructed state vectors converge to the actual state vector. If, however, one of the sensors fails, then a difference occurs in the output vector of the corresponding observer. This can be used to identify the faulty sensor. Clark (1978) also introduced a simplified IFD scheme using a single observer driven by one of the measured variables. If all sensors work precisely, there will be no difference between the reconstructed outputs and the actual instrument outputs. If one of the sensors that does not drive the observer fails there will be a reconstruction error in corresponding channels. If, however, a sensor failure occurs in the channel that drives the observer, *all* the reconstructed outputs will be erroneous. Thus, a unique detection and isolation of the faulty sensor is possible.

An analytical redundancy scheme for flight control sensors of the Lockheed L-1011 aircraft was designed by Shapiro and Decarli (1979). They employed Luenberger observers to reconstruct the signals of failed sensors from the associated unfailed sensors. To avoid a bank of observers for each failure configuration, they apply an observer that is driven by the airframe inputs and the output of the sensor with the highest reliability.

To tackle the IFD problem in the presence of random disturbances,

Clark and Setzer (1980) modified the simplified IFD scheme by using a Kalman filter with an accordingly modified detection logic. As a modification of DOS, Clark and Campbell (1982) described an IFD scheme using a dedicated Kalman filter for each of the sensors, but instead of driving each filter by a single sensor output it is driven by three sensor outputs where each filter is sensitized to faults of the corresponding sensor. A similar approach was adapted by Patton and Willcox (1985, 1986).

In the IFD schemes so far described the errors of the reconstructed states that are employed for IFD are affected by both, sensor malfunctions and variations of the process parameters. Clark (1978) already performed numerical robustness tests. Insensitivity to parameter variations as a design specification was first included in the observer design by Frank and Keller (1980). They extended the dedicated observer scheme by duplicating the observers to allow distinction between parameter variations and instrument malfunctions. One observer in each pair is designed to be insensitive to parameter variations but sensitive to instrument failures and the other one insensitive to both. The method is robust with respect to process parameter variations and applies to single output systems as well as to multiple output systems. Several other approaches to the robustness problem were proposed by Elmadbouly e.a. (1982 and 1983), Belkoura (1983) and Frank e.a. (1984). These approaches are addressed in more detail later in this paper.

A general approach of creating robustness in FDI systems was pursued over the years by the group around Willsky (see, e.g., Deckert e.a. (1977), Chow (1980), Willsky e.a. (1981), Leininger (1981), Lou (1982), Chow el.a. (1984) and Lou e.a. (1986)), and by Watanabe e.a. (1981, 1982). They look to the problem of robust residual generation from the viewpoint of analytical redundancy relations, and have introduced the concept of general parity checks. From this perspective the innovation of an observer or a Kalman filter can be considered as the most general residual containing the complete set of redundancy relations. The underlying idea of robustness generation is now to utilize only those redundancy relations for FDI that are most reliable. In the latest works of Chow e.a. (1984) and Lou e.a. (1986) two procedures for how to find the most reliable, i.e., optimally robust redundancy relations are described.

The alternative approach of increasing the robustness of observer schemes by using "robust" or "unknown input" observers was recently tackled by Wünnenberg and, Frank (1986) and Alejandro e.a. (1986), see also White e.a. (1986).

While most of the work on FDI is concerned with the application to instrument faults, lesser attention has been paid to the application to *component faults* in dynamic systems. One of the crucial points of CFD is the problem of failure *isolation* which is much more complicated than the isolation of faulty instruments. The early works of Beard (1971) and Jones (1977) on detection filters already covered this issue. If information on possible failure types exists, the hypothesis testing techniques can be applied. For the case that no information on the types of failures is available, Janssen and Frank (1984, 1985) have developed a hierarchical scheme of global and local observers to detect and isolate faulty components. Shoureshi e.a. (1986) use eigenvector sensitivity analysis for this purpose.

For the case of sparce systems that allow matrix triangulation, Singh e.a. (1983) and Hodzic and Siljak (1985) have proposed a hierarchical state estimation technique to solve the fault isolation problem.

3. RESIDUAL GENERATION USING STATE ESTIMATORS

In general, there are many ways to generate residuals using the analytical redundancy of the process, Chow and Willsky (1984). A simple method is to check the validity of the mathematical equations among the measured variables and to compare a quantitative measure of the amount of mismatch in the *actual* situation with that of the *normal* situation. For example, suppose the relation among the two measured variables y_1 and y_2 reads $\dot{y}_2 + a_1 y_2 + a_2 y_1 = 0$, then the residual may simply be taken as $r=\dot{y}_2 + a_1 y_2 + a_2 y_1$. Evidently, r=0 holds if no fault occurs and $r \neq 0$ if there is a fault (either in y_1 or y_2 or a_1 or b_1). At the other end the most general residual possible is the *estimation error* or the *innovation* of an observer or Kalmann filter of the system because it includes the complete mathematical model and, hence, all the existing relations among the measured variables.

3.1 State Estimators for FDI

State estimation is well established in modern control engineering in both theory and practice. In this section we briefly review the standard types of estimators with due regard to their application to FDI. A key question is the dependence of the estimation error on changes in the sensors, the components and the paramters, as well as on noise.

The type of estimator used depends primarily on the nature of the process and its environment. In practice, most processes can be considered to be linear, time-invariant, controllable and observable. Hence, their state equations may take the form

$$\dot{\underline{x}} = \underline{A}\,\underline{x} + \underline{B}\,\underline{u} \tag{1}$$

$$\underline{y} = \underline{K}\,\underline{x} \tag{2}$$

where $\underline{x} \in R^n$ is the state vector, $\underline{u} \in R^p$, the input vector, $\underline{y} \in R^q$ the vector of measured variables and $\underline{A}$, $\underline{B}$, $\underline{K}$ are matrices of appropriate dimensions with rank ($\underline{B}$) = p and rank ($\underline{K}$) = q. Suppose the process under consideration has a set of q sensors whose readings y_i form the output vector $\underline{y}$ in (1),(2). Let us assume that the transfer functions of the sensors can simply be modeled by gain factors $k_1,\ldots k_q$. For the reason of separating the effects of sensor faults from those of parameter variations it is useful to define the state $\underline{x}$ in (1),(2) in terms of so-called *sensor coordinates* which means that $\underline{y}$ is a subset of $\underline{x}$. This is always possible in practice. If the state is not a-priori in sensor coordinates, it can be brought there by an appropriate transformation. Then the $\underline{K}$ matrix in (2) takes the form

$$\underline{K} = \left[\mathrm{diag}(k_i),\ \underline{0}_{q,n-q}\right] \ ; \qquad i = 1, \ldots, q \ . \tag{3}$$

In general, the process under consideration may be part of a control system, i.e., be controlled by a dynamic controller

$$\dot{\underline{q}} = \underline{F}\,\underline{q} + \underline{E}\,\underline{y} + \underline{G}\,\underline{\theta} \tag{4}$$

$$\underline{u} = \underline{Z}\,\underline{q} + \underline{H}\,\underline{y} + \underline{J}\,\underline{\theta} \tag{5}$$

(with $\underline{q} \in R^i$, and $\underline{\theta} \in R^l$ the reference input) or any other possibly nonlinear or discrete control.

It can easily be seen that the mathematical model (1),(2) with the assumptions made, contains all the existing relations among the measured variables y_i (i.e., the "redundancy relations") in form of differential equations. To reconstruct the states or measured variables from other measured variables and inputs one can use linear or nonlinear full- or reduced-order state observers (in the deterministic case) or Kalman filters (when noise has to be considered).

The state equations of the *linear full-order estimator* for (1),(2) are

$$\dot{\hat{\underline{x}}} = \underline{A}\,\hat{\underline{x}} + \underline{B}\,\underline{u} + \underline{L}\,(\underline{y} - \underline{K}\,\hat{\underline{x}}) \tag{6}$$

$$\hat{\underline{y}} = \underline{K}\,\hat{\underline{x}} \ . \tag{7}$$

where $\hat{\underline{x}}$ and $\hat{\underline{y}}$ denote the estimated state or output vector, resp., and $\underline{L}$ the gain matrix by the choise of which desired dynamics of the estimator can be achieved.

Changes in the process, $\underline{\gamma}(t)$, and system noise $\underline{\nu}(t)$ can be modeled in the system state equation by

$$\dot{\underline{x}}(t) = \underline{A}\,\underline{x}(t) + \underline{B}\,\underline{u}(t) + \underline{\gamma}(t) + \underline{\nu}(t) \tag{8}$$

and sensor faults, $\Delta\underline{K}(t)$, and measurement noise, $\underline{\mu}(t)$, can be modeled in the output equation by

$$\underline{y}(t) = \underline{K}\,\underline{x}(t) + \Delta\underline{K}(t) + \underline{\mu}(t). \tag{9}$$

Substracting (6) from (8) and defining the state estimation error $\underline{\varepsilon} = \underline{x} - \hat{\underline{x}}$, the equations for the output estimation error, $\underline{e} = \underline{y} - \hat{\underline{y}}$, become

$$\dot{\underline{\varepsilon}} = (\underline{A} - \underline{L}\,\underline{K})\,\underline{\varepsilon} + \underline{v} + \underline{\nu} \tag{10}$$

$$\underline{e} = \underline{K}\,\underline{\varepsilon} + \Delta\,\underline{K} + \underline{\mu}\,. \tag{11}$$

It can be seen that if all the disturbances $\underline{v}$, ν, $\Delta\underline{K}$, $\underline{\mu}$ are zero and the matrix $\underline{A}$-$\underline{LK}$ has leftside eigenvalues the estimation error $\underline{e}$ goes to zero after any initial condition tracking error has died out. If, however, at least one of these disturbances occurs, $\underline{e}$ is driven by it. Thus its effect on $\underline{e}$ can be evaluated to discover the disturbance. Therefore $\underline{e}$ can be used as a residual for FDI.

In order to diminish the cost of implementation one often uses *reduced order (Luenberger) observers*. A typical FDI task that requires a reduced-order observer is to use one of the measured variables in (1),(2), say y_i to estimate the rest, $\underline{w}_i$ of $\underline{y}$. Let $k_i = 1$ and the $\underline{K}$ matrix be rearranged such that the i-th row takes the form

$$\underline{k}_i^T = [1, 0, \ldots, 0]\,. \tag{12}$$

Then y_i is simply given by

$$y_i = \underline{k}_i^T\,\underline{x}_i\,. \tag{13}$$

With the partition

$$\underline{x}_i^T = [y_i \,\vdots\, \underline{w}_i^T] \tag{14}$$

and appropriately defind $\underline{A}$ and $\underline{B}$ matrices the state equations of the system (1),(2) can be rewritten as

$$\dot{y}_i = A_{11i} y_i + \underline{A}_{12i}\underline{w}_i + \underline{B}_{1i}\underline{u} \tag{15}$$

$$\dot{\underline{w}}_i = \underline{A}_{21i} y_i + \underline{A}_{22i}\underline{w}_i + \underline{B}_{2i}\underline{u}\,. \tag{16}$$

Dropping the subscript i in the $\underline{A}$ and $\underline{B}$ matrices for ease of notation, the corresponding state equations of the n-1 order observer driven by y_i are given by

$$\underline{z}_i = \underline{A}_b\underline{z}_i + \underline{Q}_b\, y_i + \underline{B}_b\,\underline{u} \tag{17}$$

where

$$\underline{A}_b = \underline{A}_{22} - \underline{L}_i\underline{A}_{12}$$

$$\underline{Q}_b = \underline{A}_{22}\underline{L}_i - \underline{L}_i\underline{A}_{12}\underline{L}_i + \underline{A}_{21} - \underline{L}_i A_{11}$$

$$\underline{B}_b = \underline{B}_2 - \underline{L}_i\underline{B}_1\,,$$

and

$$\hat{\underline{w}}_i = \underline{z}_i + \underline{L}_i \, \underline{y}_i \; . \qquad (18)$$

$\underline{L}_i$ can be chosen such that $\underline{A}_b$ has prescribed eigenvalues. Note that in the case of application to FDI it is not required that the eigenvalues are placed left of those of the system. The estimated output vector $\hat{\underline{y}}$ of the reduced order observer becomes

$$\hat{\underline{y}} = K T_i^{-1} \left[\frac{\underline{y}_i}{\underline{z}_i + \underline{L}_i \underline{y}_i} \right] \; . \qquad (19)$$

As a residual for FDI the output estimation error $\underline{e} = \underline{y} - \hat{\underline{y}}$ can be used. It can be calculated from the above relations.

If FDI is to be performed in a *nonlinear process*, given by

$$\dot{\underline{x}} = \underline{f}(\underline{x},u), \qquad \underline{x}(0) = \underline{x}_0 \qquad (20)$$

$$\underline{y} = \underline{c}(\underline{x},u) \qquad (21)$$

where $\underline{x} \in R^n$, $\underline{u} \in R^p$, $\underline{y} \in R^q$ are the state, input and output vector, resp., and $\underline{f}$, $\underline{c}$ nonlinear partially differentiable functions, a *nonlinear estimator* is required that takes in general the form

$$\dot{\hat{\underline{x}}} = \hat{\underline{f}}(\hat{\underline{x}},\underline{u},\underline{y}) \; , \qquad \hat{\underline{x}}(0) = \hat{\underline{x}}_0 \; . \qquad (22)$$

Using (20) and (22) the vector differential equation for the state error $\underline{\epsilon} = \underline{x} - \hat{\underline{x}}$ becomes

$$\dot{\underline{\epsilon}} = \underline{f}(\underline{x},\underline{u}) - \hat{\underline{f}}(\hat{\underline{x}},\underline{u},\underline{y}) \; . \qquad (23)$$

To achieve asymptotic stability, i.e.

$$\lim_{t\to\infty} \underline{\epsilon}(t) = \underline{0} \qquad \forall \, \underline{u}_0 \; , \; \forall \, \underline{x}_0 \qquad (24)$$

$$||\epsilon(t)|| \leqslant \delta \qquad \forall \, t \geqslant 0, \; \delta < +\infty \qquad (25)$$

we linearize $\hat{\underline{f}}$ and $\underline{f}$ around the estimated state $\hat{\underline{x}}(t)$. This yields

$$\dot{\underline{\epsilon}} = \underline{f}(\hat{\underline{x}},\underline{u}) + \frac{\partial \underline{f}}{\partial \underline{x}} \Big|_{\hat{\underline{x}}} \underline{\epsilon} + \ldots - \hat{\underline{f}}(\hat{\underline{x}},\underline{u},\hat{\underline{y}}) - \frac{\partial \hat{\underline{f}}}{\partial \underline{y}} \frac{\partial \underline{c}}{\partial \underline{x}} \Big|_{\hat{\underline{x}}} \underline{\epsilon} - \ldots \qquad (26)$$

Choosing $\hat{\underline{f}}(\hat{\underline{x}},\underline{u},\hat{\underline{y}}) = \underline{f}(\hat{\underline{x}},\underline{u})$ and neglecting higher order terms in $\underline{\epsilon}$ the following linear error differential equation is achieved

$$\dot{\underline{\epsilon}}(t) = \left[\frac{\partial \underline{f}}{\partial \underline{x}} - H(x,u) \frac{\partial \underline{c}}{\partial \underline{x}} \right]_{\hat{\underline{x}},\underline{u}} \underline{\epsilon}(t) = \underline{F}(t) \, \underline{\epsilon}(t) \qquad (27)$$

where $\underline{H}(\hat{\underline{x}},\underline{u}) = \frac{\partial \hat{\underline{f}}}{\partial \underline{y}}\Big|_{\hat{\underline{x}},\underline{u}}$ is a time variant (nxq) feedback gain matrix.

Hence the estimator equations become

$$\dot{\hat{\underline{x}}} = \underline{f}(\hat{\underline{x}},\underline{u}) + \underline{H}(\hat{\underline{x}},u)\,(\underline{y} - \hat{\underline{y}})\ ,\quad \hat{\underline{x}}(0) = \hat{\underline{x}}_0 \tag{28}$$

$$\hat{\underline{y}} = \underline{c}(\hat{\underline{x}},\underline{u})\ . \tag{29}$$

If system noise $\underline{v}(t)$ and parameter variations $\Delta\underline{f}(t)$ occur in the plant (20), the state error equation becomes

$$\dot{\underline{\epsilon}} = \underline{F}(t)\ \underline{\epsilon}(t) + \Delta\underline{f}(t) + \underline{v}(t) \tag{30}$$

from which the output $\underline{e}(t) = \underline{y} - \hat{\underline{y}}$ can be calculated. If measurement noise $\underline{\mu}(t)$ and sensor faults $\Delta\underline{K}(t)$ have to be considered, one obtains

$$\underline{e}(t) = \underline{c}(\underline{x},\underline{u}) - \underline{c}(\hat{\underline{x}},\underline{u}) + \Delta\underline{K}(t) + \underline{\mu}(t) \tag{31}$$

After a linear approximation of $\underline{e}(t)$, the dependence of $\underline{e}$ on $\underline{\epsilon}$ becomes apparant:

$$\underline{e}(t) = \frac{\partial \underline{c}}{\partial \underline{x}}\Big|_{\hat{\underline{x}},\underline{u}}\ \underline{\epsilon}(t) + \Delta\underline{K} + u\ . \tag{32}$$

The procedure of generation of the residual $\underline{e}(t)$ using a nonlinear estimator is illustrated in Fig.2

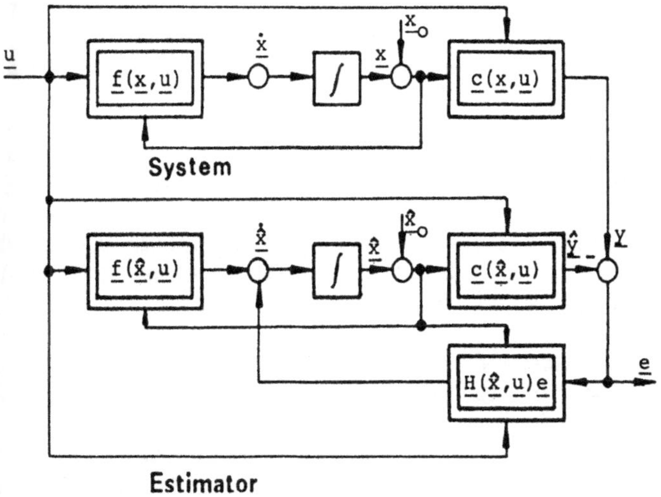

Fig.2: Residual generation by nonlinear state estimation

In many practical situations satisfactory results can be achieved by using a *simplified* version of (28),(29) by replacing the time-variant gain matrix $\underline{H}(\hat{\underline{x}},\underline{u})$ by a constant gain matrix as in the linear case. The structure of the estimator is then the same as in the linear case except that the process model in the estimator is replaced by the actual nonlinear process model.

In order to discuss the different types of estimator schemes used in the FDI system of Fig.1 for the sake of proper fault isolation according to Fig. 1 it is expedient to make a destinction between IFD and CFD.

3.2 Estimator Schemes for IFD

The goal of IFD is to detect and locate faults of the *sensors* of the process with sufficient robustness to system parameter changes and noise. There are two extremes of application, depending on the *purpose* of failure detection and the *type* of failures. One extreme is the detecting of *hard-failures* as, for example, in safety relevant systems. Typically, the failures are then large and the admissible detection time is small (e.g. seconds or fractions of seconds). The other extreme is the detecting of *soft failures* as in the case of *signal validation*, such as repeated (periodic) checking of instrumentation in a nuclear reactor. In this case the failures to be discovered are typically small (a few percent) but a large detection time is available (days or weeks). Hence, in the first case, one can get along with the deterministic approach using observers since noise (and parameter uncertainties) do not play a large role. In the second case noise cannot be neglected and hence Kalman filters and stochastic decision procedures including correlation techniques are indicated. In a concrete situation one has to decide for either (or, in poorly conditioned situations, for non) of the two approaches.

Over the years a number of different estimator schemes for IFD have been proposed in the literature. They differ in the number of estimators and the number of driving measured variables. The question of using observers or Kalman filters is not a primary matter of concern in this context. In most cases the one or the other can be used depending on the noise and failure situation of the process.

In the *dedicated observer scheme (DOS)*, Fig. 3, it is assumed that the full state or a subset of it can be observed from each measured variable of the process. Each sensor output, y_i, is used to drive a dedicated observer or Kalman filter of full or reduced order to observe as many of the remaining measured variables as possible. If the process is completely observable, a q-fold redundancy of the measurement vectors is achieved. In the no-fault case,

$$\hat{y}_{ik} = \hat{y}_{ij} = y_i \tag{33}$$

holds for all $i = 1,2,\ldots,n$ and $k,j = 1,2,\ldots,q$. If a fault occurs in

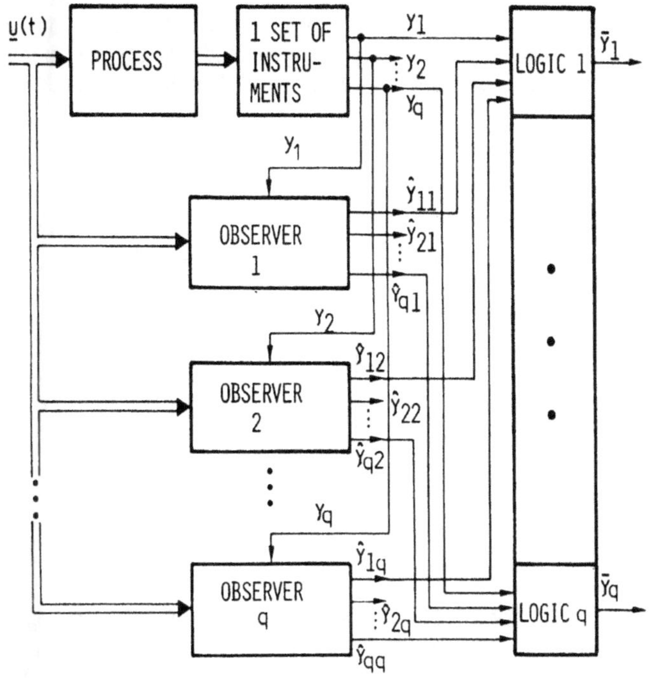

Fig. 3: Dedicated observer scheme (DOS)

the m-th instrument then

$$\hat{y}_{im} \neq \hat{y}_{ik} = y_i \tag{34}$$

holds for $i = 1,2,...,n$ and $k = 1,2,...,q$, and $k \neq m$, i.e., the observer that is driven by the faulty instrument provides a wrong estimate of all measured variables whereas the estimates of all the other observers match the corresponding y_i except that of the erroneous instrument. The output estimation error

$$\underline{e}_k = \underline{y} - \hat{\underline{y}}_k \tag{35}$$

represents now the residual vector which, in principle, allows a unique FDI even in the case of several faulty instruments at the same time.

The procedure is also applicable if only subsets of the state vectors $\underline{x}$ (or of the vector of measured variables, $\underline{y}$) are observable from the measured variables y_i. This, however, means a due loss of usable redundancy, i.e. of reliability of fault detection.

The most simple version of an observer scheme for IFD is the so-called *simplified observer scheme (SOS)* which includes only a single (full or reduced order) observer or Kalman filter driven by only one of the measured variables, see. Fig. 4 (Clark (1978)). As the driving variable one selects the most appropriate sensor output that provides full state (and output) estimation. Though only single redundancy is used this scheme allows a unique isolation of a single faulty sensor. If

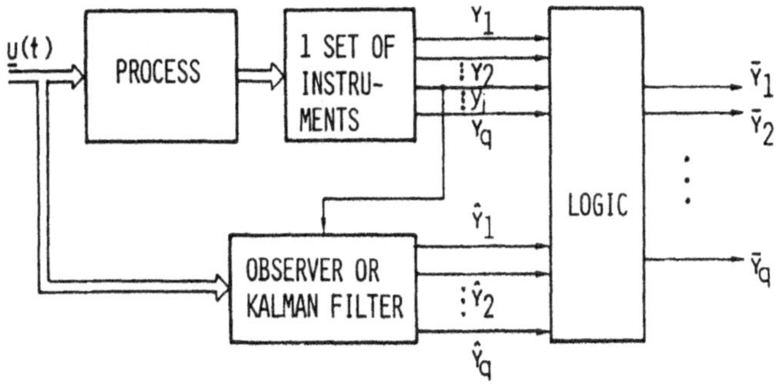

Fig. 4: Simplified observer scheme SOS

one of the sensors, m, that does not drive the observer, fails, the reconstructed output vector is correct and $y_i - \hat{y}_i = 0$ holds for all i except i = m. If the sensor s that drives the observer fails all estimates are erroneous so that $y_i - \hat{y}_i \neq 0$ for all i except i = s. An advantage of this scheme is that it applies to processes with very limited observability and that it is easy to implement. However, the price to pay is reduced redundancy and a loss of detection reliability.

At the other end, if we generalize DOS in that we drive each observer with more than one measured variable, say with overlapping subsets of $\underline{y}$, we arrive at the *generalized observer scheme GOS*, see Fig. 4. It is evident that the decision logic is of the same type as in the case of DOS or SOS. However, this scheme provides more degrees of freedom for the observer design and we shall see later that they can be used for the increase of robustness to parameter variations.

In the wide field of statistical methods for residual evaluation such as the standard tests of whiteness, mean and covariance or multiple hypothesis tests, two types of observer schemes are important. As outlined in the preceding section, the hypothesis test methods use a bank of Kalman filters. The resulting configuration of IFD scheme is then similar to the above described GOS except that an additional filter for the no-failure case has to be included. All the filters are driven by the complete measurement vector diminished by one component in the following way. In the i-th filter designed for the hypothesis of a failure in the i-th sensor, the i-th component y_i is deleted, i.e. $\underline{y}_i$ in Fig. 5 has the form

$$\underline{y}_i = [y_1 \dots y_{i-1} \; y_{i+1} \dots y_q]^T .$$

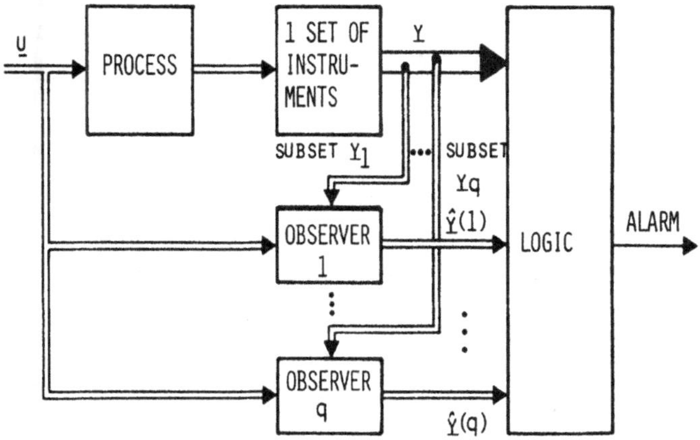

Fig. 5: Generalized observer scheme (GOS)

On the other hand, if the decision making process is based on a test of covariance combined with correlation, Belkoura (1983), then a *single Kalman filter* driven by the complete measurement vector is sufficient. In this case a failure in the i-th instrument causes a due change in the i-th row and i-th column of the time-shifted covariance matrix.

It will be seen later that it is reasonable from the point of view of insensitivity to system parameter variations to use as residuals the *differences* of estimation errors rather than the estimation errors themselves.

3.3 Estimator Schemes for CFD

The goal of CFD is to detect and locate faults of components in dynamic processes with sufficient robustness to sensor faults, system parameter changes and noise. While the IFD schemes described in section 3.2 are also suitable for the alarm task of CFD they are generally not able to solve the problem of *location* of faulty components. In this section we discuss observer schemes that allow to detect *and* locate component faults.

The most natural approach to the component fault detection and isolation problem is to decompose the system and apply a hierarchical scheme of local observers. One of the advantages of a local observer scheme is that even if the order of the overall system may be rather high, the order of each subsystem and consequently the order of the corresponding observers may be rather low. Furthermore, only local observability is required,

The problematic of this approach lies, however, in the interaction between the subsystems. If they are low (or zero) a malfunction in any of the components affects only the estimate of the corresponding local observer. It is thus possible to identify the faulty component uniquely. However, if the interactions are large and not measurable, a fault in one component propagates to observers of other components. The observer scheme would then fail to identify the faulty component. What it could achieve is to identify a fault in *the complex of components* that are largely interconnected. This is the basis for the hierarchical observer scheme introduced by Janssen and Frank (1984, 1985). The block diagram of the local observer scheme is shown in Fig.6.

The *hierarchical observer scheme*, HOS, (1984, 1985) provides to devide the overall system into two levels of components: an upper level including all components with measurable couplings and a lower level of components with unknown couplings. For each of the resulting configurations in both levels we design a scheme of local observers which includes the reconstruction of the measured outputs of the components. The observer scheme used for the upper level is called ASCOS (Available-State Coupled Observer Scheme) and the one for the lower level ESCOS (Estimated-State Coupled Observer Scheme). The difference between ASCOS and ESCOS is the way of performing the couplings in the observer part of Fig.6.

For a brief development of the main idea of ASCOS and ESCOS let us consider a system represented by its component state equations

$$\dot{\underline{x}}_i(t) = \underline{A}_{ii}\,\underline{x}_i(t) + \sum_{\substack{j=i\\ j\neq i}}^{N} \underline{A}_{ij}\,\underline{x}_j + \underline{B}_i\,\underline{u}_i \tag{36}$$

$$\underline{y}_i(t) = \underline{C}_i\,\underline{x}_i(t) \tag{37}$$

where $\underline{x}_i$, $\underline{y}_i$, $\underline{u}_i$ are the states, outputs and inputs, resp., of the i-th component and $\underline{A}_{ii}$, $\underline{A}_{ij}$, $\underline{B}_i$, $\underline{C}_i$ are matrices of appropriate dimensions. Note that $\underline{A}_{ij}$ characterizes the coupling gains with other components.

To derive ASCOS we assume that the coupling terms $\Sigma\,\underline{A}_{ij}\,\underline{x}_j$ in (36) are measurable. The state equations of the corresponding i-th observer is then given by

$$\dot{\hat{\underline{x}}}_i(t) = (\underline{A}_{ii}-\underline{H}_i\,\underline{C}_i)\hat{\underline{x}}_i(t) + \sum_{\substack{j=1\\ j\neq i}}^{N} \underline{A}_{ij}\,\underline{x}_j(t) + \underline{B}_i\,\underline{u}_i(t) + \underline{H}_i\,\underline{y}_i(t) \tag{38}$$

$$\hat{\underline{y}}_i(t) = \underline{C}_i\,\hat{\underline{x}}_i(t) \quad , \tag{39}$$

where $\underline{x}_i$, $\underline{x}_j$, $\underline{u}_i$ are measured variables and N is the total of components of the upper level. In the nominal case the state estimation

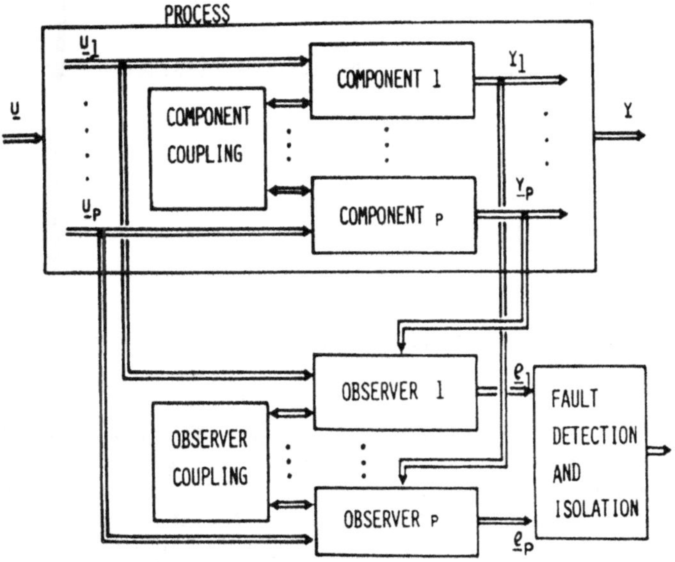

Fig. 6: Local observer scheme for component fault detection and location

error $\underline{\varepsilon}_i = \underline{x}_i - \hat{\underline{x}}_i$ obeys the homogenious vector differential equation

$$\dot{\underline{\varepsilon}}_i(t) = (\underline{A}_{ii} - \underline{H}_i \underline{C}_i)\underline{\varepsilon}_i(t) + \Delta\underline{f}_i + \underline{\nu}_i \tag{40}$$

where $\Delta\underline{f}_i$ and $\underline{\nu}_i$ denote parameter variations and component faults, resp., of the i-th component. The corresponding output estimation error $\underline{e}_i = \underline{y}_i - \hat{\underline{y}}_i$ of the i-th component then becomes

$$\underline{e}_i(t) = \underline{C}_i \underline{\varepsilon}_i(t) + \underline{N}_i \underline{n}_i + \underline{\mu}_i \tag{41}$$

where $\underline{n}_i$ and $\underline{\mu}_i$ denote sensor faults and measurement noise, resp.. Note that in the ideal case that all coupling terms are measurable, all state and output estimation errors are perfectly decoupled from each other, and $\Delta\underline{f}_i$, $\underline{\nu}_i$, $\underline{n}_i$, $\underline{\mu}_i$ only affect the error equations of the i-th compoment. This implies that a system fault in the k-th component only affects $\underline{e}_k(t)$ leaving the remaining errors $\underline{e}_i(t)$ with $i \neq k$ unchanged. This allows a unique failure isolation.

Deriving ESCOS we realize that in the second level the coupling terms $\Sigma\underline{A}_{ij}\,\underline{x}_j(t)$ are not measurable. The idea is to replace them in the observer scheme by their estimates and estimation errors which both are available from other observers. Then the state equation of the corresponding observer of the i-th component becomes

$$\dot{\hat{\underline{x}}}_i(t) = (\underline{A}_{ii} - \underline{H}_i \underline{C}_i)\hat{\underline{x}}_i(t) + \sum_{\substack{j=1 \\ j\neq i}}^{N} \Big[\underline{A}_{ij}\,\hat{\underline{x}}_j(t) +$$

$$+ \underline{H}_{ij}\,\underline{e}_j(t)\big] + \underline{B}_i\,\underline{u}_i(t) + \underline{H}_i\,\underline{y}_i(t) \quad . \tag{42}$$

This leads to the following equation for the state estimation error

$$\dot{\underline{\varepsilon}}_i(t) = (\underline{A}_{ii} - \underline{H}_i\,\underline{C}_i)\;\underline{\varepsilon}_i(t) + \sum_{\substack{j=1\\ j\neq i}}^{N} (\underline{A}_{ij} - \underline{H}_{ij}\underline{C}_j)\underline{e}_j(t) \; . \tag{43}$$

From (43) the output estimation error $\underline{e}_i$ can again be determined in the same way as above.

It is easily seen from (43) that the i-th estimation error, $\underline{e}_i$, is influenced by all other errors, $\underline{e}_k$. This implies that the estimation errors are coupled and hence the uniqueness of failure localization is lost. What can be done to restore the property of failure location is to increase the thresholds in the error evaluation logic (thresholds are anyway required because of the parameter uncertainties and the noise). However, the price to pay is a due reduction of the detection sensitivity and speed. Alternatively, in many cases the coupling terms can be made neglegibly small by a proper choice of $\underline{H}_{ij}$. This can clearly not be done without consideration of the stability of the overall system which of course complicates the design of the observer scheme considerably. The method of finding "optimal" solutions for the $\underline{H}_{ij}$ with respect to stability, Siljak (1978), is not appropriate in the present situation because it does not actually minimize the coupling terms $(\underline{A}_{ij}-\underline{H}_{ij}\underline{C}_j)$. As a remedy a systematic minimization of the coupling matrices in the sense of an Euklidian norm can be performed. This yields

$$\underline{H}_{ij} = \underline{A}_{ij}\;\underline{C}_j^T\;(\underline{C}_j\;\underline{C}_j^T)^{-1} \tag{44}$$

where $\underline{C}_j^T(\underline{C}_j\underline{C}_j^T)^{-1}$ is the pseudo-inverse of $\underline{C}_j$. Since this procedure does not include stability considerations the stability of the overall system has to be checked in addition. If the stability is violated, $\underline{H}_{ij}$ has to be changed until a satisfactory stability is attained. Due to the complexity of this task and since an analytical method is not available yet, we suggest that this should be done by CAD.

Alternative methods to degrade the interactions among the components of $\underline{e}_i$ are to utilize local observers designed for *unknown inputs* or local observers that include *coupling observers*. No work on this subject in conjunction with FDI is known so far.

In the view of practial applications one should realize that physical systems often contain *nonlinear components*. As long as the nonlinearities are weak enough the above derived local observer schemes can be successfully applied after a linearization of the component model. The crux of such a procedure is, however, that we now run into the problem of increased parameter variations due to changes of the operating points of the process. This is particularly true in the case of *essential nonlinearities*. It is then advantageous to use *nonlinear versions* of ASCOS and ESCOS (Frank and Hengy (1986)).

Let us briefly outline how to derive the estimation error and observer equations for nonlinear ASCOS and ESCOS. To start with ASCOS, consider the state equations of the i-th component to be given by

$$\dot{\underline{x}}_i = \underline{f}_i(\underline{x}_i, \bar{\underline{x}}_i, \underline{u}_i) \tag{45}$$

$$\underline{y}_i = \underline{c}_i(\underline{x}_i, \underline{u}_i) \tag{46}$$

where $i = 1,2 \ldots, N$, and N is the number of components; $\bar{\underline{x}}_i$ denotes the coupling state vector of the i-th component which is assumed to be measurable and can therefore be considered as an additional input vector. Utilizing (28),(29) the estimator equations for the i-th component can be written as

$$\dot{\hat{\underline{x}}}_i = \underline{f}_i(\hat{\underline{x}}_i, \bar{\underline{x}}_i, \underline{u}_i) + H_i(\hat{\underline{x}}_i, \bar{\underline{x}}_i, \underline{u}_i)\,(\underline{y}_i - \hat{\underline{y}}_i) \tag{47}$$

$$\hat{\underline{y}}_i = \underline{c}_i(\hat{\underline{x}}_i, \underline{u}_i)\ . \tag{48}$$

Following the derivations in section 3.1 the corresponding estimation error differential equation becomes

$$\dot{\underline{\varepsilon}}_i = F_i(\hat{\underline{x}}_i, \bar{\underline{x}}_i, \underline{u}_i)\underline{\varepsilon}_i + \Delta\underline{f}_i + \underline{\nu}_i \tag{49}$$

with

$$F_i(\hat{\underline{x}}_i, \bar{\underline{x}}_i, \underline{u}_i) = \left[\frac{\partial \underline{f}_i}{\partial \underline{x}_i} - H_i \frac{\partial \underline{c}_i}{\partial \underline{x}_i}\right]_{\hat{\underline{x}}_i, \bar{\underline{x}}_i, \underline{u}_i} \ . \tag{50}$$

One can see that as in the linear case, the estimation errors $\underline{\varepsilon}_i$ are perfectly decoupled and the parameter variations $\Delta\underline{f}_i$ as well as component faults $\underline{\nu}_i$ only affect the i-th error differential equation. The same is then true for the output estimation errors

$$\underline{e}_i = \left.\frac{\partial \underline{c}_i}{\partial \underline{x}_i}\right|_{\hat{\underline{x}}_i, \underline{u}_i} \underline{\varepsilon}_i + N_i \underline{n}_i + \underline{\mu}_i \tag{51}$$

where $\underline{n}_i$ and $\underline{\mu}_i$ denote sensor faults and measurement noise, respectively. This means that faults can be detected if the influences of $\Delta\underline{f}_i$, $\underline{n}_i$ and $\underline{\mu}_i$ are sufficiently small.

For *nonlinear ESCOS* the estimation error and observer equations can be derived in a similar way. For the sake of brevity of this presentation we only give the resulting equations and refer for more details to literature (Frank and Hengy (1986)). Suppose the state equations of the i-th component are again given by the pair (36), (37) augmented by

$$\underline{y}_i = \underline{c}_i(\underline{x}_i, \underline{u}_i) \quad . \tag{52}$$

Herein $\bar{\underline{x}}_i$ and $\bar{\underline{y}}_i$ denote, resp., the vectors of states and outputs of the i-th component that are coupled with all the other components and are not measurable. If they are replaced by their estimates the resulting nonlinear observer equations become

$$\dot{\hat{\underline{x}}}_i = \underline{f}_i(\hat{\underline{x}}_i, \hat{\bar{\underline{x}}}_i, \underline{u}_i) + \underline{H}_i(\hat{\underline{x}}_i, \hat{\bar{\underline{x}}}_i, \underline{u}_i)(\underline{y}_i - \hat{\underline{y}}_i) + $$
$$+ \bar{\underline{H}}_i(\hat{\underline{x}}_i, \hat{\bar{\underline{x}}}_i, \underline{u}_i)(\bar{\underline{y}}_i - \hat{\bar{\underline{y}}}_i) \tag{53}$$

$$\hat{\underline{y}}_i = \underline{c}_i(\hat{\underline{x}}_i, \underline{u}_i) \tag{54}$$

$$\hat{\bar{\underline{y}}}_i = \bar{\underline{c}}_i(\hat{\bar{\underline{x}}}_i, \underline{u}_i) \tag{55}$$

with

$$\underline{H}_i(\hat{\underline{x}}_i, \hat{\bar{\underline{x}}}_i, \underline{u}_i) = \left.\frac{\partial \hat{\underline{f}}_i}{\partial \underline{y}_i}\right|_{\hat{\underline{x}}_i, \hat{\bar{\underline{x}}}_i, \underline{u}_i} \tag{56}$$

$$\bar{\underline{H}}_i(\hat{\underline{x}}_i, \hat{\bar{\underline{x}}}_i, \underline{u}_i) = \left.\frac{\partial \hat{\underline{f}}_i}{\partial \bar{\underline{y}}_i}\right|_{\hat{\underline{x}}_i, \hat{\bar{\underline{x}}}_i, \underline{u}_i} \quad . \tag{57}$$

The linearized state estimation error equation becomes

$$\dot{\underline{\varepsilon}}_i = \underline{F}_{ii}\, \underline{\varepsilon}_i + \bar{\underline{F}}_{ii}\, \bar{\underline{\varepsilon}}_i + \Delta \underline{f}_i + \underline{v}_i \tag{58}$$

with $\underline{F}_{ii}$ as in (50) and

$$\bar{\underline{F}}_{ii} = \left[\frac{\partial \underline{f}_i}{\partial \bar{\underline{x}}_i} - \bar{\underline{H}}_i \frac{\partial \underline{c}_i}{\partial \bar{\underline{x}}_i}\right]_{\hat{\underline{x}}_i, \hat{\bar{\underline{x}}}_i, \underline{u}_i} \quad . \tag{59}$$

Clearly, the stability problem associated with the minimization of $\bar{\underline{F}}_{ii}\bar{\underline{\varepsilon}}_i$ is much more complicated than in the linear case, and no analytical solutions are so far known.

In many situations the system under consideration has a hierarchical structure so that its state space model

$$\dot{\underline{x}}(t) = \underline{A}\, \underline{x}(t) + \underline{B}\, \underline{v}(t) \tag{60}$$

$$\underline{y}(t) = \underline{C}\, \underline{x}(t) + \underline{\mu}(t) \tag{61}$$

can be transformed into a lower block triangular form

$$\underline{A} = \begin{bmatrix} a_{11} & & 0 \\ \vdots & \ddots & \\ a_{s1} & \cdots & a_{ss} \end{bmatrix}, \quad \underline{B} = \begin{bmatrix} b_{11} & & 0 \\ \vdots & \ddots & \\ b_{s1} & \cdots & b_{ss} \end{bmatrix}, \quad \underline{C} = \begin{bmatrix} c_{11} & & 0 \\ \vdots & \ddots & \\ c_{s1} & \cdots & c_{ss} \end{bmatrix}.$$

A system with such a well conditioned structure allows a unique failure detection and localization with the aid of a hierarchical estimator scheme, Hodzic and Siljak (1985). The estimators can be designed block by block so that their innovations carry hierarchically structured fault information. In other words, whenever a fault occurs in a certain component, this affects only the innovations on the lower level whereas the innovations in the higher level remain untouched. The change of innovations can, e.g. be evaluated with the aid of statistical tests such as mean, whiteness and covariance. The method is restricted in that it is not applicable to general patterns that do not allow the assumed triangulations of $\underline{A}$, $\underline{B}$ and $\underline{C}$ matrices.

4. DECISION LOGIC

The evaluation of the residual for detection and isolation of a fault is a problem of decision theory. Several kinds of strategies have been developed over the years.

Threshold Logic. A most simple and quite common decision strategy is to form a *decision function* in terms of the residuals and compare its value with a predetermined *threshold*. If the decision function of a certain instrument remains below the threshold the corresponding instrument is declared unfaulty, if the decision function surpasses the threshold (perhaps for longer than a defined minimum period) the instrument is declared faulty and an alarm is released. To make the supervision independent of the signals in the process the thresholds should actually be made adaptive to the signals (i.e. the inputs). Decision functions can be defined in many ways. Two simple examples are the additive and multiplicative decision function.

Additive decision functions superimpose the residuals linearly. Consider for example two (possibly partial) state vectors, $\hat{\underline{x}}_i$, $\hat{\underline{x}}_j$ of the i-th and j-th observer of DOS, respectly. Then the innovation-based residual vector can be chosen as

$$\underline{\epsilon}_{ij} = \begin{bmatrix} \epsilon_{ij1} \\ \vdots \\ \epsilon_{ijp} \end{bmatrix} \tag{62}$$

where the $\epsilon_{ij\nu}$ are the differences between of the p components of $\hat{\underline{x}}_i$ and $\hat{\underline{x}}_j$ that are common to both vectors. As a decision function we can now define the scalar

$$|\epsilon_{ij}| \triangleq \sum_{\nu=1}^{p} |\epsilon_{ij\nu}| \quad . \tag{63}$$

Furthermore, we define a threshold τ and a binary function K_{ij} (truth function) such that

$$K_{ij} = \begin{cases} 1 & \text{if } |\epsilon_{ij}| \geqslant \tau_{ij} \\ 0 & \text{if } |\epsilon_{ij}| < \tau_{ij} \end{cases} \tag{64}$$

The final decision on the occurance of a failure can now be made by entering a truth table for K_{ij}. It should be noticed that in the absence of noise and parameter variations the thresholds could be zero. In practice, however, the value of the threshold has to be increased due to the effects of noise and parameter variations in order to keep false alarms in limited range. The choise of thresholds is always a trade-off between tolerable false alarm rate and failure detection sensitivity.

Multiplicative decision functions defined by multiplication of residuals make the decision more effective since the product of small residuals (in the no-fault cases) becomes of higher order small and, vice versa, the effects of faults are enlarged when a fault occurs. Consider, e.g., the estimation error vector of one, say the i-th, observer of DOS, $\underline{e}_i = \underline{y} - \hat{\underline{y}}_i = \underline{e}_{vi}$. One may then define the decision function as

$$\Delta_i = \left|\prod_{\substack{v=1 \\ v\neq i}}^{q} e_{vi}\right| \qquad i = 1,2, \dots ,q \tag{65}$$

or

$$\Phi_i = \left|\prod_{\substack{v=1 \\ v\neq i}}^{q} e_{vi} \cdot \prod_{\substack{v=1 \\ v\neq i}}^{q} e_{iv}\right| \quad . \tag{66}$$

In the first case, all the residual components of observer i are multiplied and thus the effect of a fault in the i-th sensor is accordingly large. In the second case this effect is even enlarged by including all the remaining residuals from other observers that are affected by a failure in the i-th sensor.

More sophisticated approaches utilize *cost functions* (Frank e.a., 1980, and Saif e.a., 1986) or *statistical tests*, for example, of whiteness, mean and covariance. Statistical methods can also be used for the *evaluation* of decision functions. An example is the use of Kullback theory on the basis of which a measure, the so-called 'Kullback Discrimination Information' (KDI) was introduced by Kumamaru e.a.(1985).

Multiple hypotheses testing. This technique has become quite popular for FDI applications though it requires some expenditure of implementation. Normally for each fault that is likely to appear a Kalman filter is designed. This leads to a bank of Kalman filters. The performance of the actual process is then compared with that of the filters. The decision whether a failure has occured or not is then a question of probabilistic decision theory, and a number of techniques have been proposed for this task, for example the Bayesian decision algorithm, generalized likelihood ratio (GLR) test, orthogonal series *generalized likelihood ratio* (OSGLR) test. For details we refer to the literature, see for example Motyka and Hall (1986) or Gray (1985). Hypothesis testing has certainly gained some importance and maturity in FDI. It is clear, however, that this approach is restricted to tasks where only a limited (or even small) number of well defined faults are considered.

Knowledge-based systems.. Fault diagnosis is a typical issue for the application of knowledge-based systems. An expert system may be rather useful for the difficult task of fault isolation. The basic idea is to supervise the residuals and give alarm when information occurs from which a conclusion upon the appearance of a fault can be drawn. The knowledgebased algorithm can, for example, be combined with observers or adaptive filters and the innovations can be used as residuals. The task of the expert system is then to locate the failures (see, e.g. Oe and Tomita, (1983) and Halme and Selkäinaho (1984)). Little work has been done so far in this field.

5. SENSITIVITY TO PARAMETER VARIATIONS

The sensititvity to process parameter variations is one of the key questions for the practical applicability of FDI systems based on analytical redundancy. This is why sensitivity analysis has gained a strong consideration in this issue. On the one hand it may serve as a measure for the assessment of the usability of an FDI scheme in a certain environment, on the other hand it may be a proper tool to determine the critical parameters and find minimally sensitive solutions. A typical application is to select the least parameter sensitive sensor output to drive the observer, e.g., in a simplified observer scheme. As a quantitative measure one can use the sensitivity functions that can be found from the sensitivity equations, Frank (1978). In this section we show how the sensitivity equations for a reduced-order observer as typically used in SOS or DOS can be derived.

Let us consider the controlled process (1),(2) and suppose the $\underline{A},\underline{B},\underline{K}$ matrices are now differentiable functions of a time-invariant parameter vector $\underline{\alpha} \in R^r$ with the nominal value $\underline{\alpha}_0$, i.e., $\underline{A} = \underline{A}(\underline{\alpha})$, $\underline{B} = \underline{B}(\underline{\alpha})$, $\underline{K} = \underline{K}(\underline{\alpha})$. With the definitions $\underline{\lambda}_j = \partial\underline{x}/\partial\alpha_j$, $\underline{n}_j = \partial\underline{u}/\partial\alpha_j$, $\underline{\sigma}_j = \partial\underline{y}/\partial\alpha_j$, $\sigma_{ij} = \partial y_i/\partial\alpha_j$, $\underline{\delta}_j = \partial\underline{z}_i/\partial\alpha_j$, $\hat{\underline{\sigma}}_j = \partial\hat{\underline{y}}/\partial\alpha_j$, $\underline{A}_\alpha = \partial\underline{A}/\partial\alpha_j$ etc., all taken at nominal parameter values $\underline{\alpha}_0$, and $\underline{A}_0 = \underline{A}(0)$ etc., the *sensitivity equations* of the plant and the observer are obtained from (1),(2) and (21),(23),(24) by implicit differentation w. r. t. α_j. As a result,

$$\dot{\underline{\lambda}}_j = \underline{A}_0\underline{\lambda}_j + \underline{A}_\alpha\underline{x}_0 + \underline{B}_\alpha\underline{u}_0 + \underline{B}_0\underline{n}_j \,, \quad \underline{\lambda}_j(0) = \underline{0} \tag{67}$$

$$\underline{\sigma}_j = \underline{K}_\alpha\underline{x}_0 + \underline{K}_0\underline{\lambda}_j \tag{68}$$

$$\dot{\underline{\delta}}_j = \underline{A}_b\underline{\delta}_j + \underline{Q}_b\sigma_{ij} + \underline{B}_b\underline{n}_j \,, \quad \underline{\delta}_j(0) = \underline{0} \tag{69}$$

$$\hat{\underline{\sigma}}_j = \underline{K}_0\underline{T}_i^{-1}\begin{bmatrix} \sigma_{ij} \\ \hline \underline{\delta}_j + \underline{L}_i\sigma_{ij} \end{bmatrix} \tag{70}$$

$$\frac{\partial \underline{e}}{\partial \alpha_j} = \underline{\sigma}_j - \hat{\underline{\sigma}}_j \quad . \tag{71}$$

A set of sensitivity equations for $\underline{n}_j$ in (67),(69) can be determined in a similar way from the state equations of the controller (6),(7). The resulting equations together with (67)-(71) define the overall *sensitivity model* of the state estimation process for the control system (1),(2),(6),(7) using a reduced-order observer (21),(23),(25). It is worth noticing that $\underline{K}_\alpha = \underline{0}$ if the state $\underline{x}$ is defined in sensor coordinates (what is usually done). Then the process parameter vector $\underline{\alpha}$ does not affect the $\underline{K}$ matrix and changes in $\underline{K}$ can simply be interpreted as sensor faults whereas the changes in $\underline{A}$ and $\underline{B}$ represent the process parameter variations. This is an important fact which allows to make a clear destinction between the effect of process parameter variations and sensor faults.

6. PARAMETER SENSITIVITY REDUCTION IN FDI SCHEMES

Because of the decisive influence of the parameter variations upon the feasability of the innovation based FDI methods, there is a strong motivation for attempts to reduce the parameter sensitivity. In this section we outline several approaches to the problem of sensitivity reduction. In contrast to the work of Chow e.a. (1984) and Lou e.a. (1986) who attack this problem by utilizing only the most reliable of the redundancy relations, we adhere to the overall dynamic model of the process because the model contains the complete information about the process that can be exploited for FDI. The key to most of these approaches is the character of the estimation error which will at first be examined more closely for the effects of parameter variations that occur together with sensor faults.

6.1 Effects of Parameter Variations on the Estimation Error

In the presence of process parameter variations $\Delta\underline{\alpha}$, the state equations of the process (1),(2) can be written as

$$\dot{\underline{x}} = (\underline{A}_0 + \Delta\underline{A})\underline{x} + (\underline{B}_0 + \Delta\underline{B})\underline{u} \tag{72}$$

$$\underline{y} = (\underline{K}_0 + \Delta\underline{K})\underline{x} \tag{73}$$

where $\underline{A}_0, \underline{B}_0 \underline{K}_0$ represent the nominal values of $\underline{A}, \underline{B}, \underline{K}$. The matrices $\Delta\underline{A}$ and $\Delta\underline{B}$ model the parameter changes of the process and $\Delta\underline{K}$ the instrument faults. The separation of the influences is due to the representation of $\underline{x}$ in sensor coordinates, (3)-(5), since the matrix $\underline{K} = \underline{K}_0 + \Delta\underline{K}$ can then be characterized by

$$\underline{K}_0 = [\underline{I}, \underline{0}] \tag{74}$$

$$\Delta\underline{K} = [\mathrm{diag}\,(\Delta k_i),\ \underline{0}] \tag{75}$$

where Δk_i represents a fault of the i-th sensor. In the special case of hard-failures we have $\Delta k = -1$

We now consider an observer, say the i-th in DOS, driven by the i-th measured variable, y_i. Let y_i be given by

$$y_i = \underline{k}_i\underline{x} = [0, \ldots, 0, k_i, 0, \ldots, 0]\underline{x} \tag{76}$$

with $k_i = 1 + \Delta k_i$ the actual gain factor of the i-th sensor and Δk_i representing the sensor fault.

In the case of a full-order observer with $\hat{\underline{y}}_i$ its estimated output, the corresponding output estimation error $\underline{e}_i = \underline{y} - \hat{\underline{y}}_i$ becomes

$$\underline{e}_i = \underline{K}_0\underline{\varepsilon}_i + \Delta\underline{K}\underline{x} \tag{77}$$

$$\dot{\underline{\varepsilon}}_i = (\underline{A}_0 - \underline{L}_i)\underline{\varepsilon}_i + (\Delta\underline{A} - \underline{L}_i\Delta\underline{k}_i)\underline{x} + \Delta\underline{B}\underline{u} \tag{78}$$

The block diagram representation of (77),(78) is given in Fig.7

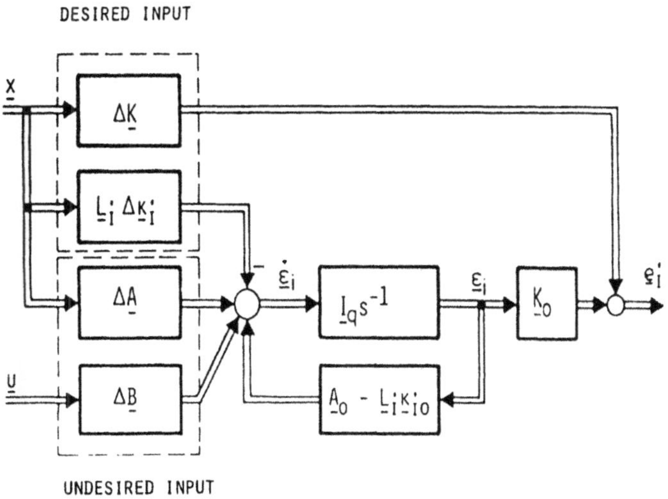

Fig.7 Estimation error of a linear full-order observer

In similar way we obtain the estimation error when a *reduced-order observer* (21),(22) is used according to the observer scheme shown in Fig.8.

In this case the resulting output estimation error $\underline{e}_i = \underline{n} - \hat{\underline{n}}$ becomes (Frank and Keller (1984))

$$\underline{e}_i = \Delta\underline{K}^i\,\underline{w}_i + \underline{K}_0^i\,\underline{\varepsilon}_i \tag{79}$$

$$\dot{\underline{\varepsilon}}_i = \underline{A}_{b0}^i\,\underline{e}_i + [\Delta\underline{Q}^i, \Delta\underline{A}_b^i, \Delta\underline{B}_b^i]\begin{bmatrix}\underline{y}_i\\ \underline{w}_i\\ \underline{u}\end{bmatrix} \tag{80}$$

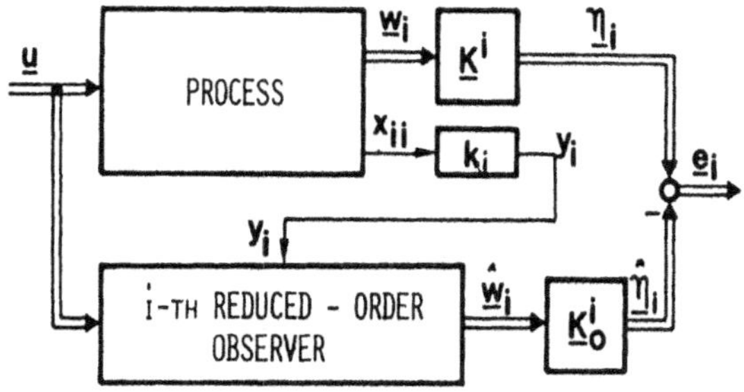

Fig. 8: State observation with a reduced-order observer

where

$$\underline{A}_{b0}^{i} = \underline{A}_{022} - \underline{L}_i\underline{A}_{022}$$

$$\Delta\underline{Q}^{i} = \Delta\underline{A}_{21}^{i} - \underline{L}_i\Delta\underline{A}_{11}^{i} - (\underline{L}_i\Delta\underline{A}_{11}^{i} + \underline{A}_{021}^{i})\Delta k_i$$

$$\Delta\underline{A}_{b}^{i} = \Delta\underline{A}_{22}^{i} - \underline{L}_i\underline{A}_{12}^{i}\Delta k_i + \underline{L}_i\Delta\underline{A}_{12}^{i}$$

$$\Delta\underline{B}_{b}^{i} = \Delta\underline{B}_2 - \underline{L}_i\Delta\underline{B}_1 - \underline{L}_i\underline{B}_1$$

The corresponding block diagram representation is given in Fig.9.

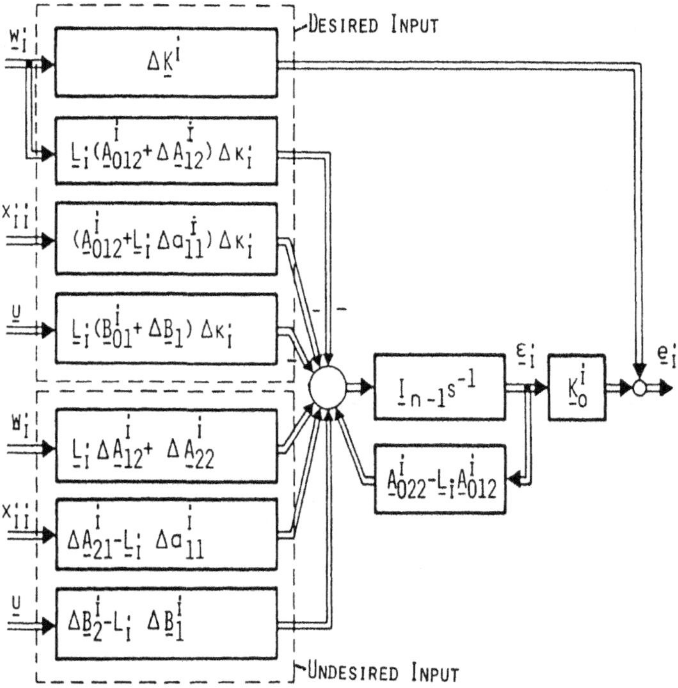

Fig. 9: Estimation error of a linear (n-1) order observer

6.2 Robust Observer Scheme ROS

Consider an observer driven by a single measured variable, as, e.g. used in DOS. The basic idea of ROS is to select the observer gain matrix $\underline{L}_i$ such that the differences of correspondent estimated variables become insensitive to parameter variations $\Delta\underline{A}$ and $\Delta\underline{B}$. This can be achieved in two steps.

(i) Determination of the *critical parameters* of the process, i.e. of the parameters with dominant influence on the system outputs

(ii) Determination of $\underline{L}_i$ so that minimal sensitivity of the estimated outputs with respect to variations of the critical parameters is achieved.

The first step can be done with the aid of the L_2-norm

$$J = \int_{t_0}^{t} \sigma_{ij}^2 dt \tag{81}$$

where

$$\sigma_{ij} = \frac{\partial y_i}{\partial \ln\alpha_j}\Big|_{\underline{\alpha}_0} \tag{82}$$

is the semirelative sensitivity function of the output y_i with respect to the j-th system parameter α_j. σ_{ij} can be found from the solution of the sensitivity equations of the process under consideration. J has to be calculated for all outputs and all parameters of interest. The critical parameters are then those with the largest L_2 norm. The procedure is illustrated in Fig. 10.

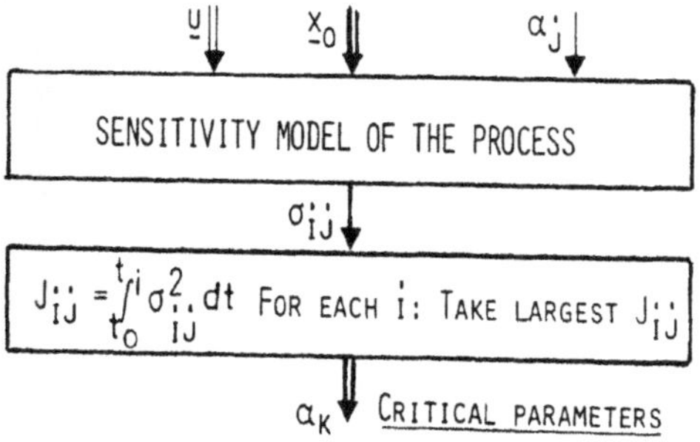

Fig. 10: Procedure for the calculation of the critical parameters

In the second step $\underline{L}_i$ is optimized according to the L_2 norm

$$J_j = \int_{t_0}^{t} \underline{\sigma}_j^T \underline{G} \underline{\sigma}_j dt \tag{83}$$

where

$$\sigma_j = \frac{\partial \hat{\underline{y}}}{\partial \alpha_j}\Big|_{\alpha_0} \tag{84}$$

is the sensitivity function of the estimated output vector of the i-th observer with respect to the most critical parameter, α_j, and $\underline{G}$ a weighting matrix. The flow chart of a digital computer program for the solution of the above minimization problem is shown in Fig. 11.

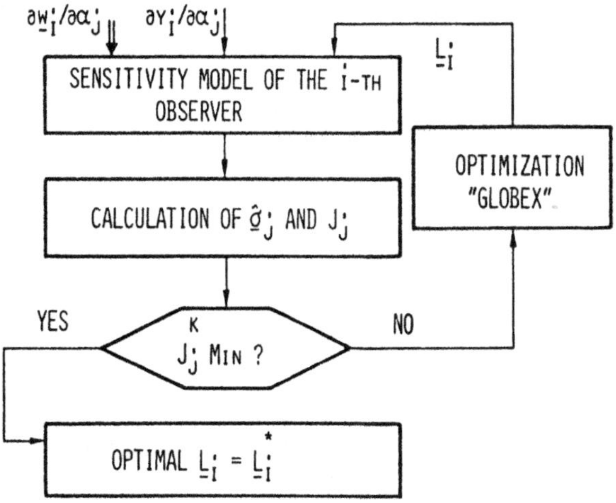

Fig. 11: Flow chart for the determination of $\underline{L}_i$ in ROS

6.3 Differential Error Observer Scheme DEOS

The basic idea of DEOS is to reduce the influences $\Delta\underline{A}^i_{11}$, $\Delta\underline{A}^i_{12}$, $\Delta\underline{B}^i_1$ in (79),(80) by making the norms of $\underline{L}_i$ and $\underline{L}_j$ small, and to reduce the influences of $\Delta\underline{A}^i_{21}$, $\Delta\underline{A}^i_{22}$, $\Delta\underline{B}^i_2$ by using the *differences of observation errors* rather than the errors themselves. Intuively, whenever the effects of parameter variations in two corresponding channels are similar or the same, the difference of estimation errors becomes insensitive to parameter variations. This fact is exploited in DEOS.

To examine the parameter influence on the differential error, let us consider the estimation error equation for the i-th observer (80) and the corresponding equation for the j-th observer. The latter reads

$$\dot{\underline{e}}_j = \underline{A}^j_{b0}\underline{e}_j + \left[\Delta\underline{Q}^j,\ \Delta\underline{A}^j_b,\ \Delta\underline{B}^j_b\right]\begin{bmatrix}\underline{y}_j\\ \underline{w}_j\\ \underline{u}\end{bmatrix} \quad . \tag{85}$$

Substracting (85) from (80) the difference $\underline{e}_{ij} = \underline{e}_i - \underline{e}_j$ becomes

$$\dot{\underline{e}}_{ij} = \underline{A}^i_{b0}\underline{e}_i - \underline{A}^j_{b0}\underline{e}_j + \Delta\underline{A}^i_{22}\underline{w}_i - \Delta\underline{A}^j_{22}\underline{w}_j +$$

$$+ \underline{L}_j \Delta \underline{A}^j_{12} \underline{w}_j - \underline{L}_i \Delta \underline{A}^i_{12} \underline{w}_i +$$
$$+ \Delta \underline{A}^i_{21} y_i - \Delta \underline{A}^j_{21} y_j + \underline{L}_j \Delta \underline{A}^j_{11} y_i - \underline{L}_i \Delta \underline{A}^i_{11} y_i$$
$$+ (\underline{L}_j \Delta \underline{B}^j_1 - \underline{L}_i \Delta \underline{B}^i_1 + \Delta \underline{B}^i_2 - \Delta \underline{B}^j_2)\underline{u} \qquad (86)$$

To become more specific we consider the case i = 1, j = 2. Then

$$\underline{w}^T_1 = [x_2 \; x_3 \; \dots \; x_n] \quad ; \quad \underline{w}^T_2 = [x_1 \; x_3 \; \dots \; x_n] \; ,$$

and $y_1 = x_1$, $y_2 = x_2$ when no fault occurs. Furthermore,

$$\Delta A^1_{11} = \Delta a_{11} \qquad \Delta A^2_{11} = \Delta a_{22}$$
$$\Delta \underline{A}^1_{12} = [\Delta a_{12} \; \Delta a_{23} \; \dots \; \Delta a_{2n}]$$
$$\Delta A^2_{12} = [\Delta a_{21} \; \Delta a_{23} \; \dots \; \Delta a_{2n}]$$
$$\Delta \underline{A}^{T1}_{21} = [\Delta a_{21} \; \Delta a_{31} \; \dots \; \Delta a_{n1}]$$
$$\Delta \underline{A}^{T2}_{21} = [\Delta a_{12} \; \Delta a_{32} \; \dots \; \Delta a_{n2}]$$

$$\Delta \underline{A}^1_{22} = \begin{bmatrix} \Delta a_{22} & \dots & \Delta a_{2n} \\ \cdot & & \cdot \\ \cdot & & \cdot \\ \cdot & & \cdot \\ \Delta a_{n2} & \dots & \Delta a_{nn} \end{bmatrix} ; \quad \Delta \underline{A}^2_{22} = \begin{bmatrix} \Delta a_{11} & \Delta a_{13} & \dots & \Delta a_{1n} \\ \Delta a_{31} & \Delta a_{33} & & \Delta a_{3n} \\ \cdot & & & \cdot \\ \cdot & & & \cdot \\ \Delta a_{n1} & \Delta a_{n3} & \dots & \Delta a_{nn} \end{bmatrix}$$

$$\Delta \underline{B}^1_1 = [\Delta b_{11} \; \dots \; \Delta b_{1p}] ; \quad \Delta \underline{B}^2_1 = [\Delta b_{21} \; \dots \; \Delta b_{2p}]$$

$$\Delta \underline{B}^1_2 = \begin{bmatrix} \Delta b_{21} & \dots & \Delta b_{2p} \\ \cdot & & \cdot \\ \cdot & & \cdot \\ \cdot & & \cdot \\ \Delta_{n1} & \dots & \Delta b_n \end{bmatrix} ; \quad \Delta \underline{B}^2_2 = \begin{bmatrix} \Delta b_{11} & \dots & \Delta b_{1p} \\ \Delta b_{31} & \dots & \Delta b_{3p} \\ \cdot & & \cdot \\ \cdot & & \cdot \\ \Delta b_{n1} & \dots & \Delta b_{np} \end{bmatrix}$$

Substituting for $\underline{e}_i = \underline{e}_{ij} + \underline{e}_j$ in the first terms of (86) with i = 1, j = 2, yields

$$\underline{A}^1_{bo} \, \underline{e}_1 - \underline{A}b^2_o \, \underline{e}_2 = \underline{A}^1_{bo} \, \underline{e}_{12} + (\underline{A}^1_{bo} - \underline{A}^2_{bo})\underline{e}_2 \quad . \qquad (87)$$

Let us first assume that the elements of $\underline{L}_1$ and $\underline{L}_2$ are sufficiently small so that we have in (87)

$$\underline{A}^1_{bo} - \underline{A}^2_{bo} \approx \underline{A}^1_{o22} - \underline{A}^2_{o22} = \begin{bmatrix} a_{22}-a_{11} & a_{23}-a_{13} & \dots & a_{2n}-a_{1n} \\ a_{32}-a_{31} & 0 & \dots & 0 \\ & \vdots & & \vdots \\ a_{n2}-a_{n1} & 0 & \dots & 0 \end{bmatrix} . \qquad (88)$$

We further assume that the differences in (88) are also sufficiently small. Then the first two terms in (86) can be approximated by $A_{b0}^{1}e_{12}$. The term $\Delta A_{22}^{1}w_1 - \Delta A_{22}^{2}w_2$ in (86) yields

$$\begin{bmatrix} \Delta\alpha \\ \Delta a_{32}x_2 - \Delta a_{31}x_1 \\ \vdots \\ \Delta a_{n2}x_2 - \Delta a_{n1}x_1 \end{bmatrix}$$

with $\Delta\alpha = \Delta a_{22}x_2 - \Delta a_{11}x_1 + (\Delta a_{23} - \Delta a_{13})x_3 + \ldots + (\Delta a_{21} - \Delta a_{1n})x_n$. The remaining terms of (86) containing L_i and L_j can be neglected due to the first assumtion. The last term in (86) is

$$(\Delta B_2^1 - \Delta B_2^2)u = \begin{bmatrix} \Delta\beta \\ 0 \\ \vdots \\ 0 \end{bmatrix}$$

with $\Delta\beta = (\Delta_{21} - \Delta b_{11})u_1 \ldots + (\Delta b_{2p} - \Delta b_{1p})u_p$.

With these notations and the two assumptions, (86) can be written as

$$\dot{e}_{12} = A_{b0}^{1}e_{12} + \begin{bmatrix} 1 \\ 0 \\ \vdots \\ 0 \end{bmatrix}\Delta y \tag{89}$$

where

$$\Delta y = \Delta\alpha + \Delta\beta + \Delta a_{21}x_1 - \Delta a_{12}x_2.$$

Including simultaneous measurement faults Δk_1 and Δk_2, the differential error equation is finally given by

$$\dot{e}_{12} = A_{b0}e_{12} + \begin{bmatrix} 1 \\ 0 \\ \vdots \\ 0 \end{bmatrix}\Delta y + \begin{bmatrix} a_{21} \\ a_{31} \\ \vdots \\ a_{n1} \end{bmatrix}\Delta k_1 x_1 - \begin{bmatrix} a_{12} \\ a_{32} \\ \vdots \\ a_{n2} \end{bmatrix}\Delta k_2 x_2 \ . \tag{90}$$

From (90) it becomes clear that undesired inputs in e_{12} are reduced compared to those in e_i or e_j.

The resulting observer scheme DEOS is illustrated in Fig.12.

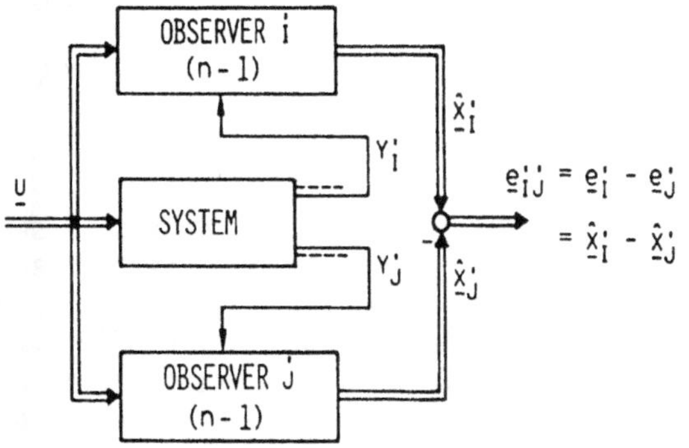

Fig. 12: Differential error observer scheme

6.4 Robust Covariance Matrix Test RCMT

This method provides robust fault detection and isolation with the aid of a single Kalman filter. To evolve the basic idea, consider the system (1),(2) in the presense of sensor faults $\Delta\underline{K}$, parameter variations $\Delta\underline{A}$, $\Delta\underline{B}$, system noise $\underline{v}(t)$ and measurement noise $\underline{w}(t)$. The *actual* system state equation in discrete time is then given by

$$\underline{x}_0(k+1)+\Delta\underline{x}(k+1) = (\underline{A}_0+\Delta\underline{A})\left[\underline{x}_0(k)+\Delta\underline{x}(k)\right]+ +(\underline{B}_0+\Delta\underline{B})\underline{u}(k)+\underline{w}(k) \tag{91}$$

$$\underline{y}_0(k)+\Delta\underline{y}(k) = (\underline{K}_0+\Delta\underline{K})\left[\underline{x}_0(k)+\Delta\underline{x}(k)\right]+\underline{v}(k) \tag{92}$$

where the subscript 0 denotes the nominal values. The standard Kalman filter used for the state estimation contains *nominal* parameters, i.e., its state equation in the common form reads

$$\underline{x}_0^*(k+1) = \underline{A}_0\hat{\underline{x}}(k)+\underline{B}_0\underline{u}(k) \tag{93}$$

$$\hat{\underline{x}}_0(k+1) = \underline{x}_0^*(k+1)+\underline{K}_f(k+1)\left[\underline{y}(k+1)-\underline{K}_0\underline{x}_0^*(k+1)\right] \tag{94}$$

where $\underline{K}_f$ means the Kalman filter gain matrix found by the usual way of minimizing the effect of noise.

Defining the state estimation error

$$\underline{\varepsilon}(k) = \underline{x}(k)-\underline{x}_0^*(k) \tag{95}$$

and the output estimation error or innovation

$$\underline{e}(k) = \underline{y}(k)-\underline{y}^*(k) = \underline{y}(k)-\underline{K}_0\underline{x}_0^*(k) \tag{96}$$

the error equation becomes

$$\underline{\varepsilon}(k+1) = \underline{A}_0\left[\underline{I}-\underline{K}_f(k+1)\underline{K}_0\right]\underline{\varepsilon}(k)-\underline{A}_0\underline{K}_f(k)\Delta\underline{K}\underline{x}(k)$$

$$- \underline{A}_0\underline{K}_f(k)\underline{v}(k)+\underline{w}(k)+\Delta\underline{A}\underline{x}(k)+\Delta\underline{B}\underline{u}(k) \tag{97}$$

$$\underline{e}(k) = K_0\epsilon_k\Delta\underline{K}\underline{x}(k) \ . \tag{98}$$

A block diagram representation of (97),(98) is given in Fig.13.

Using the innovation as the residual for FDI, the occurance of a sensor fault can be monitored with the aid of a test of the covariance matrix $\underline{P}[\underline{e}(k+1),\underline{e}(k+1)]$. The basic idea of sensitivity reduction is now to form the *shifted* covariance matrix of the innovation, $\underline{P}[\underline{e}(k+1),\underline{e}(k+1-i)]$. Not only that $\underline{P}$ allows to better *locate* a sensor fault,it also reduces the influence of parameter variations. Belkoura (1983) has developed a procedure to determine the optimal time shift such that the effects of $\Delta\underline{A}$ and $\Delta\underline{B}$ become minimal. The resulting robust FDI scheme is shown in Fig.14. A fault in the i-th sensor affects the i-th row and the i-th column of the shifted covariance matrix so that a fault isolation with a high detection sensitivity becomes possible.

6.5 IFD Scheme Using Unknown Input Observers

A fundamental approach to the problem of sensitivity reduction of state estimation is to use *robust observers* in the sense of *"unknown input ob-*

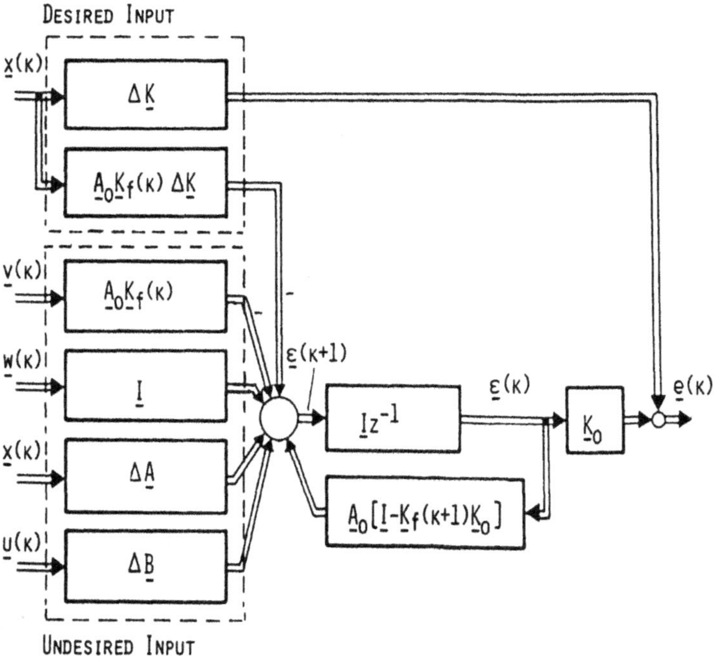

Fig.13: Block diagram representation of the estimation error of a Kalman filter

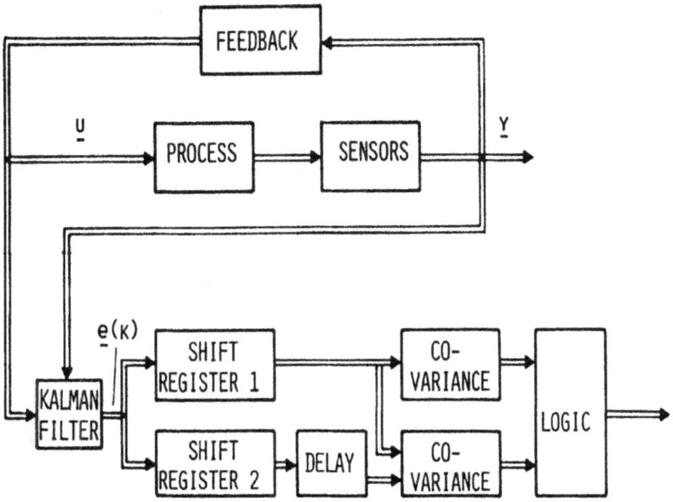

Fig.14: Robust covariance matrix test RCMT (Belkoura (1983))

servers". Suppose the system is given by

$$\dot{\underline{x}} = \underline{A}\underline{x}+\underline{B}\underline{u}+\underline{E}\underline{d} \tag{99}$$

$$\underline{y} = \underline{K}\underline{x} \,. \tag{100}$$

$\underline{E}$ is known as the distribution matrix of the unknown input signal $\underline{d}$, where $\underline{d}$ may also contain the effects of parameter variations. The idea is to choose the observer gain matrix $\underline{L}$ such that the effect of $\underline{E}\underline{d}$ in the observation error $\underline{e}$ is decoupled from the effect of the sensor or component faults, or in terms of the geometrical approach, that $\underline{E}\underline{d}$ has no influence on the subspaces into which the faults are projected. The existence of a robust observer depends upon the rank of $\underline{E}$. If rank $\underline{E} \geq$ rank $\underline{K}$ the full state vector can not be estimated with free choice in the eigenvalues of the estimation error dynamics. That means that the observer should be driven by as many measured variables as possible. In the continuous case the realizability condition may require high gains which may cause problems of practical realization. In the discrete case it is only limited anticipation that may be required. The unavoidable neglect of anticipation is problematic in a closed-loop control system, but when applied to FDI it only causes a certain delay of fault detection and may therefore be tolerated (Wünnenberg and Frank (1986)).

The resulting FDI scheme provides a bank of robust observers that are driven by all but one measured variables. The residuals are then chosen to be sensitive to all but one sensor faults. In practice, the parameter variations that are not described by the product $\underline{E}\underline{d}$ affect the residuals and cannot be distinguished from small sensor faults. Therefore thresholds have to be placed inside the detection logic as in conventional FDI schemes.

7. APPLICATIONS AND SIMULATION STUDIES

The theoretical work on fault detection and isolation by means of state estimation is essentially based on a number of idealized assumptions. Such are

- Knowledge of the mathematical model of the process (structure and parameters)
- Knowledge of the characteristic of noise
- Observability of the system
- Preciseness of the solution of the analytical equations
- In many cases: linearization of the model and assumption of white noise

The discussed methods are therefore easier to apply to well defined processes such as electrical and mechanical systems than to processes with inherent uncertainties like thermodynamic and chemical plants. As a matter of fact, a great deal of work on these methods is dedicated to applications in spacecrafts, aircrafts, hydrofoil boats and, to a minor degree, in nuclear power plants. A catalog of applications is given in the appendix.

Although theoretical investigators have paid great attention to the key question of the sensitivity with respect to violation of the idealized assumptions and, more than this, have attempted to systematically reduce these sensitivities, the question of performance under real conditions has to be answered in every particular case. The known literature shows therefore a long catalog of computer simulation studies and even some successful demonstrations at pilot plants (see the catalog in the appendix). However, only little is known about real applications in practice. In this section the results of computer simulation studies of IFD applications in nuclear reactors, CFD in a three-tank system and of an application to a real system, namely the control of an inverted pendulum, are presented.

7.1 IFD via ROS in the Primary Circuit of a Nuclear Reactor

The nuclear power station is a typical example for a process that requires high expenditure of process supervision and in which considerable parameter variations occur. The multiple purpose research reactor (MZFR) in Karlsruhe was therefore chosen for the examination of the robust observer schemes with the aid of a computer simulation study.

In the MZFR the thermal power of 200 MW is generated in the fuel rods of the reactor. This is transferred to the coolant (D_2O) which is heated from 248 ^oC to 274 ^oC. The coolant is circulated in two closed loops each consisting of a generator, a main coolant pump and the connecting pipes. The pressurizer which is half filled with water and half with steam is connected to one of the coolant loops through a surge line pipe. The pressure in the pressurizer is kept constant by a spray system and electric heating to give a value of 90 kg/cm abs. at the reactor inlet. This safely prevents boiling at the outlet of the coolant channels both in normal operation and on disturbance.

As in all heavy-water-moderated natural-uranium reactors, the moderator is separated from the coolant by shroud tubes. It has a mean temperature of 170 °C at equilibrium burn-up and is heated from 130 °C to 210 °C by neutron moderation, gama-radiation and the heat transferred from the coolant. After this it is mixed with the incoming coolant. The moderator circuit thus forms a bypass which is branched off on the discharge side of the main coolant pumps and is equiped with a separate moderator pump. The heat from the moderator coolers is used to heat the feedwater. In normal operation the feedwater is heated up to a temperature of 111 °C in the low pressure feed heaters. It is then heated further up to 200 °C in the moderator coolers.

The two steam generators which, on the secondary side, operate on the natural circulation principle, produce 310 t/h of saturated steam of 33 kg/cm^2 abs. The turbine generator produces a gross electrical output of 58 MW. On load rejection or turbine tripout, the steam is passed to the condenser through a dumping system which is capable of taking the full steam output for short periods. In these cases the reactor remains in operation of reduced power level.

Since the fuel and coolant temperatures have a negative coefficient of reactivity, the reactor is inherently stable. In view of the relatively high negative moderator temperature is changed by lowering the feedwater temperature and at partial load by increasing the feedwater flow rate. The additional feedwater is then drained behind the moderator coolers into the condenser through a flash tank.

In the first computer simulation we study ROS with the primary circuit of the reactor, Fig.15. This subsystem is of 7th order and has 3 inputs

- u_1 control rod position
- u_2 moderator inlet temperature
- u_3 live steam pressure

7 state variables

- x_1 reactor thermal power
- x_2 delayed neutron concentration
- x_3 average temperature of fuel element
- x_4 average temperature of coolant
- x_5 average temperature of moderator
- x_6, x_7 } outlet temperature of moderator at different sectors

and 5 outputs

$$y_1 = x_1 , \qquad y_2 = x_4$$
$$y_3 = -.404x_3+1.02x_4 , \qquad y_4 = .305x_3+1.43x_4$$
$$y_5 = x_5$$

The mathematical model of the process is described in detail in Frank and Keller (1984). From the experience of the operators it is known that the parameters of the process vary in a range of about 10%-40% (except the parameters of the kinematics that can vary up to 150%). Since there are 5 outputs, the application of ROS requires 5 observers each driven by one of the outputs. The overall observer scheme including 5 reduced-order observers is depicted in Fig.16.

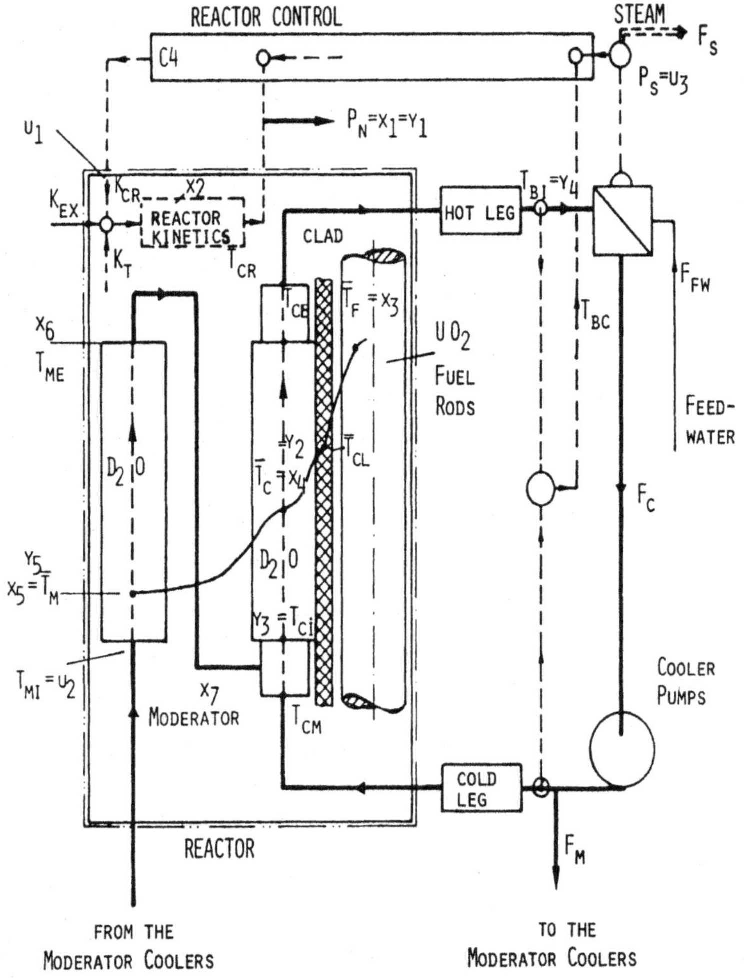

Fig.15: Schematic diagram of the primary circuit of the nuclear reactor MZFR

As a first step in applying ROS the critical parameters have to be determined by a sensitivity analysis. The result of the sensitivity analysis of the overall subsystem of 18th order has revealed that the element a_{51} of the subsystem matrix has great influence on $y_1(t)$ and $y_5(t)$, whereas element a_{43} strongly affects $y_2(t)$ and $y_4(t)$. Hence, the first and the fifth observer must be designed such that their

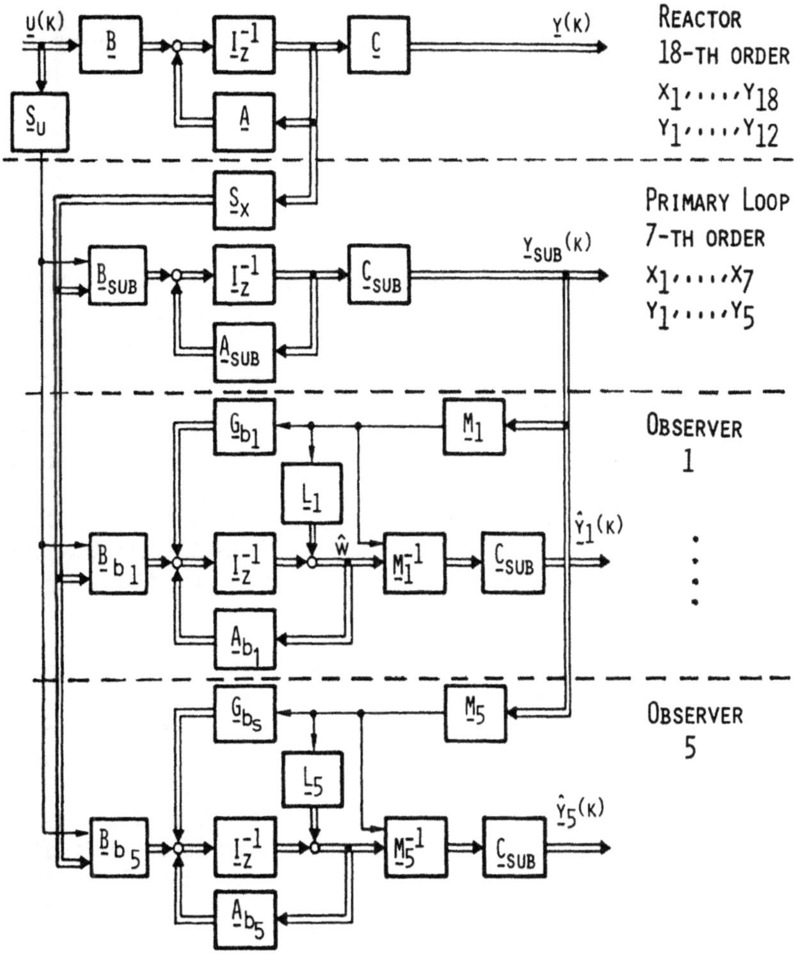

Fig. 16: ROS of the primary circuit of the MZFR power plant

estimates are insensitive versus a_{51}, the second and forth observer insensitive with respect to a_{43}.

The results of the computer simulation study are illustrated in Figures 17 and 18. The plots display the effects of abrupt instrument faults of 10% at 20 seconds on estimation errors in the presence of sudden parameter variations. It is seen that there is a remarkable difference between the effect of instrument faults on the one hand and process parameter variations on the other hand.

7.2 IFD via DEOS in the Steam Generator of the MZFR

The differential error observer scheme was tested by a computer simulation study with the steam generator of the MZFR. Fig. 19 shows the scheme of the steam generator and the block diagram of the arrangement of the observers of DEOS. The output vector of the steam generator comprises 4 components:

ΔT_{cm} coolant mixing temperature
ΔT_{WD} average coolant temperature in their downcomer
ΔY_{WD} water level in the steam generator
ΔP_s live steam pressure

Hence, 4 observers are required each estimating the full state $\hat{x}_1,\ldots,\hat{x}_4$. In order to study the net effect of the use of the differential errors instead of the estimation errors, three scale factor faults of 10% (at t = 20 sec) are simulated associated with a 20% parameter variation (coolant leakage). Results of the computer study are illustrated in Fig.20a-c. The solid lines represent the shapes of the estimation errors and the dotted lines those of the corresponding differential errors. It can be seen that the effect of a 20% change of the coolant mass (ΔM = 20% at t = 0) on the differential error is considerably smaller than on the estimation error. This illustrates the advantage of taking the differential errors as a residuals for IFD.

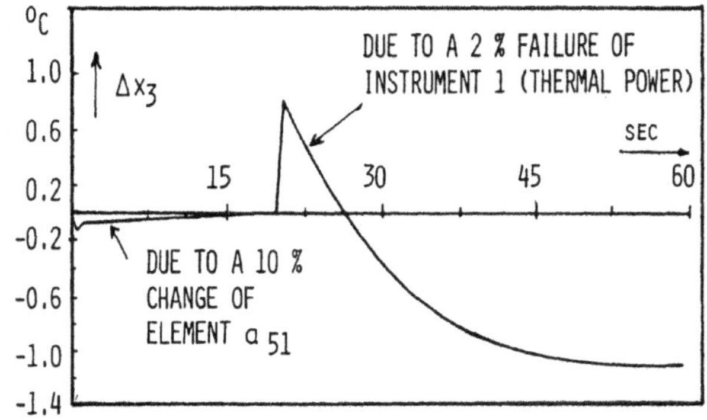

Fig.17: Estimation error of average fuel element temperature $x_3-\hat{x}_3$ for a scale factor fault of 4 MW in instrument 1 (reactor thermal power) at t=20 sec and a 10% change of a_{51} at t=0sec

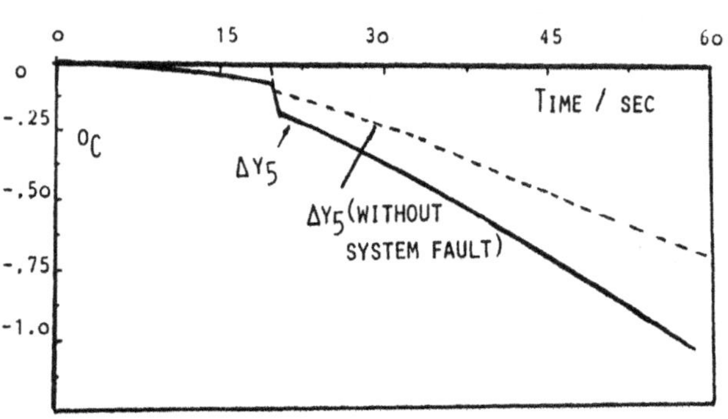

Fig.18: Average moderator temperature estimation error $x_5-\hat{x}_5$ for a scale factor fault of 10% in sensor 2 at t=20 sec and a 10%change of a_{43} at t=0sec

7.3 IFD via RCMT in the Turbo Set of a Boiling Water Reactor

As a third example we consider the application of the robust covariance matrix test (RCMT) to a 1100 Megawatt boiling water reactor of General Electric. The mathematical model of the overall plant is described in Frank and Keller (1984)(with further references). For the computer simulation study we restrict to the turbo alternator set, see Fig. 21. The mathematical model of this subsystem, linearized around the operating point, contains 4 states and 4 measured output variables (i.e. the nominal $\underline{K}$ matrix in (2) is the identity matrix): The outputs are.

$y_1 = x_1$ relative power of the turbine
$y_2 = x_2$ relative pressure at the input of the high pressure turbine
$y_3 = x_3$ relative flow rate of feed-water
$y_4 = x_4$ relative fresh water enthalpy

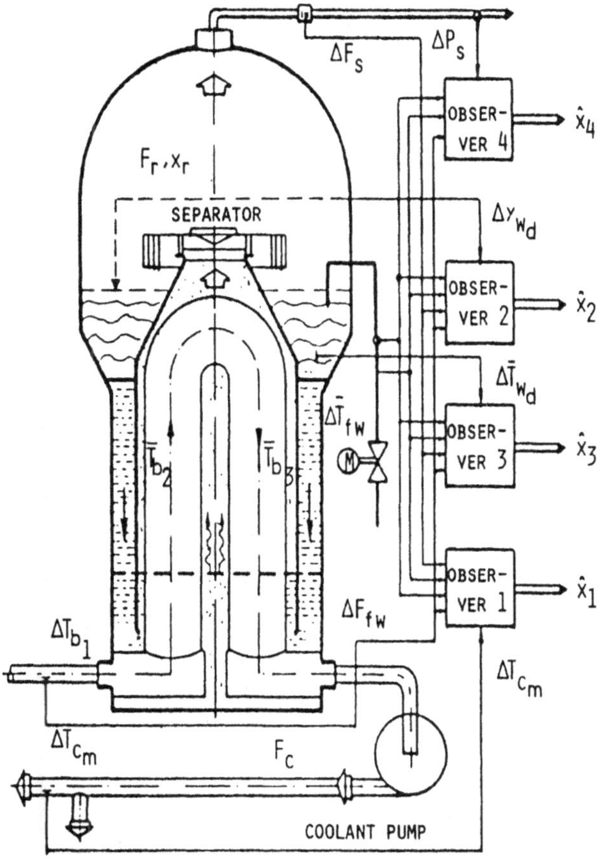

Fig. 19: Schematic diagram of the steam generator together with DEOS

The five inputs are

u_1 control of turbine valve
u_2 control of the bypass valve
u_3 control of the feedwater flow rate
u_4 control of feedwater enthalpy
u_5 pressure control in the reactor

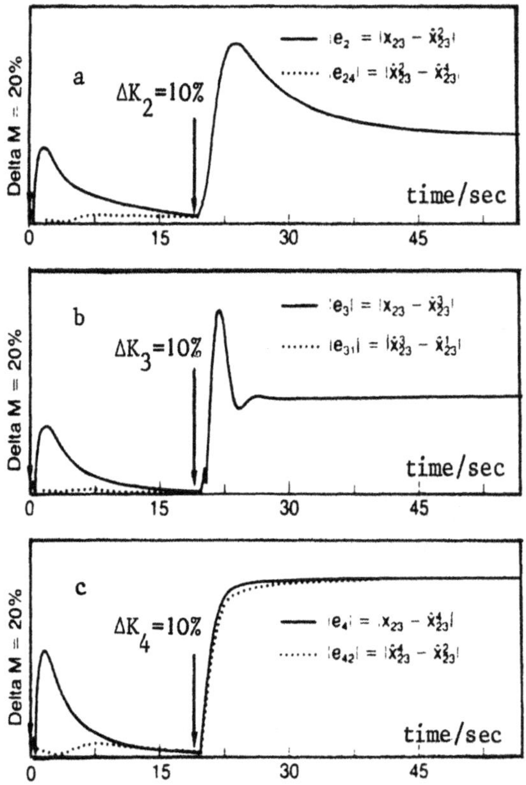

Fig. 20a-c: Typical shapes of the estimation errors (solid) and the differences of observation errors (dotted) at the occurance of a scale factor fault ΔK_i= 10% at t=20sec in the four instruments at a system parameter variation (coolant leakage) of ΔM=20% at t=0sec

The fault detection based on RCMT requires a single Kalman filter driven by the five inputs and four outputs. A computer simulation was carried out under the following conditions. System and measurement noise was considered as stationary Gaussian and white with amplitudes of about 0.1 % of the maximal values of the state variables. The optimal shifting time for the time-shifted covariance matrix was found to be 2.25 sec. The following four sensor faults were considered:

(1) Power sensor fault Δy_1 = 5 %
(2) Pressure sensor fault Δy_2 = 1 %
(3) Flow meter fault Δy_3 = 1 %
(4) Enthalpy sensor fault Δy_4 = 10 %

In Fig. 22 we show two samples of the results of the computer study, namely the shapes of the covariance matrices of the innovation with 1% failures in instruments 2. and 3.. It can bee seen that in both cases the corresponding rows and lines in the shifted covariance matrix surpass the thresholds so that a fault is displayed. The threshholds were chosen such that no failure was indicated in the nominal case and the case of a 100% change of the system parameter matrices $\underline{A}$ and $\underline{B}$. This clearly reveales the power of the RCMT approach, namely to pro-

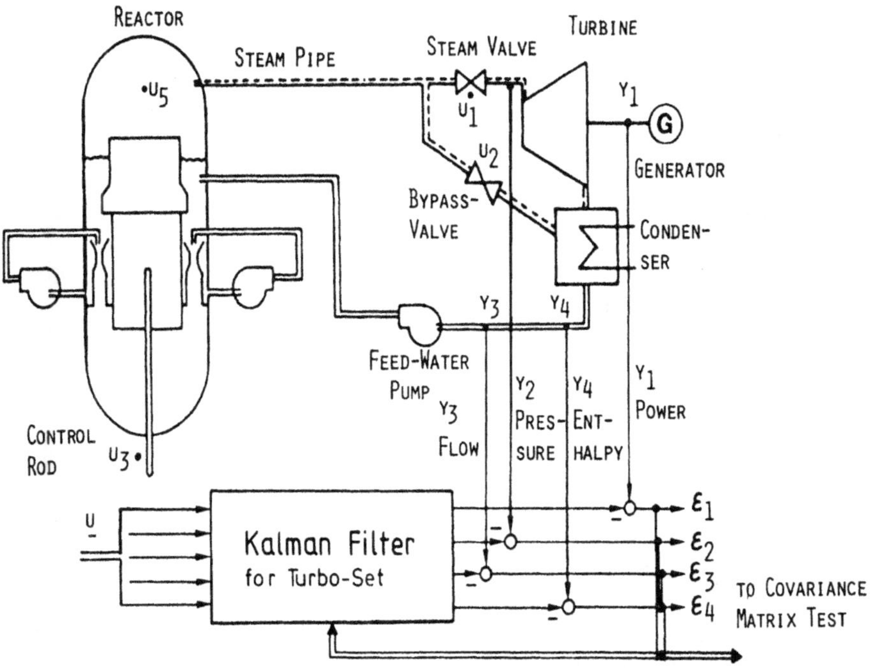

Fig.21: Schematic diagram of the boiling water nuclear reactor with RCMT applied to the turbine alternator set

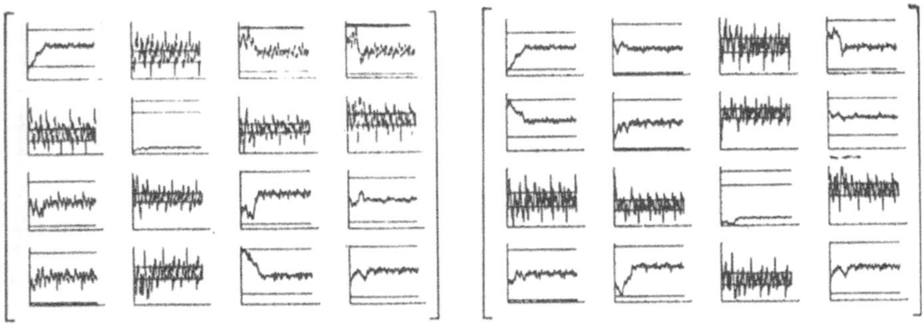

Fig.22: Covariance matrices of the innovation with time shifting of 2.5sec in the case of 1% fault in the scale of the pressure sensor, Δy_2 (left) and a 1% fault in the scale of the flow meter, Δy_3 (right)

vide large sensitivity to sensor faults and tolerate large parameter variations in the process. In conclusion, one can see that the RCMT

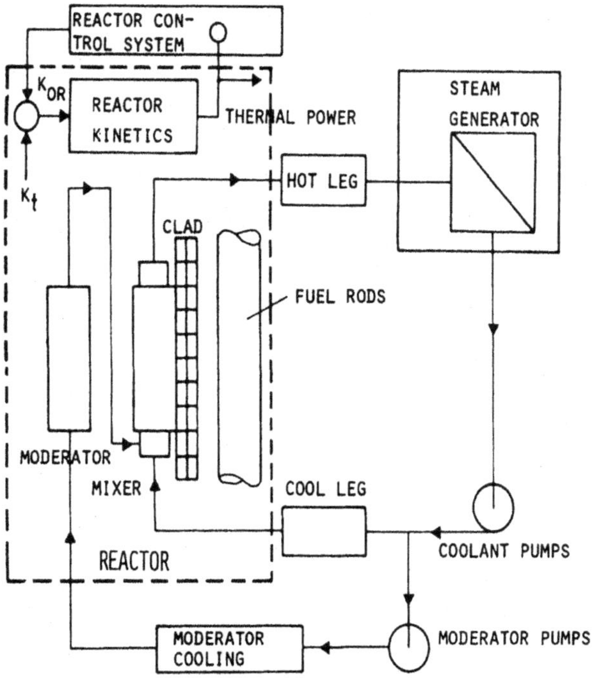

Fig.23: The primary circuit of the nuclear reactor MZFR showing the components for LOS

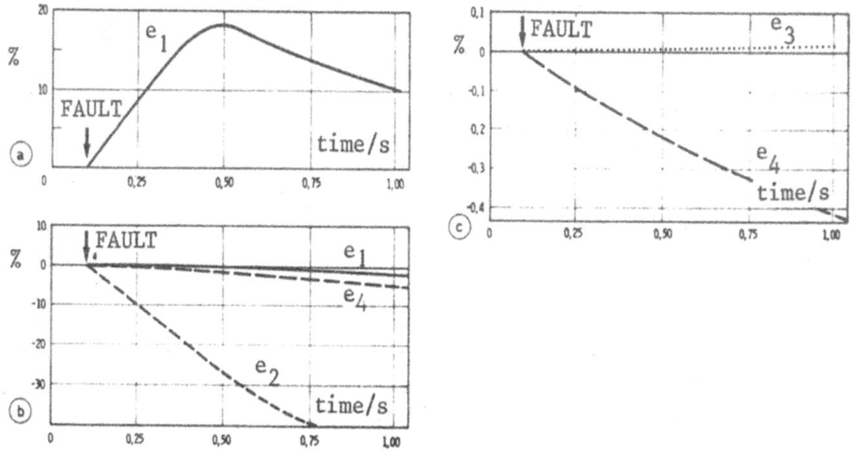

Fig.24: Time histories of the relative estimation errors $e_1, \ldots, e_4$ with ESCOS at different component failures: a) control rod ejection, b)drop-out of a coolant pump, c)drop-out of a moderator pump

approach presents itself as a rather robust method of identifying sensor faults even if they are small and under realistic conditions of substantial process uncertainty. This may largely encourage to apply this IFD scheme to real systems.

7.4 CFD in the Primary Circuit of the MZFR using the Local Observer Scheme HOS

As an example of component fault detection, let us consider the application of the hierarchical observer scheme HOS to the primary circuit of the nuclear reactor MZFR, Fig. 15. In a first step the process model has to be linearized around the operating point. Then the process has to be decomposed with respect to the type of interconnections. It turns out that the primary circuit can be decomposed into four components with unobservable interconnections

- reactor kinetics
- cooling in the core
- moderator
- mixer .

This means that ESCOS has to be applied without hierarchical structuring of the observer scheme. The schematic diagram of the process under consideration showing all the four components is depicted in Fig.23. In a digital simulation study with ESGOS the following accidents are simulated

a) Control rod ejection : malfunction in component 1
b) Drop-out of a coolant pump: malfunction in components 1,3,4
c) Drop-out of a moderator pump: malfunction in components 4,5

The plots in Fig. 24 show the resulting time histories of the four local estimation errors (e_1 thermal power, e_2 average moderator temperature, e_3 average coolant temperature, e_4 temperature at mixer outlet) at an abrupt change of the input vector.

From these shapes of the residuals it is possible to detect and isolate the three component failures precisely by the application of a proper threshold logic.

7.5 Component Fault Detection in a Three-Tank System using LOS

To study the efficiency of the local observer scheme when using linear observers in the case of a weakly nonlinear system, LOS was applied to a three tank system shown in Fig.25. The observer scheme was implemented on a micro-computer and the process was actually modeled by an analog computer (including noise). The failure detection was carried out on-line. Thus the results can be considered to be almost as realistic as in a real application.

Defining the three tanks as the components of interest, the three levels h_1,h_2,h_3 can be considered as the state variables. Then the nonlinear state equation reads

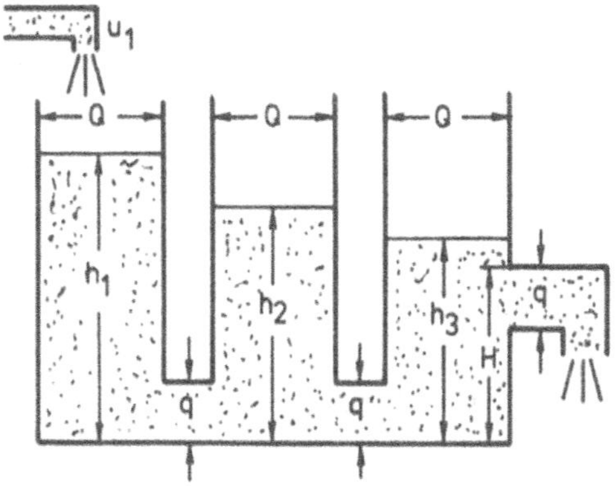

Fig.25: Schematic diagram of the three-tank system

$$h_1 = b\sqrt{h_1-h_2} + \frac{1}{Q}u_1$$

$$h_2 = b\sqrt{h_1-h_2} - b\sqrt{h_2-h_3}$$

$$h_3 = b\sqrt{h_2-h_3} - b\sqrt{h_3-H}$$

with $b = q\sqrt{2g}/Q$ and g the gravitational constant. Note that in this case the components have no unavailable interconnections, so that ASCOS can be applied. The nonlinear state equations can easily be linearized around an operating point. The linearized equations were used for the design of the observer scheme whereas the nonlinear equations were modeled on the analog computer. As component failures for the investigation we considered a leakage in tank 2, a clogging (reduction to 0,25q) in outlet 3 and a clogging (reduction to 0,25q) in the pipe between tank 1 and 2.

The resulting time histories of the three local estimation errors under abrupt faults in the components are shown in Fig.26. It can be seen that all faults can be detected by an appropriate threshold logic before the CFD scheme comes out of the region of validity caused by mismatches between the nonlinear system models and the linear models used in the observers. Though a unique fault detection is possible, the models in the observers can easily be replaced by nonlinear models, so that the region of validity becomes infinite. The improvements with nonlinear models are described in Frank and Hengy, 1986.

7.6 A Real FDIA Application to an Inverted Pendulum Control

While robustness of the FDI scheme plays a great role in chemical or thermodynamical systems it is less critical in mechanical systems

Leakage in tank 2 at 13.5s

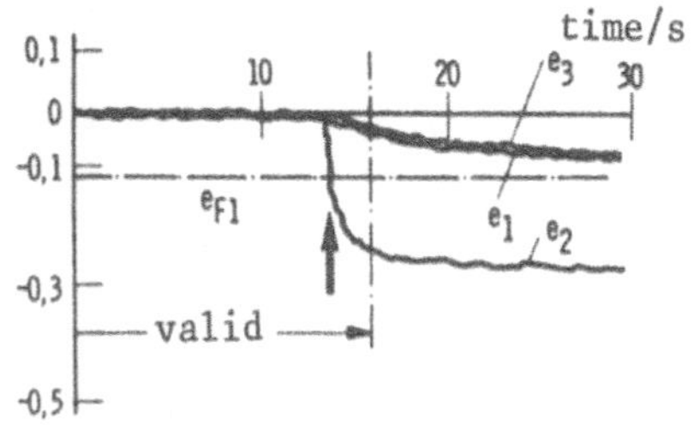

Diagnosis:

. Failure in tank 2, detected at 14.1s

. Tank 1 and 3 o.k.

. Model valid up to 15,7s

Clogging in the outlet of tank 3 at 17.6s

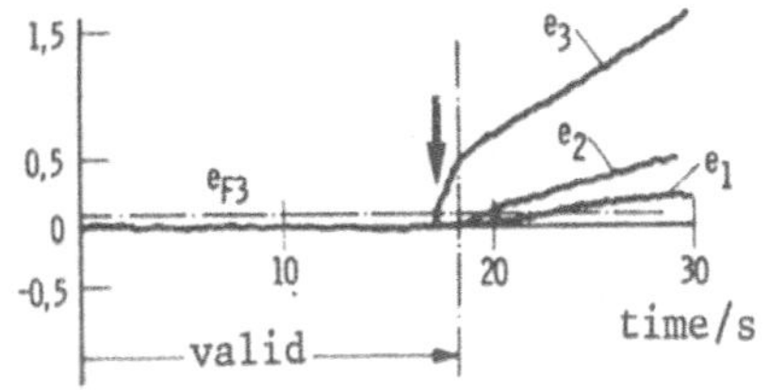

Diagnosis:

. Failure in tank 3 detected at 17.8s

. Tank 1 and 2 o.k.

. Model valid up to 18.4s

Clogging in the pipe between tank 1 and 2 at 16.4s

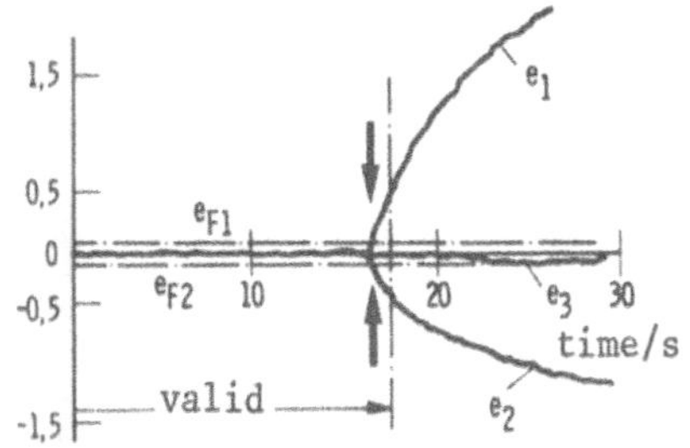

Diagnosis:

. Failure in tank 1 detected at 16.7s

. Failure in tank 2 detected at 16.6s

. Tank 3 o.k.

. Model valid up to 17.4s

Fig. 26: Time histories of the local estimation errors with the local observer scheme applied to a three tank system

because of the better knowledge and constancy of their parameters. This is for example true for an inverted pendulum control that will be addressed in this section. Actually, two IFD schemes similar to DOS and SOS are alternatively implemented on a micro-computer or mini-computer, resp., and applied to a laboratory set up of a servomechanism which balances a pendulum in upside position (Wünnenberg e.a.(1985)). Note that the pendulum is an unstable process, hence the failure detection has to be fast enough so that a failure accommodation can be accomplished before the pendulum collapses.

The schematic diagram of the inverted pendulum and its driving base is shown in Fig.27. Fig.28 shows the block diagram of the control loop to

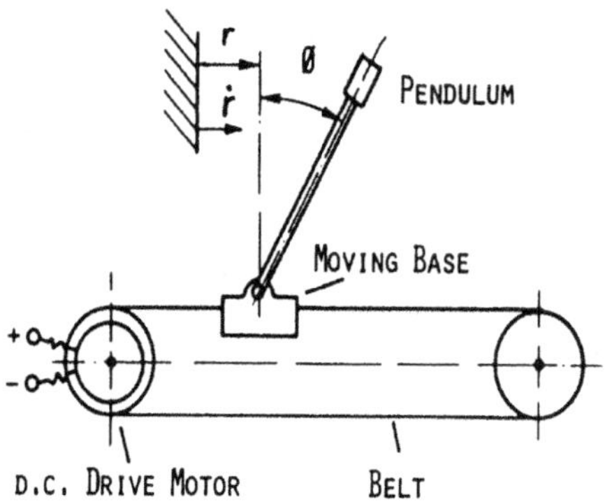

Fig.27: Schematic diagram of the inverted pendulum

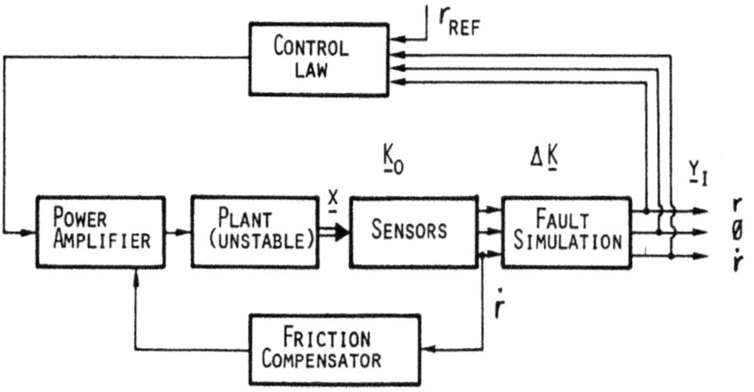

Fig. 28: Block diagram of the pendulum control

stabilize the pendulum. The linearized mathematical model of the plant including its output sensors is

$$\dot{\underline{x}} = \underline{A}\underline{x} + \underline{B}u \tag{101}$$

$$\underline{y}_I = \underline{K}\underline{x} + \underline{y}_F \tag{102}$$

where

$$\underline{x} = \begin{bmatrix} r \\ \Phi \\ \dot{r} \\ \dot{\Phi} \end{bmatrix} = \begin{bmatrix} \text{position of base} \\ \text{angular position of pendulum} \\ \text{velocity of base} \\ \text{angular velocity of pendulum} \end{bmatrix} . \tag{103}$$

u is the DC voltage that drives the motor. The measured output vector is

$$\underline{y}_I = \begin{bmatrix} y_{I1} \\ y_{I2} \\ y_{I3} \end{bmatrix} = \begin{bmatrix} r \\ \Phi \\ \dot{r} \end{bmatrix} + \begin{bmatrix} y_{F1} \\ y_{F2} \\ y_{F3} \end{bmatrix} = \begin{bmatrix} r+\Delta r \\ \Phi+\Delta\Phi \\ \dot{r}+\Delta\dot{r} \end{bmatrix} \tag{104}$$

where $y_F = [y_{F1}\ y_{F2}\ y_{F3}]^T$ represents the vector of sensor faults. The three output signals $r, \Phi, \dot{r}$ are used for state feedback. The signal Φ is reconstructed in the controller and fed back in addition to $r, \Phi, \dot{r}$.

In order to detect hard-failures of $r, \Phi, \dot{r}$, an IFD scheme similar to DOS can be applied, Fig.29. The total of the seven estimated signals

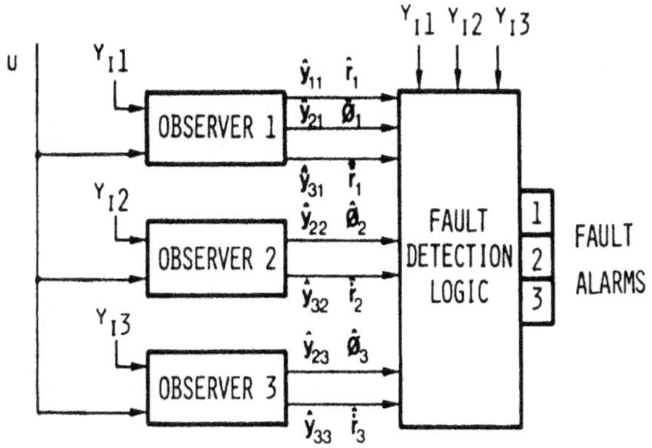

Fig.29: Block diagram of the IFD scheme implemented on the micropro- cess or INTEL 8086/87

$$\hat{y}_1 = \begin{bmatrix} \hat{y}_{11} \\ \hat{y}_{12} \\ \hat{y}_{13} \end{bmatrix} = \begin{bmatrix} \hat{r}_1 \\ \hat{\Phi} \\ \hat{\dot{r}}_1 \end{bmatrix} \quad \text{estimates of observer 1} \tag{105}$$

$$\hat{y}_2 = \begin{bmatrix} \hat{y}_{22} \\ \hat{y}_{23} \end{bmatrix} = \begin{bmatrix} \hat{\Phi}_2 \\ \hat{\dot{r}}_2 \end{bmatrix} \quad \text{estimates of observer 2} \tag{106}$$

$$\hat{y}_3 = \begin{bmatrix} \hat{y}_{32} \\ \hat{y}_{33} \end{bmatrix} = \begin{bmatrix} \hat{\Phi}_3 \\ \hat{\dot{r}}_3 \end{bmatrix} \quad \text{estimates of observer 3} \tag{107}$$

is then used to form the residuals

$$e_i^j = y_{Ii} - \hat{y}_{ji} \ . \tag{108}$$

Defining the decision functions

$$D_1 = | e_1^1\ e_2^1\ e_2^1 | \qquad \text{from observer 1} \tag{109}$$

$$D_2 = | e_2^2\ e_3^2 | \qquad \text{from observer 2} \tag{110}$$

$$D_3 = \left| e_2^3 \; e_3^3 \right| \qquad \text{from observer 3} \qquad (111)$$

the decision can be made due to appropriately defined thresholds S_i.

The threshold S_i can be determined such that no alarm is given under normal conditions. When a fault is displayed, it can be accommodated either by using the estimates from one of the observer variables instead of the faulty measured ones or by switching to a stand-by sensor. In the reported case all the described functions, namely the feedback control law, the state reconstruction in the controller, the observer scheme for IFD, the descision logic and the accommodation are implemented on a single microcomputer on the basis of an Intel 8086/87 processor with a sampling period of 25m sec.

Figures 30-32 show the time histories of the decision functions D_1, D_2, D_3 at a change of the reference input Δr_{ref} = 40cm (at t=0) and 100% scale factor faults (at t=5sec) in the base position sensor y_{I1}, the pendulum angle sensor y_{I2} and the base velocity sensor y_{I3}, respectively. Added to these plots are the corresponding thresholds.

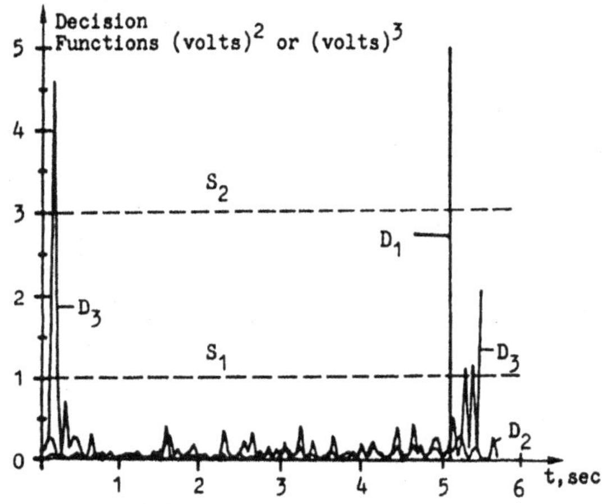

Fig.30: Time response of the decision functions $D_1, \ldots, D_3$ to a step input Δr_{ref}= 40cm at t=0 and a scale factor fault (100% change to zero) at t=5sec in the base position sensor

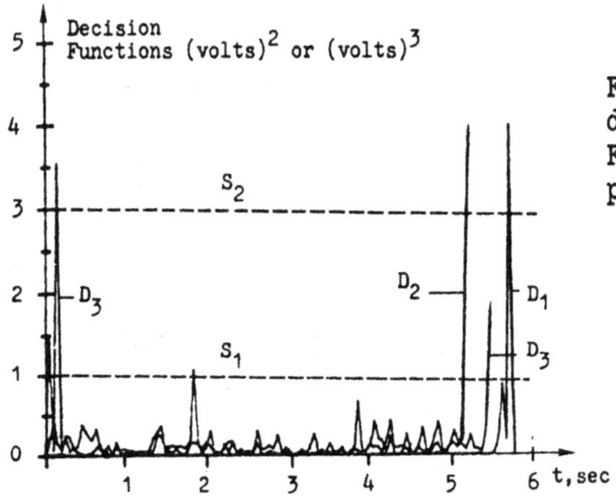

Fig.31: Time responses of decision functions as in Fig.30 but for fault in pendulum angle sensor

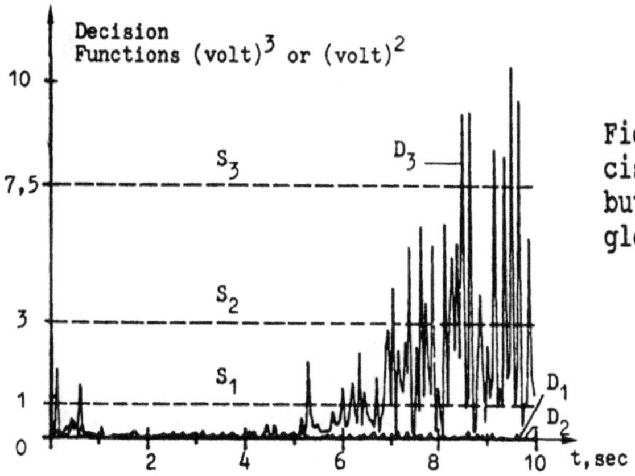

Fig.32: Time responses of decision functions as in Fig.30 but for fault in pendulum angle sensor

These three figures show instances of correct detection and isolation of sensor faults without false alarms being caused by either system inputs or by failures in falsely identified sensors. Moreover the detection time is short enough so that the pendulum can be stabilized by a proper failure accommodation after a failure has occured. Detection, isolation and accommodation was accomplished and tested for a long period including changes in the pendulum mass (up to 300%) and changes of environmental conditions. Despite these changes the device has proved to work perfectly in the case of abrupt hard-failures. There are of course limits in the detectability due to failure size and type

Alternatively, the simplified observer scheme SOS with a single observer can be applied. The schematic diagram including the block diagram of a possible observer scheme is shown in Fig. 33. The observer was

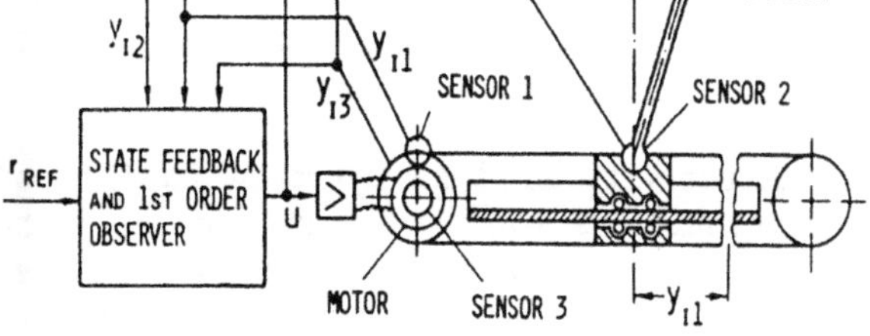

Fig.33: IFD in the inverted pendulum control using SOS

implemented on a hp 9825 table computer with a sampling period of 96 msec. The time responses of the observation errors for the same faults as in the case of DOS are shown in Fig.34. Though the performance of the hard-failures could be detected and isolated in due time, the superior performance of DOS pays out in the case of soft failures. This statement holds quite generally in the application of DOS and SOS.

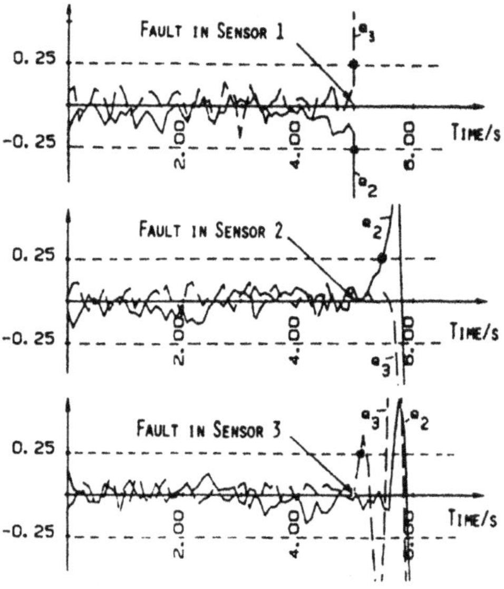

Fig.34: Time histories of estimation errors e_2 and e_3 of the measured variables y_2 (angle of pendulum Φ) and y_3 (speed of the base r) at a drop-out of sensor 1 (position r) after 5 seconds

8. CONCLUSIONS

The problem of FDI has been discussed from the viewpoint of residual generation with the aid of state estimation. This approach to FDI belongs to the group of model-based methods, since a model of the process is needed in some form in the state estimators. The neuralgic point of all model-based FDI techniques is the sensitivity to parameter variations. Emphasis has therefore been given to the question of parameter sensitivity and the design and examination of parameter insensitive or robust estimator schemes.

The advantage of the state estimation approach to FDI lies in the fact that the FDI task can completely be transfered into the control room and, since only available data have to be processed, there is no interaction with the process that could cause any risk. Moreover, once a faulty sensor is identified, the method allows to directly replace its output by the output of one of the estimators for the task of accommodation.

Despite the unquestionable attractivity of this approach to FDI and the profound work done by theoreticians over the past decade, there are still many open questions. The main problem is of course concerned with the necessary knowledge of the process model and its environment. This is the crucial point with respect to practical applications. It is therefore not surprising that applications have been discussed since long in transport systems and mechanical devices of power systems where rather well defined models exist. However, the situation is much different in chemical plants where the models are not only largely unprecise but also highly nonlinear and substantially changing with time. This is the main reason that only few attempts have been made to apply analytical FDI methods to chemical or thermodynamical processes. Another problem of practical relevance is that no general answer to the question of feasability can be given. Each task and given situation asks for an individual answer since noise and parameter variations normally considered as side effects are the key problem in this issue.

However, these problems that are somehow typical for all model-based methods, may not lead to the conclusion that FDI methods based on state estimation have little chance for practical applications. Their success will be closely tied to the advances in mathematical modeling and the treatment of uncertain systems. The experiences so far made with pilot plants are rather encouraging. They reveal that in special fields of application such as for the detection of well defined faults in well defined processes, a certain degree of maturity has been reached so that a practical application is possible.

ACKNOWLEDGEMENT

The author is pleased to acknowledge the financial support of the Ministerium für Wissenschaft und Forschung Nordrhein-Westfalen and the Deutsche Forschungsgemeinschaft for parts of the work reported in this paper. Much gratitude is owed to the co-workers cited in the references for their successful assistance.

REFERENCES

Beard, R.V. (1971). 'Failure Accommodation in Linear Systems through Self-Reorganization'. *Rept. MVT-71-1, Man Vehicle Laboratory*, Cambridge, Massachusetts

Belkoura, M. (1983). 'Entdeckung von Instrumentenfehlern mittels Kalman-Filter'. *Ph.D. Thesis*, University of Duisburg

Bonnice, W.F., Wagner E., Motyka, P. and Hall, S.R. (1985). 'The Application of the Detection Filter to Aircraft Control Surface and Actuator Failure Detection and Isolation'.*AIAA Guidance, Navigation and Control Conf.*, Snowmass, Co.

Chow, E.Y. (1982). 'A Failure Detection System Design Methodology'. *Sc.D. Thesis, Lab. Inform. and Decision Syst.*, M.I.T., Cambridge, Massachusetts

Chow, E.Y. and Willsky, A.S. (1984). 'Analytical Redundancy and the Design of Robust Failure Detection Systems'. *IEEE Trans. Aut. Contr.*, Vol. AC-29, No.7

Clark, R.N., Fosth, D.C. and Walton, V.M. (1975). 'Detecting Instrument Malfunctions in Control Systems', *IEEE, Trans. Aerospace Electron. Syst.*, AES-11, pp 465-473

Clark, R.N. (1978). 'A Simplified Instrument Failure Detection Scheme'. *IEEE Transactions on Aerospace and Electronic Systems*, Vol. AES-14, No.4

Clark, R.N. (1978). 'Instrument Fault Detection'. *IEEE Trans. on Aerospace and Electronic Systems'*, AES-14, 3, pp 456-465

Clark, R.N. (1979). 'The Dedicated Observer Approach to Instrument Failure Detection'. *Proc. 18th IEEE CDC*, pp 237-241

Clark, R.N. and Setzer, W. (1980). 'Sensor Fault Detection in a System with Random Disturbances'. *IEEE Trans. on Aerospace and Electronic Systems*, AES-16, 4, pp 468-473

Clark, R.N. and Campbell, B. (1982). 'Instrument Fault Detection in a Pressurized Water Reactor Pressurizer'. *J. on Nuclear Technology*, Vol.56,

Cunningham, T.B. and Poyneer, R.D. (1977). 'Sensor Failure Detection using Analytical Redundancy'. *Proc. of the 1977 JACC, IEEE*, New York,

Deckert, J.C., Desai, M.N., Deyst, J.J. and Willsky, A.S. (1977). 'DFBW Sensor Failure Identification using Analytic Redundancy'. *IEE Trans. Aut. Contr.*, Vol. AC-22, No.5, pp 795-803

Elmadbouly, E., Keller, L. and Frank, P.M. (1982). 'Application of Observers to Instrument Failure Detection'. *IFAC Symposium on Theory and Applications of Digital Control*, New Delhi

Elmadbouly, E. and Frank, P.M. (1983). 'Robust Analytical Instrument Failure Detection of Nuclear Power Plants'. *IFAC Symposium on Control Applications for Power System Security'*, Florence

Frank, P.M. (1978). 'Introduction to System Sensitivity Theory'. *Academic Press*, N.Y., San Francisco, London

Frank, P.M. and Keller, L. (1980). 'Sensitivity Discriminating Observer Design for Instrument Failure Detection'. *IEEE Trans. Aerosp. Electr. Syst.*, Vol. AES-16, pp 460-467

Frank, P.M. and Keller, L. (1984). 'Entdeckung von Instrumentenfehlanzeigen mittels Zustandsschätzung in technischen Regelungssystemen'. *Fortschrittberichte VDI-Z.*, Reihe 8, Nr. 80, VDI-Verlag Düsseldorf

Frank, P.M. and Hengy, D. (1986). 'Component failure detection via nonlinear state observers'. *IFAC Workshop on Fault Detection and Safety in Chemical Plants*, Kyoto, Japan

Frank, P.M. and Hengy, D. (1986). 'Fehlererkennung und -lokalisierung in komplexen Systemen'. *Abschlußbericht DFG-Projekt Fr 365/8, Universität Duisburg*, Fachgebiet Meß- und Regelungstechnik, Duisburg GFR

Garton, P. and Gray, J.O. (1985). 'The Classification of Sample Defects in Electromagnetic Non Destructive Testing'. *Proc. IFAC Identification and System Parameter Estimation*, York, 1631-1634

Halme, A. and Selkäinaho, J. (1984). 'Instrument Fault Detection using an Adaptive Filtering Method'. *Proc. 9th IFAC Congr.*, Budapest, 1765-1770

Hodzic, M. and Siljak, D.D. (1985). 'Failure Detection in Large Sparce Systems'. *Proc. of the 7th IFAC/IFORS Symp. on Identification and Process Parameter Estimation*, York, U.K., pp 273-279

Isermann, R. (1984). 'Process Fault detection based on Modeling and Estimation Methods - A Survey'. *Automatica*, 20, pp 387-404

Janssen, K. and Frank, P.M. (1984). 'Component Failure Detection via State Estimation'. *Proc. of the 9th IFAC congress*, Budapest

Janssen, K. and Frank, P.M. (1986). 'Entdeckung von Komponentenfehlern durch hierarchische Zustandsschätzung'. *Automatisierungstechnik at, 34. Jahrgang, Heft 1*, pp 23-31

Jones, H.L. (1973). 'Failure Detection in Linear Systems'. *Ph.D. Thesis*, Dept. of Aeronautics, M.I.T., Cambridge, Massachusetts

Leininger, G.G. (1981). 'Sensor Failure Detection with Model Mis-match'. *J.A.C.C.*, Charlottesville, Virginia

Lou, X.C. (1982). 'A System Failure Detection Method: The Failure Projection Method'. *Rep; LIDS-TH-1203, M.I.T.* Lab. for Inf. and Decision Syst., Cambridge, Massachusetts

Lou, X.C., Willsky A.S. and Verghese, G.L. (1986). 'Optimally Robust Redundancy Relations for Failure Detection in Uncertain Systems'. *Automatica*, Vol. 22, No. 3, pp 333-344

Mehra, R.K. and Peschon, J. (1971). 'An Innovations Approach to Fault Detection and Diagnosis in Dynamic Systems'. *Automatica*, Vol. 7, pp 637-640

Meier, L., Ross, D.W. and Glaser, M.B. (1971). 'Evaluation of the Feasibility of using Internal Redundancy to Detect and Isolate On-Bord Control Data Instrumentation Failures'. *Technical Report* AFFDL-TR-70-172

Merril, W.C. (1985). 'Sensor Failure Detection for Jet Engines using Analytical Redundancy'. *J. Guidance Control and Dynamics*, Vol. 8, No. 6, pp 673-681

Montgomery, R.C. and Caglayan, A.K. (1974). 'A Self-Reorganizing Digital Flight Control System for Aircraft'. *AIAA*, 12th Aerospace Meeting, Washington

Montgomery, R.C. and Caglayan, A.K. (1976). 'Failure Accommodation in Digital Flight Control Systems by Bayesian Decision Theory'. *J. Aircraft*, 13(2), pp 69-75

Montgomery, R.C. and Tabak, D. (1979). 'Application of Analytical Redundancy Management to Shuttle Crafts'. *Proc. IEEE Conf.* on Decision and Control, San Diego, pp 442-448

Motyka, P. and Hall, S.R. (1986). 'A Comparison of Aircraft Control Surface Failure Detection and Isolation Algorithm'. *ACC, Seattle, Washington (late paper)*

Oe, S. and Tomita, Y. (1983).'Failure detection and Prediction System by using Adaptive Digital Filter'. *Proc. IFAC/IFIP Symp. Realtime Digital Control Application*, Guadalajara, Mexico, pp 57-63

Onken, R. and Stuckenberg, N. (1979). 'Failure Detection in Signal Processing and Sensing in Flight Control Systems'. *Proc. IEEE Conf.* on Decision and Control, San Diego, pp 449-454

Patton, R.J. and Willcox, S.W. (1985). 'Comparison of two Techniques of IFD based on a Nonlinear Stochastic Model of an Aircraft'. *Proc. of the 7th IFAC/IFORS Symp. on Identification and Process Parameter Estimation*, York, U.K., pp 711-717

Patton, R.J., Willcox, S.W. and Winter, J.S. (1986). 'A Parameter Insensitive Technique for Aircraft Sensor Fault Diagnosis using Eigenstructure Assignment and Analytical Redundancy'. *Proc. A.I.A.A., Navigation, Guidance and Control Conf.*, Williamsburg, U.S.A.

Saif, M. and Villaseca, F.E. (1986). 'Sensor Failure Detection in Optimally controlled systems'. *Proc. ACC, Seattle*, Washington, pp 2104-2110

San Martin, A.M. and Vander Velde, W.E. (1986). 'Design of Robust Failure Detection Filters'. *Proc. ACC, Seattle*, Washington, pp 1052-1059

Shapiro, E.Y. and Decarli, H.E. (1979). 'Analytical Redundancy for Flight Control Sensors on the Lockheed L-1011 Aircraft'. *Proc. IEEE Conf.* on Decision and Control, San Diego, pp 371-376

Shoureshi, R. and Hostein, R.F. (1986). 'Failure Detection and Isolation Using System Structure Knowledge and Eigenvalue Sensitivity'. *Procc. ACC, Seattle*, Washington, pp 1933-1938

Siljak, D.D. (1978). 'Large Scale Dynamic Systems', *North Holland Book Comp.*, N.Y., Amsterdam, Oxford

Walker, B.K. (1983). 'Recent Developments in FAult Diagnosis and Accommodation'. AIAA Guidance and Control Conf., *AIAA Paper* No. 83-2258-CP

Watanabe, K., Yoshimura, T. and Sodea, T. (1981). 'A Diagnosis Method for Linear Stochastic Systems with Parametric Failures'. *Trans. of the ASME*, Vol. 103, pp 28-35

Watanabe K. and Himmelblau, D.M. (1982). 'Instrument Fault Detection in Systems with Uncertainties'. *Int. J. Systems Sci*, Vol. 13, No. 2, pp 137-158

Weiss, J.L., Pattipati, K.R., Willsky, A.S., Eterno, J.S. and Crawford, J.T. (1985). 'Robust Detection/Isolation/Accommodation for Sensor Failures'. *NASA Contr. Rep.* 174797, Lewis Res. Cent. NAS 3-24078, Alphatech Inc., Burlington, Massachusetts

White, J.E. and Speyer, J.L. (1986). 'Detection Filter Design: Spectral Theory and Algorithms'. *Proc. ACC, Seattle,* Washington, pp 1475-1481

Willsky, A.S., Deyst, J.J. and Crawford, B.S. (1974). 'Adaptive Filtering and Self-Test Methods for Failure Detection and Compensation'. *Proc. of the 1974 JACC*, Austin, Texas

Willsky, A.S., Deyst, J.J. and Crawford, B.S. (1975). 'Two Self-Test Methods applied to an Internal System Problem'. *J. Spacecrafts and Rockets*, Vol. 12, No. 7, pp 434-437

Willsky, A.S. (1976). 'A Survey of Design Methods for Failure Detection'. *Automatica*, 12, pp 601-611

Willsky, A.S. and Jones, H.L. (1976). 'A Generalized Likelyhood Ratio Approach to the Detection and Estimation of Jumps in Linear Systems'. *IEEE Trans. on Aut. Contr.*, AC-21, pp 108-112

Wünnenberg, J., Clark, R.N. and Frank P.M. (1985). 'An Application of Instrument Fault Detection'. *Proc. of the 7th IFAC/IFORS Symp. on Identification and Process Parameter Estimation,* York, U.K., pp 699

Wünnenberg, J. and Frank, P.M. (1986). 'Sensor Fault Detection via Robust Observers'. *First European Workshop on Fault Diagnostics, Reliability and Related Knowledge - Based Approaches,* Rhodes, Greece

APPENDIX: APPLICATIONS OF INNOVATION BASED FDI SCHEMES

EXAMPLE	DETECTION PROCEDURE	REFERENCE	
Airplane F-4	2 Kalman filters, consistency test of measurements and estimates	Meier, Ross, Glaser	1971
Satellite, autopilot for the lateral axis dynamics of an airplane	Detection filter design for deterministic systems and systems with stochastic disturbances	Jones	1971
F8-C	State estimators with nonlinear one step ahead prediction - several extended Kalman filters - decision logic with cost minimization by hypothesis testing	Montgomery, Price	1974
Inertial navigation systems	Definition of confidence areas for the deviation of the trajectories of the system and a state estimator. One Kalman filter has to be designed for every instrument	Kerr	1974
TF-30	Hard failures are detected via rate limit thesholds	Hrach et.al.	1975
Hydrofoil boat	One observer is designed for every system output that estimates the full state of the system	Clark, Fosth	1975
Turbofan	Threshold check of weighted average estimates	Ellis	1976
Space shuttle	Extended Kalman filter, generalized likelyhood test	Deyst, Deckert	1976
Inertial navigation system	Kalman filter and extended Kalman filter, statistical fault evaluation for a window of measurements	Maybeck	1976
Space shuttle	Kalman filter bank, filter design for hypothetical errors and for the nominal state	Montgomery, Caglayan	1976
Traffic on freeways	Kalman filter and generalized likelyhood test for the detection of jumps in linear systems	Willsky, Jones	1976

Airplane A-JD	For each measurement 2 instruments and 2 Kalman filters. Detection of scale factor and bias-type errors by hypothesis testing. Bandpass-Kalman filters considered	Cun-ningham, Poyneer	1977
Jet engines	Conditioned Kalman filters, Bayes estimator for hypothesis tests	Wells, de Silva	1977
NASA F8-DFBW Airplane	For each measurement 2 instruments. Detection of bias-type errors by sequential probability ratio test. Fault identification by Kalman filters	Deckert, Deyst Desai, Willsky	1977
Single spool	Maximum likelyhood testing	De Hoff, Hall	1978
Navigation system	Kalman filter bank, where each instrument is driven by a subset of all sensors. Comparison of the estimates and calculation of the covariance matrices	Smestad Orpen	1978
Hydrofoil boat	DOS-scheme, reduced order observers, decision logic, robustness test	Clark	1978
Hydrofoil boat	Simplified DOS-scheme, one reduced order observer that is driven by one system output signal, robustness test	Clark	1978
Space shuttle	UD-factorized filter with word limited digital processing of the measurements. Minimization of a Bayes-cost function	Mont-gomery, Tabak	1979 1979
Third-order system with system and measurement noise	Fault diagnosis for systems in discrete time in the presence of parametric failures. Kalman filter for the normal operating condition, sequential probability ratio test, adaptive extended Kalman filter estimates the state and the parameters under parameter variations	Yoshi-mura, Watanabe, Konishi Sodea	1979
Lockeed L-1011 Airplane	One observer of full order for one flight condition in the closed loop to drive the autopilot	Shapiro, Decarli	1979

HFB 320 Airplane	Fault safety concept for a double sensor configuration. One observer is designed for each sensor	Onken, 1979 Stuckenberg
Internal navigation system	One Kalman filter for each instrument and the nominal system model. If the confidence areas of the estimated and the nominal states do no longer overlap a failure is concluded	Kerr 1980
Nuclear reactor	5 Kalman filters are designed for 5 instruments that are each driven by 3 instrument readings	Campbell 1980
Power water reactor	Extension of the DOS-scheme (Clark, 1978) for a system with measurement noise, Kalman filter bank	Kitamura 1980
Single input - single output system	2 observers with different sensitivity	Frank, 1980 Keller
Hydrofoil boat	One Kalman filter is driven by one system output signal. Exponential smoothing of the output estimation error	Clark, 1980 Setzer
F 100 Nonlinear simulation	Detection filter	Meserole 1981
Pendulum, hydrofoil boat, airplane	Fixing of directions in the state space that are unsensitive against parameter variations. Extension of the DOS-scheme (Clark) according to (Frank/Keller) to a dual DOS-scheme	Chen 1981
Discrete single input, single output system	Kalman filter for the nominal state, extended Kalman filter for the state and parameter estimation. Hypothesis test of the filter innovations	Watanabe, 1981 Yoshimura, Sodea
Turbojet	Bayesian hypothesis testing	de Silva 1982
Chemical reactor with heat exchanger	Detection of instrument faults in the presence of parameter uncertainties, system- and measurement noise for a class of nonlinear systems with unsensitive state estimators	Watanabe, 1982 Himmelblau

Nuclear reactor	Robust observer scheme based on DOS (Clark)	Elmad-bouly, Keller, Frank	1982
A-J Airplane	Considering several output signals to drive an observer, logic design	Clark, Hertel	1982
Nuclear reactor	One Kalman filter, calculation of the corvariance innovation for a time window of estimated and sampled values. Parameter variations are considered	Bel-koura	1983
Nuclear reactor	Difference of the estimation error of two observer that are driven by different measurements (TEOS). Parameter invariant observer scheme (PIOS)	Elmad-bouly, Frank	1983
Power water reactor	One Kalman filter for each instrument. Consistency checks between outputs of the filters	Tylee	1983
Pressurized water reactor	Fault detection by decentralized state estimation	Janssen, Frank	1984
Pulsed extraction column	Kalman filtering and sequential probability testing	Voni-vento, Tonielli	1984
F 100 Linear simulation	Sequential probability ratio test	Weiss et.al.	1985
F 100 Linear simulation	Hypothesis testing	Emami-Naeini et.al	1985
F 100 Nonlinear simulation	Threshold checks	Brown et.al.	1985
Nuclear reactor	DOS-scheme (Clark) applied to detect instrument and component faults	Noccoli et.al.	1985
Inverted pendulum	DOS-scheme applied to detect instrument faults	Wünnen-berg et.al.	1985
Gas-pipeline	Adaptive state estimation, fault detection by fault models and cross-correlation	Bill-mann	1985
Nonlinear air-	Detection filter design by eigen-	Patton	1986

craft model	structure assignment	et.al.
Nuclear reactor	Sequential decision making procedure that seek out the largest consistent subset for use in estimation of the measured variable	Polenta 1986 et.al.
Nuclear reactor	Kalman filter bank each filter dedicated to one instrument, combinatory logic	Tylee, 1986 Purviance
F 100 Turbine engine	Accommodation filter to detect and isolate "hard failures", hypothesis filters to detect and isolate "soft failures", adaptive decision thresholds	Merrill,1986 Delaat

REFERENCES

Meier, L., Ross, D.W. Glaser, M.B.:
Evaluation of the Feasibility of Using Internal Redundancy to Detect and Isolate Onboard Control Data Instrumentation Failures. AFFDL-TR-70-172-AF Flight Dynamics Laboratory, W-PAFB, Ohio, 1971

Jones, H.L.:
Failure Detection in Linear Systems. PH.D. Thesis, Dept. of Aeronautics and Astronautics, M.I.T., Cambridge, Mass., 1973

Montgomery, R.C., Price, D.B.:
Management of Analytical Redundancy in Digital Flight Control Systems for Aircraft. AIAA Mechanics and Control of Flight Conference, Anaheim, Calif., 1974

Kerr, T.H.:
A Two Ellipsoid Overlap Test for Real Time Failure Detection and Isolation by Confidence Regions. Proc. of the 5th Conference on Decision and Control, Phoenix, Ariz., 1974

Hrach, F.J.; Arpasi, D.J., and Bruton, W.M.:
Design and Evaluation of a Sensor Fail-Operational Control System for a Digitally Controlled Turbofan Engine. NASA TM-X 3260, 1975

Clark, R.N., Fosth, D.C, Walton, V.M.:
Detecting Instrument Malfuncitons in Control Systems. IEEE Trans. Aerosp. Electron. Syst., Vol. AES-11, pp 465-473, 1975

Ellis, S.H.:
Self-Correcting Control for a Turbofan Engine. Proceedings, 3rd International Symposium on Air Breathing Engines, Munich, FRG, pp 171-186, 1976

Deyst, J.J., Deckert, J.C.:
Maximum Likelihood Failure Detection Techniques Applied to the Shuttle RCS Jets. Journal of Spacecraft and Rockets, Vol. 13, No. 2, pp 65-74, 1976

Maybeck, P.:
Failure Detection without Excessive Hardware Redundancy. Aerospace and Electronics Conf. (Naecon 1976), Dayton, Ohio, 1976

Montgomery, R.C., Caglayan, A.K.:
Failure Accommodation in Digital Flight Control Systems by Bayesian Decision Theory. J. Aircraft, Vol. 13, No. 3, 1976

Willsky, A.S., Jones, H.L.:
A Generalized Likelihood Ratio Approach to the detection and Estimation of Jumps in Linear Systems. IEEE Transactions on Automatic Control, 1976

Cunningham, T.B., Poyneer, R.D.:
Sensor Failure Detection Using Analytical Redundancy. Proc. Joint Automatic Control Conference, New York, 1977

Wells, W.R., de Silva, C.W.:
Failure State Detection of Aircraft Engine Output Sensors. Proceedings of the 1977 Joint Automatic Control Conference, Vol. 2, San Francisco, CA, pp 1493-1497, 1977

Deckert, J.C., Deyst, J.J., Desai, M.N., Willsky, A.S.:
F-8 DF BW Sensor Failure Identification Using Analytic Redundancy. IEEE Trans. on Automatic Control, Vol. AC-22, No. 5, pp 795-803, 1977

De Hoff, R.L., Hall, W.E.Jr:
Advanced Fault Detection and Isolation Methods for Aircraft Turbine Engines. System Control Inc., ONR-CR-215-245-1 (AD-A058891). 1978

Smestad, T, Orpen:
Application of Parallel Filters for Malfunction Detection and Alternative Mode Capability. AGARD Symposium, Sandefjord, Norway, 1978

Clark, R.N.:
Instrument Fault Detection. IEEE Trans. Aerosp. Electron. Systems, Vol. AES-14, pp 456-465, 1978

Clark, R.N.:
A Simplified Failure Detection Scheme. IEEE Trans. Aerosp. Electron. Systems, Vol. AES-14, pp 558-563, 1978

Montgomery, R.C., Tabak, D:
Application of Analytical Redunadancy Management to Shuttle Crafts. IEEE Conf. on Decision and Control. San Diego, 1979

Yoshimura, T., Watanabe, K., Konishi, K., Sodea, T.:
A Sequential Failure Detection Approach and the Identification of Failure Parameters. Int. J. Systems Sci., Vol. 10, No. 7, pp 827-836, 1979

Shapiro, E.Y., Decarli, H.E.:
Analytic Redundancy for Flight Control Sensors on the Lockheed L-1011 Aircraft. IEEE Conf. on Decision and Control, San Diego, 1979

Onken, R., Stuckenberg, N.:
Failure Detection in Signal Processing and Sensing in Flight Control Systems. Proc. of the 1978 IEEE Conf. on Decision and Control, San Diego, Cal., 1979

Kerr, T.H.:
Statistical Analysis of Two-Ellipsoid Overlap Test for Real-Time Failure Detection.IEEE Trans.on Automatic Control, Vol.AC-25, No.4, pp 762-772, 1980

Campbell, B.:
Instrument Failure Detection Using Functional Redundancy Applied to a Nuclear Pressurizer. Master Thesis, Univ. of Washington, U.S.A., 1980

Kitamura, M.:
Detection of Sensor Failures in Nuclear Plants Using Analytic Redundancy. Trans. Am. Nucl. Soc., 34, pp 581-583, 1980

Frank, P.M., Keller, L.:
Sensitivity Discriminating Observer Design for Instrument Failure Detection. IEEE Trans. Aerosp. Electron. Systems, Vol. AES-16, pp 460-467,1980

Clark, R.N., Setzer, W.:
Sensor Fault Detection in a System with Random Disturbances. IEEE Trans. on Aerospace and Electronic Systems, Vol. AES-16, No. 4, pp 468-473, 1980

Meserole, J.S.:
Detection Filters for Fault Tolerant Control of Turbofan Engines. Ph.D. Thesis, Massachusetts Institute of Technology, Cambridge, MA, 1981

Chen, W.-K.:
Instrument Failure Detection for a Linear Uncertain System. Dissertation, University of Washington, U.S.A.; 1981

Watanabe, K., Yoshimura, T., Sodea, T.:
A Diagnosis Method for Linear Stochastic Systems with parametric Failures. Transactions of the ASME, Vol. 103, pp 28-35, 1981

de Silva, C.W.:
Real-Time Failure Detection of Aircraft Engine Output Sensors. Arabian Journal for Science and Engineering, Vol. 7, pp 45-53, 1982

Watanabe, K., Himmelblau, D.M.:
Instrument Fault Detection in Systems with Uncertainties. Int. J. Systems Sci., Vol. 13, No. 2, pp 137-158, 1982

Elmadbouly, E., Keller, L., Frank, P.M.:
Application of Observers to Instrument Failure Detection. IFAC-Symposium on Theory and Applications of Digital Control, New Dehli, 1982

Hertel, J.E., Clark, R.N.:
Instrument Failure Detection in Partially Observable Systems. IEEE Trans. on Aerospace and Electronic Systems, Vol. AES-18, No. 3, pp 310-317, 1982

Belkoura, M.:
Entdeckung von Instrumentenfehlanzeigen mittels Kalman Filter. Dissertation Universität -GH- Duisburg, FRG, 1983

Elmadbouly, E., Frank, P.M.:
Robust Instrument Failure Detection via Luenberger Observers in Nuclear Power Plants. Cigre Symposium on Control Application for Power System Sensitivity, Florence, 1983

Tylee, J.L.:
On-Line Failure Detection in Nuclear Power Plant Instrumentation. IEEE Trans. on Automatic Control, Vol. AC-28, No. 3, 1983

Janssen, K., Frank, P.M.:
Component Failure detection via State Estimation. 9-th IFAC-World Congress of the International Federation of Automatic Control, Budapest, 1984

Bonivento, C., Tonielli, A.:
A Detection-Estimation Multifilter Approach with Nuclear Application. 9-th IFAC-World Congress of the International Federation of Automatic Control, Budapest, 1984

Weiss, J.L., Pattipati, K.R., Willsky, A.S., Eterno, J.S., Crawford, J.T.:
Robust Detection/Isolation/Accommodation for Sensor Failures. Alphatec Inc., Burlington, MA, TR-213, 1985, also NASA CR-17479, 1985

Emami-Naeini, A., Akhter, M.M., Rock, S.M.:
Robust Detection Isolation, and Accomodation for Sensor Failures. NASA CR-174825, 1985

Brown, H., Corley, R.C., Elgin, J.A., Spang, H.A.:
Sensor Failure Detection and Isolation in Multiengine Aircraft. General Eletric Co., Aircraft Engine Business Group, Cincinnati, OH, NASA CR-174846, 1985

Niccoli, L.G., Wilburn, N.P., Colley, R.W., Alexandro, F., Clark, R.N., Bouzerdoum, S.:
Detection of Instrument or Component Failures in a Nuclear Plant by Luenberger Observers. International Topical Meeting on Computer Applications for Nuclear Plant Operation and Control Pasco, WA, 1985

Wünnenberg, J., Clark, R.N., Frank, P.M.:
An Application of Instrument Fault Detection. IFAC Identification and System Parameter Estimation, York, U.K., 1985

Billmann, L.:
Methoden zur Lecküberwachung und Regelung von Gasfernleitungen. Fortschrittberichte VDI, Reihe 8, Nr. 85, 1985

Patton, R.J., Willcox, S.W., Winter, S.J.:
A Parameter Insensitive Technique for Aircraft Sensor Fault Analysis Using Eigenstructure Assignment and Analytical Redundancy. AIAA Navigation, Guidance and Control Conference, Williamsburg, Virginia, U.S.A., 1986

Polenta, H.P., Ray, A., Menadier, P.T., Benard, J.A., Lanning, D.D.:
Implementaion of a Fault Detection Procedure. Proc. 1986 American Control Conference, Seattle, WA., U.S.A., 1986

Tylee, J.L., Purviance, J.E.:
Providing Nuclear Reactor Control Information in the Presence of Instrument Failures. Proc. 1986 American Control Conference, Seattle, WA., U.S.A., 1986

Merrill, W.C., Delaat, J.C.:
A Real-Time Simulation Evaluation of an Advanced Detection, Isolation and Accommodation Algorithm for Sensor Failures in Turbine Engines. Proc. 1986 American Control Conference, Seattle, WA., U.S.A., 1986

A DISTRIBUTED HYPOTHESIS-TESTING TEAM DECISION PROBLEM WITH COMMUNICATIONS COST*

Jason Papastavrou**
Michael Athans**

ABSTRACT

In this paper we formulate and solve a distributed binary hypothesis-testing problem. We consider a cooperative team that consists of two decision makers (DM's); one is refered to as the primary DM and the other as the consulting DM. The team objective is to carry out binary hypothesis testing based upon uncertain measurements. The primary DM can declare his decision based only on its own measurements; however, in ambiguous situations the primary DM can ask the consulting DM for an opinion and it incurs a communications cost. Then the consulting DM transmits either a definite recommendation or pleads ignorance. The primary DM has the responsibility of making a final definitive decision. The team objective is the minimization of the probability of error, taking into account different costs for hypothesis misclassification and communication costs. Numerical results are included to demonstrate the dependence of the different decision thresholds on the problem parameters, including different perceptions of the prior information.

---

* Research conducted at the MIT Laboratory for Information and Decision Systems with support provided by the Office of Naval Research under contract ONR/N00014-84-k-0519 (NR 649-003)
** Room 35-406/LIDS, MIT, Cambridge, MA 02139.

This paper has been submitted to the 25th IEEE Conference on Decision and Control.

S. Tzafestas et al. (eds.), System Fault Diagnostics, Reliability and Related Knowledge-Based Approaches, Vol. 1, 99–130.

1. Introduction and Motivation.

In this paper we formulate, solve, and analyze a distributed hypothesis-testing problem which is an abstraction of a wide class of team decision problems. It represents a normative version of the "second-opinion" problem in which a primary decision maker (DM) has the option of soliciting, at a cost, the opinion of a consulting DM when faced with an ambiguous interpretation of uncertain evidence.

1.1 Motivating Examples.

Our major motivation for this research is provided by generic hypothesis-testing problems in the field of Command and Control. To be specific, consider the problem of target detection formalized as a binary hypothesis testing problem (H_0 means no target, while H_1 denotes the presense of a target). Suppose that independent noisy measurements are obtained by two geographically distributed sensors (Figure 1). One sensor, the primary DM, has final responsibility for declaring the presense or absence of a target, with different costs associated with the probability of false alarm versus the probability of missed detection. If the primary DM relied only on the measurements of his own sensor, then we have a classical centralized detection problem that has been extensively analyzed; see, for example, Van Trees [1]. If the actual measurements of the second sensor were communicated to the primary DM, we have once more a classical centralized detection problem in which we have two independent measurements on the same hypothesis; in this case, we require communication of raw data and this is expensive both from a channel bandwidth point of view and, perhaps more importantly, because radio or acoustic communication can be intercepted by the enemy.

Continuing with the target detection problem, we can arrive at the model that we shall use in the sequel by making the following assumptions which model the desire to communicate as little as possible. The primary DM can look at the data from his own sensor and attempt to arrive at a decision using a likelihood-ratio test (lrt), which yields a threshold test in the linear-Gaussian case. Quite often the primary DM can be confident about the quality of his decision. However, we can imagine that there will be instances that the data will be close to the decision threshold,

corresponding to an ambiguous situation for the primary DM. In such cases it may pay off to incur a communications cost and seek some information from the other available sensor. It remains to establish what is the nature of the information to be transmitted back to the primary DM.
In our model, we assume the existence of a consulting DM having access to the data from the other sensor. We assume that the consulting DM has the ability to map the raw data from his sensor into decisions. The consulting DM is "activated" only at the request of the primary DM. It is natural to speculate that its advise will be ternary in nature: YES, I think there is a target; NO, I do not think there is a target; and, SORRY, NOT SURE MYSELF. Note that these transmitted decisions in general require less bits than the raw sensor data, hence the communication is cheap and more likely to escape enemy interception. Then, the primary DM based upon the message received from the consulting DM has the responsibility of making the final binary team decision on whether the target is present or absent.

The need for communicating with small-bit messages can be appreciated if we think of detecting an enemy submarine using passive sonar(Figure 2). We associate the primary DM with an attack submarine, and the consulting DM with a surface destroyer. Both have towed-array sonar capable of long-range enemy submarine detection. Request for information from the submarine to the destroyer can be initiated by having the sub eject a slot-buoy with a prerecorded low-power radio message. A short active sonar pulse can be used to transmit the recommendation from the destroyer to the submarine. Thus, the submarine has the choice of obtaining a "second opinion" with minimal compromise of its covert mission.

Of course, target detection is only an example of more general binary hypothesis-testing problems. Hence, one can readily extend the basic distributed team decision problem setup to other situations. For example, in the area of medical diagnosis we imagine a primary physician interpreting the outcomes of several tests. In case of doubt, he sends the patient to another consulting physician for other tests (at a dollar cost), and seeks his recommendation. However, the primary physician has the final diagnostic responsibility. Similar scenarios occur in the intelligence field where the "compartmentalization" of sensitive data, or the protection of a spy, dictate infrequent and low-bit communications. In more general military Command and Control problems, we seek insight on formalizing the need to break EMCON, and at what cost, to resolve tactical situation assessment ambiguities.

1.2 Literature Review.

The solution of distributed decision problems is quite a bit different, and much more difficult, as compared to their centralized counterparts. Indeed there is only a handful of papers that deal with solutions to distributed hypothesis-testing problems. The first attempt to illustrate the difficulties of dealing with distributed hypothesis-testing problems was published by Tenney and Sandell [2]; they point out that the decision thresholds are in general coupled. Ekchian [3] and Ekchian and Tenney [4] deal with detection networks in which downstream DM's make decisions based upon their local measurements and upstream DM decisions. Kushner and Pacut [5] introduced a delay cost (somewhat similar to the communications cost in our model) in the case that the observations have exponential distributions, and performed a simulation study. Recently, Chair and Varshney [6] have pointed out how the results in [2] can be extended in more general settings. Boettcher [7] and Boettcher and Tenney [8], [9], have shown how to modify the normative solutions in [4] to reflect human limitation constraints, and arrive in at normative/descriptive model that captures the constraints of human implementation in the presense of decision deadlines and increasing human workload; experiments using human subjects showed close agreement with the predictions of their normative/descriptive model. Finally, Tsitsiklis [10] and Tsitsiklis and Athans [11] demonstrate that such distributed hypothesis-testing problems are NP-complete; their research provides theoretical evidence regarding the inherent complexity of solving optimal distributed decision problems as compared to their centralized counterparts (which are trivially solvable).

1.3 Contributions of this Research.

The main contribution of this paper relates to the formulation and optimal solution of the team decision problem described above. Under the assumption that the measurements are conditionally independent, we show that the optimal decision rules for both the primary and the consulting DM are deterministic and are expressed as likelihood-ratio tests with constant thresholds which are tightly coupled (see Section 3 and the Appendix).

When we specialize the general results to the case that the observations are linear and the statistics are Gaussian, then we are able to derive explicit expressions for the decision thresholds for both the primary and consulting DM's (see Section 4). These threshold equations are tightly coupled, thereby necessitating an iterative solution. They provide clear-cut evidence that the DM's indeed operate as team members; their optimal thresholds are very different from those that they would use in isolation, i.e. in a non-team setting. This, of course, was the case in other versions of the distributed hypothesis-testing problem, e.g. [2].

The numerical sensitivity results (summarized in Section 5) for the linear-Gaussian case provide much needed intuitive understanding of the problem and concrete evidence that the team members operate in a more-or-less intuitive manner, especially after the fact. We study the impact of changing the communications cost and the measurement accuracy of each DM upon the decision thresholds and the overall team performance. In this manner we can obtain valuable insight on the optimal communication frequency between the DM's. As to be expected, as the communication cost increases, the frequency of communication (and asking for a second opinion) decreases, and the team performance approaches that of the primary DM operating in isolation. In addition, we compare the overall distributed team performance to the centralized version of the problem in which the primary DM had access, at no cost, to both sets of observations. In this manner, we can study the degree of inherent performance degradation to be expected as a consequence of enforcing the distributed decision architecture in the overall decision making process.

Finally, we study the team performance degradation when one of the team members, either the primary or the consulting DM, has an erroneous estimate of the hypotheses prior probabilities. This corresponds to mildly different mental models of the prior situation assesment; see Athans [12]. As expected the team performance is much more sensitive to misperceptions by the primary DM as compared to similar misperceptions by the consulting DM. This implies that, if team training reduces misperceptions on the part of the DM's, the greatest payoff is obtained in training the primary DM.

2. Problem Definition

The problem is one of hypothesis testing. The team has to choose among two alternative hypotheses H_0 and H_1, with a priori probabilities

$$P(H_0)=p_0 \qquad\qquad P(H_1)=p_1 \tag{1}$$

Each of two DM's, one called primary (DM A) and one consulting (DM B), receives an uncertain measurement y_α and y_β respectively (Figure 1), distributed with known joint probability density functions

$$P(y_\alpha, y_\beta \mid H_i) \quad ; \quad i=0,1. \tag{2}$$

The final decision of the team u_f (0 or 1, indicating H_0 or H_1 to be true) is the responsibility of the primary DM. DM A initially makes a preliminary decision u_α where it can either decide (0 or 1) on the basis of its own data (ie y_α), or at a cost (C≥0) can solicit DM B's opinion (u_α=I), prior to making the commital decision.

The consulting DM's decision u_β consists of three distinct messages (call them : x,v and z) and is activated only when asked. We decided to assign three messages to DM B, because we wanted to have one message indicating each of the two hypotheses and one message indicating that the consulting DM is 'not sure.' In fact, we proved that the optimal content for the messages of DM B is the one mentioned above.

When the message from DM B is received,the burden shifts back to the primary DM, which is called to make the commital decision of the team based on his own data and the information from the consulting DM.

We now define the following cost function :

$$J : \{0,1\} \times \{H_0, H_1\} \rightarrow \mathbf{R} \tag{3}$$

with $J(u_f, H_i)$ being the cost incurred by the team choosing u_f, when H_i is true.

Then, the optimality criterion for the team is a function

$$J^*: \{0,1,I\} \times \{0,1\} \times \{H_0,H_1\} \rightarrow \mathbb{R} \qquad (4)$$

with :

$$J^*(u_\alpha,u_f,H_i)= \begin{cases} J(u_f,H_i)+C & ; \quad u_\alpha=I \text{ (information requested)} \\ J(u_f,H_i) & ; \quad \text{otherwise} \end{cases} \qquad (5)$$

The cost structure of the problem is the usual cost structure used in Hypothesis Testing problems, but also includes the non-negative communication cost, which the team incurs when the DM A decides to obtain the consulting DM's information.

<u>Remark</u> : According to the rules of the problem, when the preliminary decision u_α of the primary DM is 0 or 1, then the final team decision is 0 or 1 respectively (ie $P(u_f=i \mid u_\alpha=i)=1$ for $i=0,1$).

The objective of the decision strategies will be to minimize the expected cost incurred

$$\min E[J^*(u_\alpha,u_f,H)] \qquad (6)$$

where the minimization is over the decision rules of the two DMs. Note that the decision rule of the consulting DM is implicitly included in the cost function, through the final team decision u_f (which is a function of the decision of the consulting DM).

All the prior information is known to both DMs. The only information they do not share is their observations. Each DM knows only its own observation and, because of the conditional independence assumption, nothing about the other DM's observation.

The problem can now be stated as follows :

Problem : Given p_0, p_1, the distributions $P(y_\alpha,y_\beta \mid H_i)$ for $i=0,1$ with $y_\alpha \in Y_\alpha$, $y_\beta \in Y_\beta$, and the cost function J^*, find the decision rules u_α,u_β and u_f as functions

$$\gamma_\alpha : Y_\alpha \rightarrow \{0,1,I\} \qquad (7)$$

$$\gamma_\beta : Y_\beta \rightarrow \{x,y,z\} \tag{8}$$

and

$$\gamma_f : Y_\alpha \times \{x,y,z\} \rightarrow \{0,1\} \tag{9}$$

(subject to : $P(u_f=i \mid u_\alpha=i)=1$ for $i=0,1$), which minimize the expected cost.

NOTE : The centralized counterpart of the problem, where a single DM receives both observations is a well known problem. The solution is deterministic and given by a likelihood ratio test (lrt). That is :

$$\gamma_c : Y_\alpha \times Y_\beta \rightarrow \{0,1\} \tag{10}$$

with

$$\gamma_c(y_\alpha,y_\beta)=\begin{cases} 0 & ; \quad \Lambda(y_\alpha,y_\beta) \geq t \\ 1 & ; \quad \text{otherwise} \end{cases} \tag{11}$$

where

$$\Lambda(y_\alpha,y_\beta) = [P(y_\alpha,y_\beta \mid H_0)p_0]/[P(y_\alpha,y_\beta \mid H_1)p_1]$$

$$= P(H_0 \mid y_\alpha,y_\beta)/P(H_1 \mid y_\alpha,y_\beta) \tag{12}$$

and t is a precomputed threshold

$$t = [J(0,H_1)-J(1,H_1)]/[J(1,H_0)-J(0,H_0)] \tag{13}$$

provided $J(1,H_0) > J(0,H_0)$. Thus, the difficulty of our problem arises because of its decentralized nature.

We will show that, under certain assumptions, the most restrictive of which is conditional independence of the observations, the optimal decision rules for the *Problem* are deterministic and given by lrt's with constant thresholds. The thresholds of the two DMs are coupled, indicating that the DMs work as a team rather than individuals.

3. About the Solution to the General Problem

In order to be able to solve the *Problem* , we make the following assumptions.

ASSUMPTION 1 : $J(1,H_0) > J(0,H_0)$; $J(0,H_1) > J(1,H_1)$ (14)

or it is more costly for the team to err than to be correct.

This logical assumption is made in order to motivate the team members to avoid erring and in order to enable us to algebraically put the optimal decisions in lrt form.

ASSUMPTION 2 : $P(y_\alpha | y_\beta, H_i) = P(y_\alpha | H_i)$; $P(y_\beta | y_\alpha, H_i) = P(y_\beta | H_i)$; $i=0,1$ (15)

or the observations y_α and y_β are conditionally independent.

This assumption removes the dependence of the one observation on the other and thus allows us to write the optimal decision rules as lrt's with <u>constant</u> thresholds.

ASSUMPTION 3 : <u>Without loss of generality</u> assume that :

$$\frac{P(u_\beta=x \,|u_\alpha=I,H_0)}{P(u_\beta=x \,|u_\alpha=I,H_1)} \geq \frac{P(u_\beta=v \,|u_\alpha=I,H_0)}{P(u_\beta=v \,|u_\alpha=I,H_1)} \geq \frac{P(u_\beta=z \,|u_\alpha=I,H_0)}{P(u_\beta=z \,|u_\alpha=I,H_1)} \tag{16}$$

This assumption is made in order to distinguish between the messages of DM B.

As shown in detail in the Appendix, the optimal decision rules for all three decisions of our problem (u_α, u_β, u_γ) are given by *deterministic* functions which are expressed as *likelihood ratio tests* with *constant* thresholds. The three thresholds of the primary DM (two for u_α and one for u_γ) and the two thresholds of the consulting DM (for u_β) can not be obtained in closed form. They are <u>coupled</u>, that is the thresholds of one DM are given as functions of the thresholds of the other DM.

Another important result is that, when the optimal decision rules are employed and the consulting DM's decision is x (or z), then the optimal final decision rule of the primary DM is always 0 (or 1):

$$P(u_f=0 \mid u_\alpha=I, u_\beta=x, y_\alpha)=1 \quad \text{for all } y_\alpha \in \{y_\alpha \mid P(u_\alpha=I \mid y_\alpha)=1,\ y_\alpha \in Y_\alpha\} \qquad (17)$$

and

$$P(u_f=1 \mid u_\alpha=I, u_\beta=z, y_\alpha)=1 \quad \text{for all } y_\alpha \in \{y_\alpha \mid P(u_\alpha=I \mid y_\alpha)=1,\ y_\alpha \in Y_\alpha\} \qquad (18)$$

Thus, we can simplify our notation by changing the DM B decisions from x to 0, from z to 1 and from v to ? (which is interpreted as :"I am not sure"). The team's decision process can be now described as follows : Each of the two DMs receives an observation. Then, the primary DM can either make the final decision (0 or 1) or can decide to incur the communication cost (u_α=I) and pass the responsibility of the final decision to the consulting DM. When called upon, the consulting DM can either make the final decision or shift the burden back to DM A (u_β=?), in which case the primary DM is forced to make the final decision, based on its own observation (y_α) and the fact that DM B decided u_β=? .

A detailed presentation of the facts discussed above can be found in the Appendix.

4. A Gaussian Example

We now present detailed threshold equations for the case where the probability distributions of the two observations are Gaussian. We selected the Gaussian distribution, despite its cumbersome algebraic formulae, because of its generality. Our objective is to perform numerical sensitivity analysis to the solution of this example, in order to gain information on the team 'activities.'

We assume that the observations are distributed with the following Gaussian distributions :

$$y_\alpha \sim N(\mu, \sigma_\alpha^2) \quad ; \quad y_\beta \sim N(\mu, \sigma_\beta^2) \qquad (19)$$

The two alternative hypotheses are :

$$H_0 : \mu=\mu_0 \quad \text{or} \quad H_1 : \mu=\mu_1 \tag{20}$$

Without loss of generality, assume that :

$$\mu_0 < \mu_1 \tag{21}$$

The rest of the notation is the same as in the general problem described above.

We can show that the optimum decision rules for this example are given by thresholds on the *observation* axes, as shown in Figure 3. Before presenting the equations of the thresholds, we define some variables.

Ψ_α^l : lower threshold of DM A Ψ_α^u : upper threshold of DM A

Ψ_α^f : threshold for the final decision of DM A

Ψ_β^l : lower threshold of DM B Ψ_β^u : upper threshold of DM B

$$\Phi_i^j(k) = \int_{-\infty}^{\frac{\Psi_i^j - \mu_k}{\sigma_i}} (2\pi)^{-0.5} \exp\left(-0.5\,x^2\right) dx \qquad \text{for } i=\alpha,\beta \;;\; j=l,f,u \;;\; k=0,1$$

Note that the above function is the well-known error function, presented with notational modifications to fit the purposes of the problem.

$$W^1 = 0.5\,[\Phi_\beta^u(0)-\Phi_\beta^u(1)] \tag{22}$$

$$W^2 = \frac{\Phi_\beta^u(0)-\Phi_\beta^l(0)-\Phi_\beta^u(1)+\Phi_\beta^l(1)+\Phi_\beta^l(0)\Phi_\beta^u(1)-\Phi_\beta^u(0)\Phi_\beta^l(1)}{\Phi_\beta^u(0)-\Phi_\beta^l(0)+\Phi_\beta^u(1)-\Phi_\beta^l(1)} \tag{23}$$

$$W^3 = \frac{\Phi_\beta^l(0)\Phi_\beta^u(1)-\Phi_\beta^l(1)\Phi_\beta^u(0)}{\Phi_\beta^u(0)-\Phi_\beta^l(0)+\Phi_\beta^u(1)-\Phi_\beta^l(1)} \tag{24}$$

$$W^4 = 0.5\,[\Phi_\beta^u(0)-\Phi_\beta^u(1)] \tag{25}$$

$$\psi_\alpha^* = (\mu_0+\mu_1)/2 + [\sigma_\alpha^2/(\mu_1-\mu_0)]\ln[p_0/(1-p_0)] \tag{26}$$

$$\psi_\beta^* = (\mu_0+\mu_1)/2 + [\sigma_\beta^2/(\mu_1-\mu_0)]\ln[p_0/(1-p_0)] \tag{27}$$

In (26) and (27), the (centralized) maximum likelihood estimators for each DM are defined.

COROLLARY 1 : If $P(u_\alpha=I)>0$ (i.e. information is requested for some y_α) and if $P(u_\beta=?\,|\,u_\alpha=I)>0$ (i.e. "I am not sure" is returned for some y_β, when information is requested), then the optimal final decision rule of the primary DM is a deterministic function defined by :

$$\gamma_f(y_\alpha) = \begin{cases} 0 & \text{if} \quad y_\alpha \le \psi_\alpha^f \\ 1 & \text{if} \quad y_\alpha > \psi_\alpha^f \end{cases} \tag{28}$$

where :

$$\psi_\alpha^f = \psi_\alpha^* + \frac{\sigma_\alpha^2}{\mu_1-\mu_0}\ln\left(\frac{\Phi_\beta^u(0)-\Phi_\beta^l(0)}{\Phi_\beta^u(1)-\Phi_\beta^l(1)}\right) \tag{29}$$

COROLLARY 2 : If $P(u_\alpha=I)>0$ (i.e. information is requested for some y_α) and the primary DM's final decision rule is the one given by Corollary 1, then the optimal decision rule of the consulting DM is a deterministic function defined by :

$$\gamma_\beta(y_\beta) = \begin{cases} 0 & \text{if} \quad y_\beta \le \psi_\beta^l \\ ? & \text{if} \quad \psi_\beta^l < y_\beta \le \psi_\beta^u \\ 1 & \text{if} \quad \psi_\beta^u < y_\beta \end{cases} \tag{30}$$

where :

$$\psi_\beta^l = \min\left\{\psi_\beta^* + \frac{\sigma_\beta^2}{\mu_1-\mu_0}\ln\left(\frac{\Phi_\alpha^u(0)-\Phi_\alpha^f(0)}{\Phi_\alpha^u(1)-\Phi_\alpha^f(1)}\right),\ \psi_\beta^* + \frac{\sigma_\beta^2}{\mu_1-\mu_0}\ln\left(\frac{\Phi_\alpha^u(0)-\Phi_\alpha^l(0)}{\Phi_\alpha^u(1)-\Phi_\alpha^l(1)}\right)\right\} \tag{31}$$

and :

$$\psi_\beta^u = \max\left\{\psi_\beta^* + \frac{\sigma_\beta^2}{\mu_1-\mu_0}\ln\left(\frac{\Phi_\alpha^u(0)-\Phi_\alpha^l(0)}{\Phi_\alpha^u(1)-\Phi_\alpha^l(1)}\right), \psi_\beta^* + \frac{\sigma_\beta^2}{\mu_1-\mu_0}\ln\left(\frac{\Phi_\alpha^f(0)-\Phi_\alpha^l(0)}{\Phi_\alpha^f(1)-\Phi_\alpha^l(1)}\right)\right\} \quad (32)$$

COROLLARY 3 : Given that the final decision rule employed by the primary DM is the one of Corollary 1 and that the decision rule employed by the consulting DM is the one of Corollary 2, then the optimal decision rule for the preliminary decision u_α of the primary DM is a deterministic function defined by :

$$\gamma_\alpha(y_\alpha) = \begin{cases} 0 & \text{if} \quad y_\alpha \le \psi_\alpha^l \\ I & \text{if} \quad \psi_\alpha^l < y_\alpha \le \psi_\alpha^u \\ 1 & \text{if} \quad \psi_\alpha^u < y_\alpha \end{cases} \quad (33)$$

where :

$$\psi_\alpha^l = \begin{cases} \psi_\alpha^* + \dfrac{\sigma_\alpha^2}{\mu_1-\mu_0}\ln\left(\dfrac{1-\Phi_\beta^l(0)+C}{1-\Phi_\beta^l(1)-C}\right) & ; \quad 0 \le C < \min\{W^1, W^2\} \\ \psi_\alpha^* + \dfrac{\sigma_\alpha^2}{\mu_1-\mu_0}\ln\left(\dfrac{1-\Phi_\beta^u(0)+C}{1-\Phi_\beta^u(1)-C}\right) & ; \quad W^2 < C \le W^4 \\ \psi_\alpha^* & ; \quad \text{otherwise} \end{cases} \quad (34)$$

and :

$$\psi_\alpha^u = \begin{cases} \psi_\alpha^* + \dfrac{\sigma_\alpha^2}{\mu_1-\mu_0}\ln\left(\dfrac{\Phi_\beta^l(0)-C}{\Phi_\beta^l(1)+C}\right) & ; \quad 0 \le C < \min\{W^3, W^4\} \\ \psi_\alpha^* + \dfrac{\sigma_\alpha^2}{\mu_1-\mu_0}\ln\left(\dfrac{\Phi_\beta^u(0)-C}{\Phi_\beta^u(1)+C}\right) & ; \quad W^3 < C \le W^1 \\ \psi_\alpha^* & ; \quad \text{otherwise} \end{cases} \quad (35)$$

<u>REMARK</u> : Observe that the equations of all the thresholds include (and possibly reduce to) a "centralized" part (ψ_i^*) indicating the relation of our problem to its centralized counterpart.

5. NUMERICAL SENSITIVITY ANALYSIS

We now perform sensitivity analysis to the solution of the Gaussian example. Our objective is to analyze the effects on the team performance from varying the parameters of our problem, in order to obtain better understanding of the decentralized team decision mechanism. We vary the quality of the observations of each DM (the variance of each DM), the a priori likelihood of the hypotheses and the communication cost. Finally, we study the effects of different a priori knowledge for each DM.

We use the following 'minimum error' cost function:

$$J(u_f,H_1) = \begin{cases} 0 & ; \quad u_f = 1 \\ 1 & ; \quad u_f \neq 1 \end{cases} \tag{36}$$

We do not need to vary the cost function, because this would be mathematically equivalent to varying the a priori probabilities of the two hypotheses.

5.1 Effects of varying the quality of the observations of the Primary DM

Denote :

$C1^*$ = cost incured if the consulting DM makes the decision alone .

We distinguish two cases depending on the cost associated with the information (ie the of quality of information)

CASE 1: $\min(P_0, 1-P_0) \leq C1^* + C$

As the variance of the primary DM increases, it becomes less costly for the team to have the primary DM always decide the more likely hypothesis, than request for information. This occurs becayse the observation of DM A becomes increasingly worthless. Thus, the primary DM progressively ignores its observation and in order to minimize cost has to choose between "de facto" deciding the more likely hypothesis (and incuring cost equal to the probability of the least likely hypothesis) or "de facto" requesting for information (and thus incuring the communication cost plus the cost of the consulting DM). In this case, the prior is less than the latter and so the optimum decision of the primary DM, as its variance tends to infinity is to always decide the more likely hypothesis (Figure 4, $P_0 = .8$). Thus :

$$\lim_{\sigma_0^2 \to \infty} P(u_\alpha = I) \to 0$$

Moreover, the percentage gain in cost achieved by the team of DMs, relative to the cost incured by a single DM obtaining a single observation, assymptotically goes to 0, as the variance of the primary DM goes to infinity (Figure 5, $P_0 = .8$).

An interesting insight can be obtained from Figure 5 ($P_0 = .8$). As the variance of the primary DM increases the percentage improvement in cost (defined above) is <u>initially increasing</u> and then decreasing assymptotically to zero. The reason for this is that for very small variances, the observations of the primary DM are so good that it does not need the information of the consulting DM. As the variance increases, the primary DM makes better use of the information and so the percentage improvement increases. But, at a certain point as the quality of the observations worsens, the primary DM finds less costly to start declaring more often the <u>more</u> likely hypothesis (ie to bias its decision towards the more likely hypothesis) than requesting for information, for reasons mentioned above, and so the percentage improvement from then on decreases.

<u>CASE 2 :</u> $\min(P_0, 1-P_0) > C1^* + C$

With reasoning similar to the above, we obtain that (Figure 4, $P_0 = .5$) :

$$\lim_{\sigma_0^2 \to \infty} P(u_\alpha = I) \to 1$$

Moreover, the percentage improvement is strictly increasing (and keeps incresing to a precomptutable limit ; Figure 5, $P_0 = .5$). This reinforces the last point we made in (Case 1) above. Since in the present case it is always less costly for the primary DM to request and use the information than to bias its decision towards the more likely hypothesis, the percentage improvement curve does not exhibit the non-monotonic behavior observed in (Case 1) above (where $P_0 = .8$).

5.2 Effects of varying the quality of the observations of the Consulting DM

As the variance of the consulting DM's observations increases, less information is requested by the primary DM, that is the primary DM's upper and lower thresholds move closer to each other (Figure 6). This is something we expected, since information of lesser quality is less profitable (more costly) to the team of DMs.

We should note here that the thresholds of a DM is an alternative way of representing the probabilities of the DM's decisions, since the decision regions are characterized by the thresholds. For example :

$$P(u_\alpha=I) = \sum_H \int_{y_\alpha:\, \psi_\alpha^l < y_\alpha < \psi_\alpha^u} P(y_\alpha \mid H)\, P(H)$$

The thresholds of the consulting DM demonstrate some interesting points of the team behavior (Figure 7). For small values of the variance they are very close together, as the quality of the observations is very good and so the consulting DM is willing to make the final team decision. As the variance increases, DM B becomes more willing to return u_β=? (i.e. "I am not sure") and let DM A make the final team decision. As the variance continues to increase, the thresholds of the consulting DM converge again. It might seem counter-intuitive, but there is a simple explanation. The consulting DM recognizes that the primary DM, despite knowing that the quality of the consulting DM's information is bad, is willing to incur the communication cost to obtain the information. This indicates that the primary DM is 'confused', that is, the a posteriori probabilities of the two hypotheses (given its observation) are very close together. Hense, the consulting DM becomes more willing to make the final decision. After a certain point ($\sigma_\beta^2 \approx 62.4$) the primary DM does find it worthwhile to request for information at all.

REMARK : Note in Figure 7 that the thresholds of the consulting DM converge to 1 which would have been the maximum likelihood threshold had the a priori probabilities of the two hypotheses been equal. But, the a priori probabilities *which the consulting DM uses in its calculations* are functions of the given a priori probabilities (ie p_1) and the fact that the primary DM requested for information (ie $P(u_\alpha=I \mid H_i)$). In fact, the consulting DM uses as its a priori probabilities its own estimates of the primary DM's a posteriori probabilities. That is :

$$P(H_0 \mid u_\alpha=I) = \frac{P(H_0)\, P(u_\alpha=I \mid H_0)}{\sum_H P(H)\, P(u_\alpha=I \mid H)} \tag{37}$$

From the above, we deduce that for large variances ($\sigma_\beta^2 \approx 62$) the estimates, of the consulting DM, for the a posteriori probabilities of the primary DM are very close to .5, reenforcing our point about the primary DM "being confused."

Finally, it is clear, that as the variance of the consulting DM increases, the percentage gain in cost, achieved by the team of DMs, decreases to 0, since the primary DM eventually makes all the decisions alone (centralized).

5.3 Effects of varying the Communication Cost

Increasing the communication cost is very similar to increasing the variance of the consulting DM, since in both cases the team "gets less for its money" (because the team has to incur an increased cost, either in the form of an increased communication cost, or in the form of the final cost, because of the worse performance of the consulting DM).

The thresholds of the primary DM, exhibit the same behavior as in 5.2 above (converging together at $C \approx .35$). The thresholds of the consulting DM (Figure 8) converge together for the same reasons as in 5.2 above. Of course, the thresholds do not start together for small values of the communication cost (as in 5.2), because low communication cost does not imply ability for the consulting DM to make accurate decisions. In fact, for small values of the communication cost, DM A is compelled to request for information more often than what is really needed and so the consulting DM returns more often $u_\beta=?$ (ie "I am not sure") and lets DM A make the team final decision.

Again it is clear that, as the communication cost increases, the percentage gain achieved by the team of the DMs decreases to zero (as the communication becomes more costly and less frequent, until we reach the centralized case).

5.4 Effects of varying the a priori probabilities of the hypotheses

This case does not present many interesting points. As expected, there is symmetry in the performance of the team around the line $p_0 = 0.5$. The closer p_0 is to 0.5 the more often information is requested by DM A (Figure 9) and the more often "I am not sure" is returned by DM B (Figure 10). This is understandable, because the closer p_0 is to 0.5, the bigger the a priori uncertainty. Consequently, the percentage improvement achieved by the team of the DMs is monotonically increasing with p_0 from 0 to 0.5 and monotonically decreasing from 0.5 to 1.

5.5 Effects of imperfect a priori information

CASE 1: Only the consulting DM knows the true p_0

From Figure 11, where the true p_0 is 0.8, we deduce that our model is relatively robust. If the primary DM's erroneous p_0 is anywhere between 0.7 and 0.9, performance of the team will be not more than 10% away from the optimum.

CASE 2: Only the primary DM knows the true p_0

As we see in Figure 12, where the true p_0 is 0.8, our model exhibits remarkable robustness qualities. If the consulting DM's erroneous p_0 is as far out as 0.01, the performance of the team will not be further than 7% away from the optimal. This can be explained by looking at the consulting DM's thresholds as functions of p_0 (Figure 13). We observe that for values of p_0 between 0.01 and 0.99, the thresholds do not change by much. This occurs because, as explained in detail in 5.2 above, the consulting DM knows that the primary DM requests for information when its a posteriori probabilities of the two hypotheses are roughly equal, which is the case indeed. As already stated, the consulting DM uses as its a priori probabilities its estimates of the a posteriori probabilities of the primary DM. Therefore, the consulting DM's estimates of the primary DM's a posteriori probabilities are good, besides the discrepancy in p_0, and the team's performance is not tampered by much.

Acknowledgment

The authors would like to thank Professor John N. Tsitsiklis for his valuable suggestions.

6. REFERENCES

[1] Van Trees, H.L. (1969)
Detection, Estimation and Modulation Theory, vol. I.
New York : Wiley, 1969

[2] Tenney, R.R., and Sandell, N.R. (1981)
Detection with Distributed Sensors
IEEE Trans. of Aerospace and Electronic Systems
17, 4 (July 1981), pp. 501-509

[3] Ekchian, L.K. (1982)
Optimal Design of Distributed Detection Networks
Ph.D. Dissertation, Dept. of Elect. Eng. and Computer Science
Mass. Inst. of Technology

[4] Ekchian, L.K., and Tenney, R.R. (1982)
Detection Networks
Proceedings of the 21st IEEE Conference on Decision and Control
pp. 686-691

[5] Kushner, H.J., and Pacut, A. (1982)
A Simulation Study of a Decentralized Detection Problem
IEEE Trans. on Automatic Control, 27, 5 (Oct. 1982)
pp. 1116-1119

[6] Chair, Z., and Varshney, P.K. (1986)
Optimal Data Fusion in Multiple Sensor Detection Systems
IEEE Trans. on Aerospace and Electronic Systems
21, 1, (January 1986), pp. 98-101

[7] Boettcher, K.L. (1985)
A Methodology for the Analysis and Design of Human Information Processing Organizations
Ph.D. Dissertation, Dept. of Elect. Eng. and Computer Science
Mass. Inst. of Technology

[8] Boettcher, K.L., and Tenney, R.R. (1985)
On the Analysis and Design of Human Information Processing Organizations
Proceedings of the 8th MIT/ONR Workshop on C^3 Systems
pp. 69-74

[9] Boettcher, K.L., and Tenney, R.R. (1985)
Distributed Decisionmaking with Constrained Decision Makers A Case Study
Proceedings of the 8th MIT/ONR Workshop on C^3 Systems
pp. 75-79

[10] Tsitsiklis, J.N. (1984)
Problems in Decentralized Decision Making and Computation
Ph.D. Dissertation, Dept. of Elec. Eng. and Computer Science
Mass. Inst. of Technology

[11] Tsitsiklis, J.N., and Athans, M. (1985)
On the Complexity of Decentralized Decision Making and Detection Problems
IEEE Trans. on Automatic Control, 30, 5, (May 1985)
pp. 440-446

APPENDIX : Solution to the General Problem

ASSUMPTION 1 : $J(1,H_0) > J(0,H_0) \quad ; \quad J(0,H_1) > J(1,H_1)$ (38)

or it is more costly for the team to err than to be correct.

This logical assumption is made in order to motivate the team members to avoid erring and in order to enable us to put the optinal decisions in lrt form.

ASSUMPTION 2 : $P(y_\alpha | y_\beta, H_i) = P(y_\alpha | H_i) \quad ; \quad P(y_\beta | y_\alpha, H_i) = P(y_\beta | H_i) \quad ; \quad i=0,1$ (39)

or the observations y_α and y_β are conditionally independent.

This assumption removes the dependence of the one observation on the other and thus allows us, as we are about to show, to write the optimal decision rules as lrt's with <u>constant</u> thresholds.

ASSUMPTION 3 : <u>Without loss of generality</u> assume that :

$$\frac{P(u_\beta=x \mid u_\alpha=I,H_0)}{P(u_\beta=x \mid u_\alpha=I,H_1)} \geq \frac{P(u_\beta=v \mid u_\alpha=I,H_0)}{P(u_\beta=v \mid u_\alpha=I,H_1)} \geq \frac{P(u_\beta=z \mid u_\alpha=I,H_0)}{P(u_\beta=z \mid u_\alpha=I,H_1)} \qquad (40)$$

This assumption is made in order to be able to distinguish between the messages of DM B.

THEOREM 1 : Given decision rules u_α and u_β, and that information is requested by the primary DM for some $y_\alpha \in Y_\alpha$ (i.e. $P(u_\alpha=I)>0$), then the optimal final decision of the primary DM after the information has been received, can be expressed as a deterministic function

$$\gamma_f : Y_\alpha \times \{x,v,z\} \rightarrow \{0,1\}$$

which is defined as likelihood ratio tests :

$$\gamma_f(y_\alpha,u_\beta)=\begin{cases} 0 & ; \text{ if } u_\beta=i \text{ and } \Lambda_\alpha(y_\alpha) \geq \alpha_i \\ 1 & ; \text{ otherwise} \end{cases} \qquad \text{for } i=x,v,z \quad (41)$$

where :

$$\Lambda_\alpha(y_\alpha) = \frac{p_0 P(y_\alpha \mid H_0)}{p_1 P(y_\alpha \mid H_1)} \qquad (42)$$

and :

$$\alpha_i = \frac{P(u_\beta=i \mid u_\alpha=I, H_1)\,[J(1,H_1)-J(0,H_1)]}{P(u_\beta=i \mid u_\alpha=I, H_0)\,[J(1,H_0)-J(0,H_0)]} \quad ; \quad i=x,v,z \qquad (43)$$

<u>Remark</u> : (29) is the equation for the corresponding threshold of the Gaussian example.

THEOREM 2 : Given the optimal decision rule u_f (derived in Theorem 1), a decision rule u_α and that information is requested for some $y_\alpha \in Y_\alpha$ (i.e. $P(u_\alpha=I)>0$),the optimal decision rule of the consulting DM is a deterministic

function

$$\gamma_\beta : Y_\beta \rightarrow \{x,v,z\}$$

defined as the following likelihood ratio tests :

$$\gamma_\beta(y_\beta) = \begin{cases} x & \text{if } \Lambda_\beta(y_\beta) \geq b_1 \text{ and } \Lambda_\beta(y_\beta) \geq b_2 \\ v & \text{if } \Lambda_\beta(y_\beta) < b_1 \text{ and } \Lambda_\beta(y_\beta) \geq b_3 \\ z & \text{if } \Lambda_\beta(y_\beta) < b_2 \text{ and } \Lambda_\beta(y_\beta) < b_3 \end{cases} \tag{44}$$

where :

$$\Lambda_\beta(y_\beta) = \frac{p_0\, P(y_\beta \mid H_0)}{p_1\, P(y_\beta \mid H_1)} \tag{45}$$

and :

$$b_1 = \frac{P(u_\alpha=I \mid H_1) \sum_{u_f} [P(u_f \mid u_\alpha=I, u_\beta=v, H_1) - P(u_f \mid u_\alpha=I, u_\beta=x, H_1)] J(u_f, H_1)}{P(u_\alpha=I \mid H_0) \sum_{u_f} [P(u_f \mid u_\alpha=I, u_\beta=x, H_0) - P(u_f \mid u_\alpha=I, u_\beta=v, H_0)] J(u_f, H_0)} \tag{46}$$

$$b_2 = \frac{P(u_\alpha=I \mid H_1) \sum_{u_f} [P(u_f \mid u_\alpha=I, u_\beta=z, H_1) - P(u_f \mid u_\alpha=I, u_\beta=x, H_1)] J(u_f, H_1)}{P(u_\alpha=I \mid H_0) \sum_{u_f} [P(u_f \mid u_\alpha=I, u_\beta=x, H_0) - P(u_f \mid u_\alpha=I, u_\beta=z, H_0)] J(u_f, H_0)} \tag{47}$$

$$b_3 = \frac{P(u_\alpha=I \mid H_1) \sum_{u_f} [P(u_f \mid u_\alpha=I, u_\beta=z, H_1) - P(u_f \mid u_\alpha=I, u_\beta=v, H_1)] J(u_f, H_1)}{P(u_\alpha=I \mid H_0) \sum_{u_f} [P(u_f \mid u_\alpha=I, u_\beta=v, H_0) - P(u_f \mid u_\alpha=I, u_\beta=z, H_0)] J(u_f, H_0)} \tag{48}$$

Equivalently, we can write :

$$\gamma_\beta(y_\beta) = \begin{cases} x & \text{if } \Lambda_\beta(y_\beta) \geq \beta_1 \\ v & \text{if } \Lambda_\beta(y_\beta) < \beta_1 \text{ and } \Lambda_\beta(y_\beta) \geq \beta_2 \\ z & \text{if } \Lambda_\beta(y_\beta) < \beta_2 \end{cases} \tag{49}$$

where :

$$\beta_1 = \max\{b_1, b_2\} \tag{50}$$

and :

$$\beta_2 = \min\{b_2, b_3\}. \tag{51}$$

Remark : (31) and (32) are the equations of the corresponding thresholds of the Gaussian example.

LEMMA 1 : Given the decision rule u_β of the consulting DM and the final decision rule u_f of the primary DM, the preliminary decision rule u_α of the primary DM can be expressed as a deterministic function

$$\gamma_\alpha : Y_\alpha \to \{0,1,I\}$$

defined as the following *degenerate* (because the thresholds are functions of y_α) lrts :

$$\gamma_\alpha(y_\alpha) = \begin{cases} 0 & \text{if } \Lambda_\alpha(y_\alpha) \geq a_1 \text{ and } \Lambda_\alpha(y_\alpha) \geq a_2 \\ I & \text{if } \Lambda_\alpha(y_\alpha) < a_2 \text{ and } 1/\Lambda_\alpha(y_\alpha) < 1/a_3 \\ 1 & \text{if } \Lambda_\alpha(y_\alpha) < a_1 \text{ and } 1/\Lambda_\alpha(y_\alpha) \geq 1/a_3 \end{cases} \tag{52}$$

where $\Lambda_\alpha(y_\alpha)$ is defined in (42) and :

$$a_1 = \frac{J(0,H_1) - J(1,H_1)}{J(1,H_0) - J(0,H_0)} \tag{53}$$

$$a_2 = \frac{\sum_{u_f,u_\beta} P(u_f \mid u_\alpha=I,u_\beta,y_\alpha)\, P(u_\beta \mid u_\alpha=I,H_1)[J(u_f,H_1) + C] - J(0,H_1)}{J(0,H_0) - \sum_{u_f,u_\beta} P(u_f \mid u_\alpha=I,u_\beta,y_\alpha)\, P(u_\beta \mid u_\alpha=I,H_0)[J(u_f,H_0) + C]} \tag{54}$$

$$a_3 = \frac{\sum_{u_f,u_\beta} P(u_f \mid u_\alpha=I,u_\beta,y_\alpha)\, P(u_\beta \mid u_\alpha=I,H_1)[J(u_f,H_1) + C] - J(1,H_1)}{J(1,H_0) - \sum_{u_f,u_\beta} P(u_f \mid u_\alpha=I,u_\beta,y_\alpha)\, P(u_\beta \mid u_\alpha=I,H_0)[J(u_f,H_0) + C]} \tag{55}$$

We proceed to show that the thresholds derived above are *independent* of y_α.

COROLLARY 1 : *If for some y_α information is requested, according to the rule of Lemma 1 and u_β=x (or z) is returned, then the optimal final decision u_f of the primary DM is always 0 (or 1); that is :*

$$P(u_f=0 \mid u_\alpha=I, u_\beta=x, y_\alpha) = 1 \quad \text{for all } y_\alpha \in \{y_\alpha \mid P(u_\alpha=I \mid y_\alpha)=1,\ y_\alpha \in Y_\alpha\} \tag{56}$$

and

$$P(u_f=1 \mid u_\alpha=I, u_\beta=z, y_\alpha) = 1 \quad \text{for all } y_\alpha \in \{y_\alpha \mid P(u_\alpha=I \mid y_\alpha)=1,\ y_\alpha \in Y_\alpha\} \tag{57}$$

Remark : From Corrolary 1 we can now give another interpretation to the team procedure: the primary DM can decide 0 or 1 using his own observation or can decide,because of uncertainty, to incur the communication cost (C) and shift the burden of the decision to the consulting DM. Then it is the consulting DM's turn to choose between deciding 0 or 1, or, because of uncertainty, shifting the burden back (at no cost) to the primary DM, which is required to make the final decision given his observation and the fact that the consulting DM's observation is not good enough for the consulting DM to make the final decision.

According to the above, we can simplify our notation of the consulting DM's messages by changing x to 0, z to 1 and v to ? (which is interpreted as the consulting DM saying "I am not sure").

Define the following secondary variables :

$$\Delta J_0 = J(1,H_0) - J(0,H_0) \tag{58}$$

$$\Delta J_1 = J(0,H_1) - J(1,H_1) \tag{59}$$

$$W^1 = \frac{\Delta J_0\, \Delta J_1\, [P(u_\beta=1 \mid H_1) - P(u_\beta=1 \mid H_0)]}{\Delta J_0 + \Delta J_1} \tag{60}$$

$$W^2 = \frac{\Delta J_0\, \Delta J_1 [P(u_\beta=?\mid H_0)P(u_\beta=1\mid H_1) - P(u_\beta=?\mid H_1)P(u_\beta=1\mid H_0)]}{\Delta J_0\, P(u_\beta=?\mid H_0) + \Delta J_1\, P(u_\beta=?\mid H_1)} \tag{61}$$

$$W^3 = \frac{\Delta J_0\, \Delta J_1 [P(u_\beta=?\mid H_1)P(u_\beta=0\mid H_0) - P(u_\beta=?\mid H_0)P(u_\beta=0\mid H_1)]}{\Delta J_0\, P(u_\beta=?\mid H_0) + \Delta J_1\, P(u_\beta=?\mid H_1)} \tag{62}$$

$$W^4 = \frac{\Delta J_0\, \Delta J_1\, [P(u_\beta=0\mid H_0) - P(u_\beta=0\mid H_1)]}{\Delta J_0 + \Delta J_1} \tag{63}$$

$$a_{2,1} = \frac{P(u_\beta=1\mid H_1)\, \Delta J_1 - C}{P(u_\beta=1\mid H_0)\, \Delta J_0 + C} \tag{64}$$

$$a_{2,2} = \frac{[P(u_\beta=1\mid H_1) + P(u_\beta=?\mid H_1)]\, \Delta J_1 - C}{[P(u_\beta=1\mid H_0) + P(u_\beta=?\mid H_0)]\, \Delta J_0 + C} \tag{65}$$

$$a_{3,1} = \frac{P(u_\beta=0\mid H_1)\, \Delta J_1 + C}{P(u_\beta=0\mid H_0)\, \Delta J_0 - C} \tag{66}$$

$$a_{3,2} = \frac{[P(u_\beta=0\mid H_1)+P(u_\beta=?\mid H_1)]\, \Delta J_1 + C}{[P(u_\beta=0\mid H_0)+P(u_\beta=?\mid H_0)]\, \Delta J_0 - C} \tag{67}$$

THEOREM 3 : Given the optimum final decision rule u_f of the primary DM (derived in Theorem 1) and the optimum decision rule u_β of the consulting DM, the optimum decision rule for the preliminary decision of the primary DM is given by a deterministic function

$$\gamma_\alpha : Y_\alpha \to \{0,1,I\}$$

defined by the following likelihood ratio tests :

$$\gamma_\alpha(y_\alpha) = \begin{cases} 0 & \text{if } \Lambda_\alpha(y_\alpha) \geq \alpha_1 \\ I & \text{if } \Lambda_\alpha(y_\alpha) < \alpha_1 \text{ and } \Lambda_\alpha(y_\alpha) \geq \alpha_2 \\ 1 & \text{if } \Lambda_\alpha(y_\alpha) < \alpha_2 \end{cases} \tag{68}$$

where :

$$\alpha_1 = \begin{cases} a_{2,1} & \text{if} \quad 0 \le C \le \min\{W^1, W^2\} \\ a_{2,2} & \text{if} \quad W^2 < C \le W^4 \\ a_1 & \text{otherwise} \end{cases} \tag{69}$$

$$\alpha_2 = \begin{cases} a_{3,1} & \text{if} \quad 0 \le C \le \min\{W^3, W^4\} \\ a_{3,2} & \text{if} \quad W^3 < C \le W^1 \\ a_1 & \text{otherwise} \end{cases} \tag{70}$$

<u>Remark</u> : (34) and (35) are the equations of the corresponding thresholds for the Gaussian example.

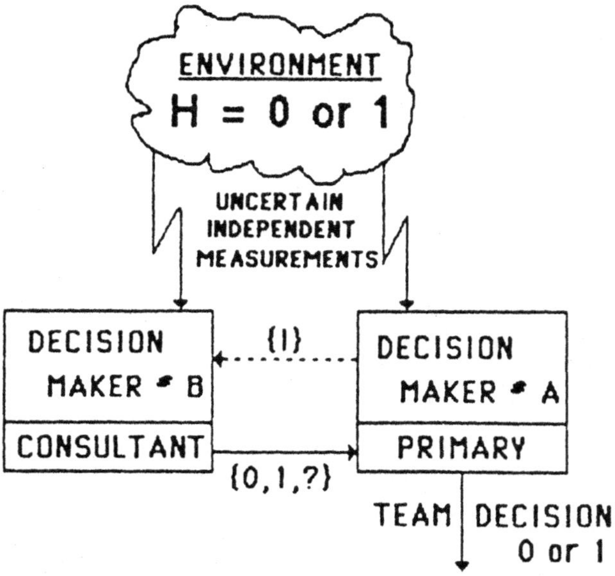

Figure 1 Problem Formulation

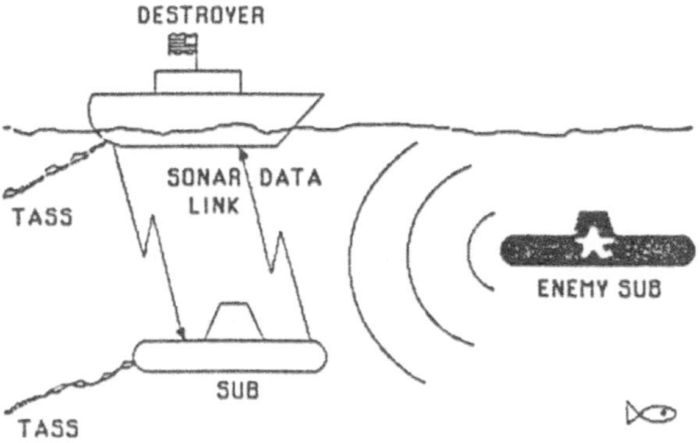

Figure 2. Anti-Submarine Warfare (ASW) Example

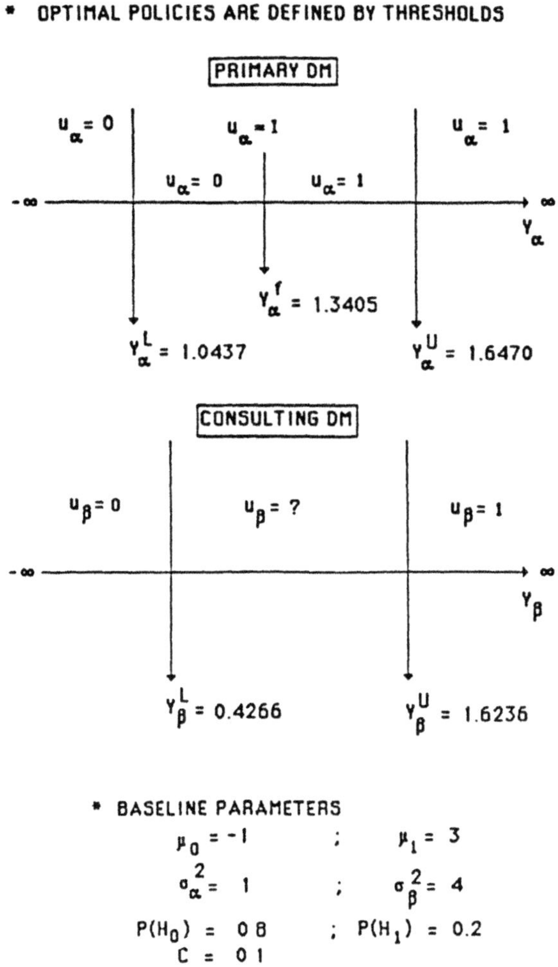

Figure 3. The Gaussian Case

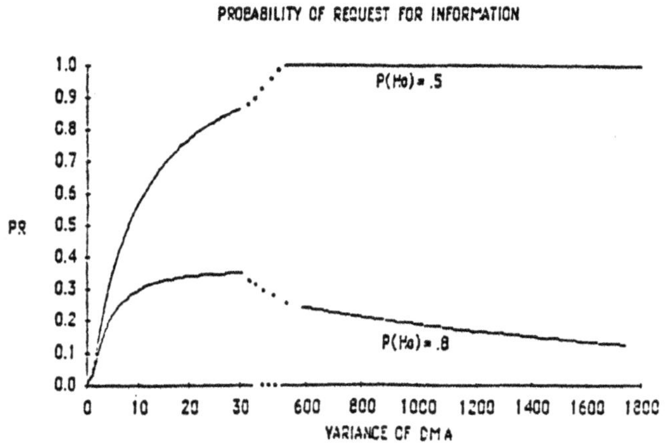

FIGURE 4:

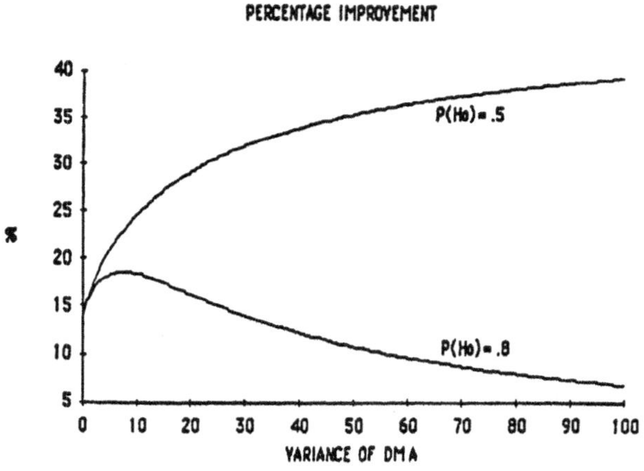

FIGURE 5

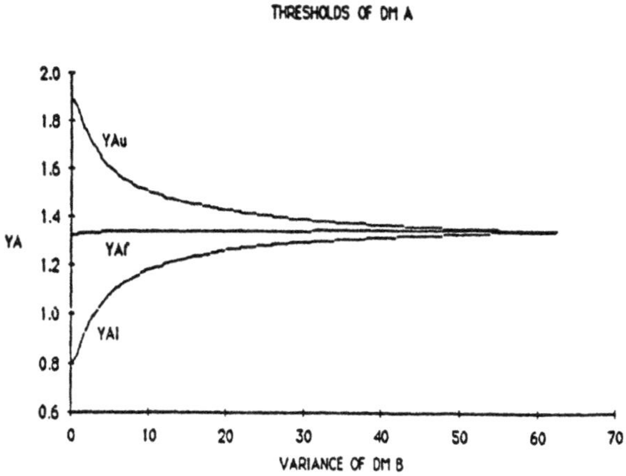

FIGURE 6

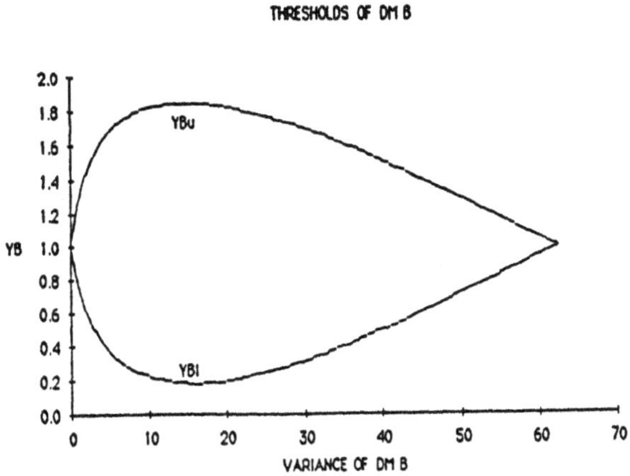

FIGURE 7

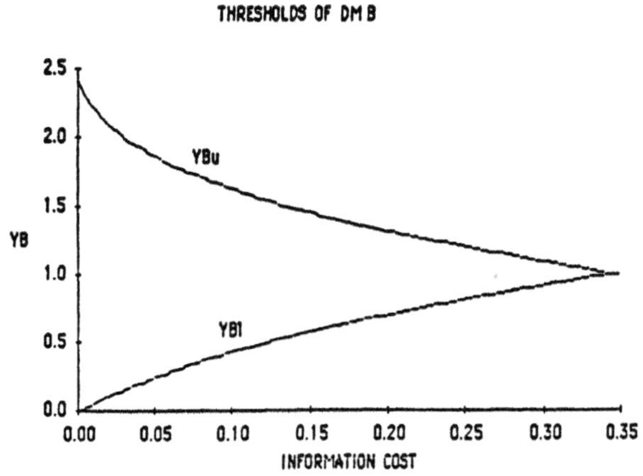

FIGURE 8

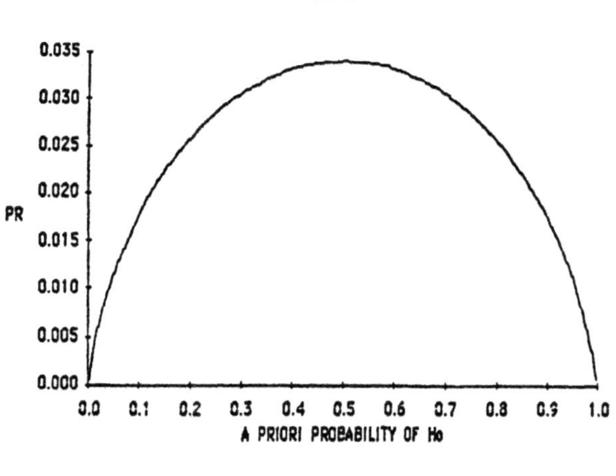

FIGURE 9

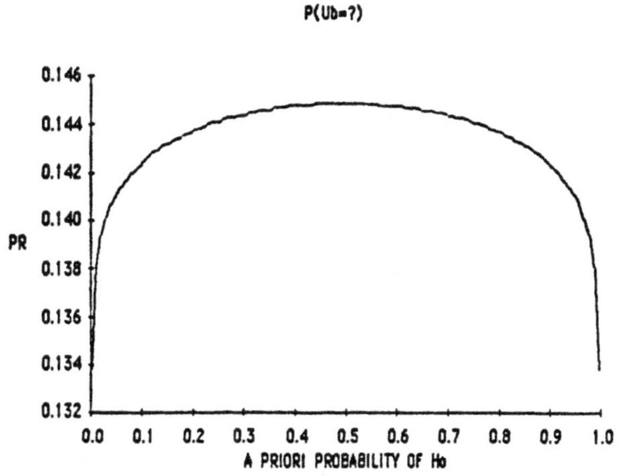

FIGURE 10

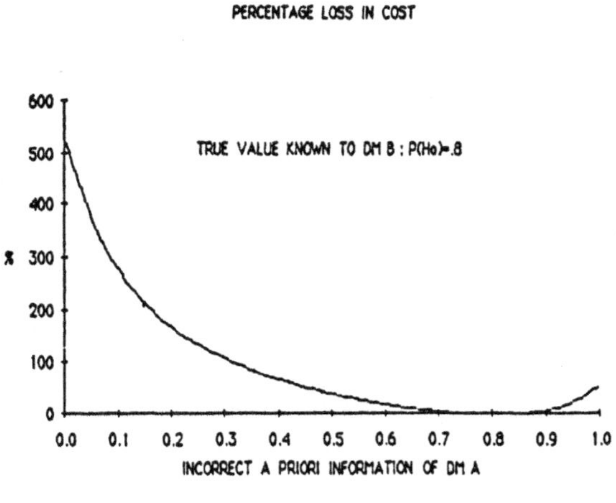

FIGURE 11

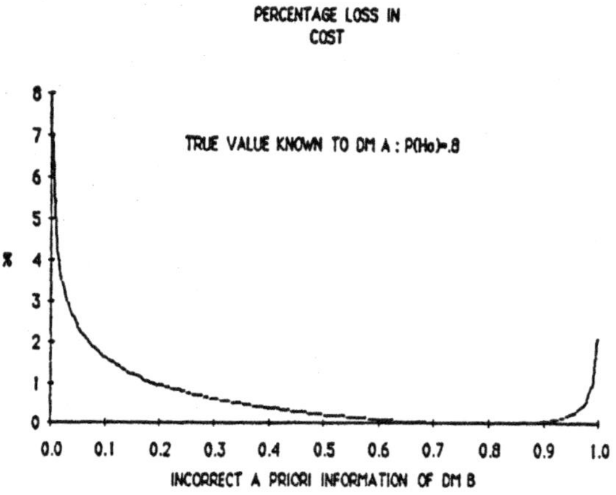

FIGURE 12

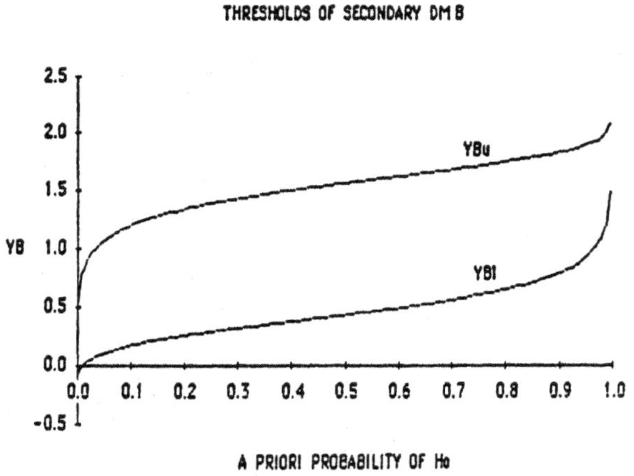

FIGURE 13

PART 2: FAULT DIAGNOSIS AND FAIL-SAFE/ROBUST CONTROL OF DYNAMIC SYSTEMS

COMPOSITE OBSERVERS FOR FAILURE DETECTION IN MULTIVARIABLE MULTI-TIME SCALE SYSTEMS

G. Dauphin-Tanguy, P. Borne
Laboratoire d'Automatique et d'Informatique Industrielle
Institut Industriel du Nord
B.P. 48
59651 Villeneuve d'Ascq Cédex - France

ABSTRACT. We propose a method for failure detection in complex multivariable multi-time scale systems. It consists in the introduction of redundancies in the global system, with overlapping transformations on the slow variables. This decomposition applied simultaneously with the singular perturbations method allows to build composite decentralised observers and to identify malfunctioning sensors. A comparison is made on an example between failure detection results obtained with full-rank composite observers and those deduced from the design of slow local subobservers.

INTRODUCTION

Most of the methods described in the literature (Willsky, 1976) for solving the problem of sensor fault detection are based on the idea of functional, or analytical redundancy, where, by making use of state estimations techniques, redundant signals are generated and treated in order to detect and isolate faulty instruments. In these methods, one state observer or several parallel observers are used to generate redundancy. They are of practically the same order as the overall system, and it is assumed that all the states of the system are completely observable from any one of the sensors. The condition of complete observability is highly restrictive and limits the range of applicability of these approachs. For large scale systems, the test for observability as well as the design of the observers would require a prohibitive computational effort due to the high dimension of these systems.

We propose a sensor fault detection method which overcomes these problems and can therefore be applied to large scale systems. Our purpose is to build redundant state estimators through simplifications in modelling by use of decentralisation and singular perturbations methods. A comparison on an example between results obtained with full rank composite observers and local slow subobservers is made.

S. Tzafestas et al. (eds.), System Fault Diagnostics, Reliability and Related Knowledge-Based Approaches, Vol. 1, 133–145.

1. FAILURE DETECTION BY DECENTRALISED OBSERVERS

The method proposed by (Singh & al, 1983) for the designing of decentralised observers uses overlapping decomposition technique of (Ikeda & Siljak, 1980).

Consider a large scale linear interconnected system (S) described by the following state and output equations :

$$\left| \begin{array}{l} \overset{\circ}{X} = A\,X + B\,u \qquad X \in \mathbb{R}^n \qquad u \in \mathbb{R}^m \qquad y \in \mathbb{R}^q \\ y = C\,X \end{array} \right. \tag{1}$$

and composed of N two-time-scale interconnected subsystems (S_i) associated to state vector $X_i \in \mathbb{R}^{n_i}$ and modelled under a singularly perturbed form as :

$$\left| \begin{array}{l} \begin{bmatrix} \overset{\circ}{x}_i \\ \mu_i\, \overset{\circ}{z}_i \end{bmatrix} = A^{(i)} \begin{bmatrix} x_i \\ z_i \end{bmatrix} + B^{(i)}\, u_i + \sum\limits_{\substack{j=1 \\ j\neq i}}^{N} H^{(ij)} \begin{bmatrix} x_j \\ z_j \end{bmatrix} \\ y_i = C_i \begin{bmatrix} x_i \\ z_i \end{bmatrix} \end{array} \right. \tag{2}$$

with :

$$x_i \in \mathbb{R}^{n_{i1}} \qquad z_i \in \mathbb{R}^{n_{i2}} \qquad A^{(i)} = \{A_{k\ell}^{(i)} \;/\; k,\ell = 1,2\}$$

$$B^{(i)} = \{B_j^{(i)} \;/\; j = 1,2\} \qquad H^{(ij)} = \{H_{k\ell}^{(ij)} \;/\; k,\ell = 1,2\}$$

In order to get an overlapping decomposition of the overall system and to construct the augmented system, we expand each subsystem by including the slow state variables in the preceding subsystem.

The overlapped subsystem $\tilde{S}$ is defined as :

$$\tilde{X} = \left[\tilde{X}_1^T\,,\,\dots\,,\,\tilde{X}_N^T \right]^T \qquad \tilde{X} \in \mathbb{R}^{\tilde{n}} \qquad \tilde{n} = n + \sum_{i=1}^{N-1} n_{i1} \tag{3}$$

with :

$$\left| \begin{array}{l} \tilde{X}_i = \left[x_i^T\,,\, z_i^T\,,\, x_{i+1}^T \right]^T \qquad i = 1,\,\dots,\,N-1 \\ \tilde{X}_N = X_N \end{array} \right. \tag{4}$$

The overlapped state vector is deduced from the original one X by a transformation T such as :

$$\tilde{X} = T\,X \qquad (5)$$

where T is a rectangular block-diagonal matrix.

The state and output equations of the expanded overall system $\tilde{S}$ are :

$$\left|\begin{array}{l} \mathring{\tilde{X}} = \tilde{A}\,\tilde{X} + \tilde{B}\,u \\ y = \tilde{C}\,\tilde{X} \\ \text{where :} \\ \tilde{A} = T\,A\,T^{I} + M \qquad \tilde{B} = T\,B + N \qquad \tilde{C} = C\,T^{I} + L \end{array}\right. \qquad (6)$$

M, N, L are complementary matrices of appropriate dimension, T^I is the generalized inverse of T.

In order to estimate the state $\tilde{X}_i$, we construct subsystem observers of the form :

$$\phi_i : \left|\begin{array}{l} \mathring{\hat{X}}_i = F^{(i)}\,\hat{X}_i + G_i\,y_i + \sum\limits_{\substack{j=1 \\ j\neq i}}^{N} L^{(ij)}\,\hat{X}_j + \tilde{B}^{(i)}\,u \\ \text{with :} \\ \left.\begin{array}{l} F^{(i)} = \tilde{A}^{(i)} - G_i\,\tilde{C}^{(i)} \\ L^{(ij)} = \tilde{H}^{(ij)} \end{array}\right| \quad i = 1, \ldots, N \end{array}\right. \qquad (7)$$

The output sensors failure detection is made at each time by comparison between the estimations of a same state by two different observers.

Because the redundancy is made on the slow part, then only the slow variables x_{i+1} in vectors $\tilde{X}_i$ and $\tilde{X}_{i+1}$ can be estimated simultaneously by the corresponding observers ϕ_i and ϕ_{i+1} $(i=1,\ldots,N-1)$.

Consider $(x_{i+1})_{\phi_i}$ and $(x_{i+1})_{\phi_{i+1}}$ the same state estimated by two different observers. In normal operation, we have :

$$\left| (x_{i+1})_{\phi_i} - (x_{i+1})_{\phi_{i+1}} \right| < \varepsilon_i \qquad i = 1, \ldots, N-1 \qquad (8)$$

When the ith sensor has failed, we have :

$$\left| \begin{array}{l} \left| (x_{i+1})_{\phi_i} - (x_{i+1})_{\phi_{i+1}} \right| > \varepsilon_i \\ \\ \left| (x_{i+2})_{\phi_{i+1}} - (x_{i+2})_{\phi_{i+2}} \right| < \varepsilon_{i+1} \end{array} \right. \tag{9}$$

When the $(i+2)^{th}$ sensor has failed, we have :

$$\left| \begin{array}{l} \left| (x_{i+1})_{\phi_i} - (x_{i+1})_{\phi_{i+1}} \right| < \varepsilon_i \\ \\ \left| (x_{i+2})_{\phi_{i+1}} - (x_{i+2})_{\phi_{i+2}} \right| > \varepsilon_{i+1} \end{array} \right. \tag{10}$$

When the $(i+1)^{th}$ sensor has failed, we have :

$$\left| \begin{array}{l} \left| (x_{i+1})_{\phi_i} - (x_{i+1})_{\phi_{i+1}} \right| > \varepsilon_i \\ \\ \left| (x_{i+2})_{\phi_{i+1}} - (x_{i+2})_{\phi_{i+2}} \right| > \varepsilon_{i+1} \end{array} \right. \tag{11}$$

Vectors ε_i and ε_{i+1} are fixed by the user according to sensors sensitivity, admissible tolerances, and imprecisions due to simplifications in modelling (as presented in the following section).

2. DESIGN OF DECENTRALISED OBSERVERS

The comparison between slow and fast redundancies has been developed in (Dauphin-Tanguy, 1986).

The initial subsystems S_i modelled by (2) become after expansion the augmented subsystems $\tilde{S}_i$ modelled by (12) as :

$$\left| \begin{array}{l} \begin{array}{ll} \tilde{S}_i : & \left| \begin{array}{l} \mathring{\tilde{X}}_i = \tilde{A}^{(i)} \tilde{X}_i + \tilde{B}^{(i)} \begin{bmatrix} u_i \\ u_{i+1} \end{bmatrix} + \sum_{\substack{j=1 \\ j\neq i}}^{N} \tilde{H}^{(ij)} \tilde{X}_j \\ \\ y_i = \tilde{C}^{(i)} \tilde{X}_i \end{array} \right. \\ i=1,\dots,N-1 & \end{array} \\ \\ \text{with :} \\ \\ \tilde{X}_i = \left[x_i^T, z_i^T, x_{i+1}^T \right]^T \\ \vdots \end{array} \right.$$

$$\tilde{A}^{(i)} = \left[\begin{array}{cc:c} \multicolumn{2}{c:}{A^{(i)}} & \begin{array}{c} H_{11}^{(i,i+1)} \\ H_{21}^{(i,i+1)} \end{array} \\ \hdashline H_{11}^{(i+1,i)} & H_{12}^{(i+1,i)} & A_{11}^{(i+1)} \end{array}\right]$$

$$\tilde{B}^{(i)} = \begin{bmatrix} B^{(i)} & 0^{(i)} \\ 0^{(i+1)} & B_1^{(i+1)} \end{bmatrix} \tag{12}$$

$$\tilde{H}^{(i,i+1)} = \left[\begin{array}{ccc} 0 & H_{12}^{(i,i+1)} & 0 \\ 0 & H_{22}^{(i,i+1)} & 0 \\ \hdashline 0 & A_{12}^{(i+1)} & 0 \end{array}\right]$$

$$\tilde{H}^{(i,j)} = \begin{bmatrix} 0 & H_{12}^{(i,j)} & 0 \\ 0 & H_{22}^{(i,j)} & 0 \\ 0 & H_{12}^{(i+1,j)} & 0 \end{bmatrix} \qquad \begin{array}{l} j \neq i+1 \\ \\ j \neq i \end{array}$$

$$\tilde{C}^{(i)} = \begin{bmatrix} C^{(i)} & 0^{(i+1)} \end{bmatrix}$$

($0^{(i)}$ and $0^{(i+1)}$ are zero-matrices of appropriate dimensions). $\tilde{S}_N$ is the same as S_N.

The state subvector $\tilde{X}_i$ has the two-time-scale property and this particularity will appear in the corresponding subsystem observer $\hat{X}_i$ and the estimation error vector e_i :

$$e_i = \tilde{X}_i - \hat{X}_i$$

where :

$$e_i = \left[e_{i,1}^T , e_{i,2}^T \right]^T$$

$$\text{with : } e_{i,1} = \begin{bmatrix} x_i - \hat{x}_i \\ x_{i+1} - \hat{x}_{i+1} \end{bmatrix} \qquad \text{slow part}$$

$$e_{i,2} = \left[z_i - \hat{z}_i \right]$$

The design of the state observer, defined in (7), needs the determination of gain matrices G_i leading to an asymptotically stable estimation error vector e_i.

Two steps of simplification in modelling will be used in this paper.

The error vector e_i satisfying the following equation :

$$\overset{\circ}{e}_i = (\tilde{A}^{(i)} - G_i \, \tilde{C}^{(i)}) \, e_i + \sum_{\substack{j=1 \\ j \neq i}}^{N} \tilde{H}^{(ij)} \, e_j, \tag{13}$$

a first step of simplification consists in the assumption that the interconnection terms $\Sigma \, \tilde{H}^{(ij)} e_j$ in (13) are neglectible (we assume that the conditions of validity of such an hypothesis given in (Singh & al, 1983) are satisfied).

A second step is the application of the singular perturbations technique in order to obtain a composite observer, by separated study of the slow and the fast parts (O'Reilly, 1979).

After permutation between state components, the slow variables and the fast ones are grouped and decoupled as :

$$\left| \begin{array}{l} \tilde{X}_s^{(i)} = \begin{bmatrix} x_i \\ x_{i+1} \end{bmatrix}_s \\ \overset{\circ}{\tilde{X}}{}_s^{(i)} = \tilde{A}_s^{(i)} \, \tilde{X}_s^{(i)} + \tilde{B}_s^{(i)} \begin{bmatrix} u_i \\ u_{i+1} \end{bmatrix}_s \\ y_{i_s} = \tilde{C}_s^{(i)} \, \tilde{X}_s^{(i)} + \tilde{D}_s^{(i)} \begin{bmatrix} u_i \\ u_{i+1} \end{bmatrix}_s \end{array} \right. \tag{14}$$

with :

$$\tilde{A}_s^{(i)} = \begin{bmatrix} A_{11}^{(i)} & H_{11}^{(i,i+1)} \\ H_{11}^{(i+1,i)} & A_{11}^{(i+1)} \end{bmatrix} - \begin{bmatrix} A_{12}^{(i)} \\ H_{12}^{(i+1,i)} \end{bmatrix} \begin{bmatrix} A_{22}^{(i)} \end{bmatrix}^{-1} \begin{bmatrix} A_{21}^{(i)} & H_{21}^{(i,i+1)} \end{bmatrix}$$

$$\tilde{B}_s^{(i)} = \begin{bmatrix} B_1^{(i)} & 0 \\ 0 & B_1^{(i+1)} \end{bmatrix} - \begin{bmatrix} A_{12}^{(i)} \\ H_{12}^{(i+1,i)} \end{bmatrix} \begin{bmatrix} A_{22}^{(i)} \end{bmatrix}^{-1} B_2^{(i)}$$

$$\tilde{C}_s^{(i)} = \begin{bmatrix} C_1^{(i)} & 0 \end{bmatrix} - C_2^{(i)} \begin{bmatrix} A_{22}^{(i)} \end{bmatrix}^{-1} \begin{bmatrix} A_{21}^{(i)} & H_{21}^{(i,i+1)} \end{bmatrix}$$

$$\tilde{D}_s^{(i)} = - C_2^{(i)} \begin{bmatrix} A_{22}^{(i)} \end{bmatrix}^{-1} B_2^{(i)}$$

and :

$$\left| \begin{array}{l} \tilde{X}_f^{(i)} = (z_i)_f \\ \overset{\circ}{\tilde{X}}{}_f^{(i)} = A_{22}^{(i)} z_{i_f} + B_2^{(i)} u_{i_f} \\ y_{i_f} = C_2^{(i)} z_{i_f} \end{array} \right. \tag{15}$$

To equation (14) is associated a slow state observer $\hat{X}_s^{(i)}$ with an estimation error defined by :

$$e_{i_s} = \hat{X}_s^{(i)} - \tilde{X}_s^{(i)} \tag{16}$$

satisfying :

$$\overset{\circ}{e}_{i_s} = \begin{bmatrix} \tilde{A}_s^{(i)} - G_{i_s} \tilde{C}_s^{(i)} \end{bmatrix} e_{i_s} \tag{17}$$

To equation (15) is associated a fast state observer $\hat{X}_{i,f}$ with an estimation error defined by :

$$e_{i_f} = \hat{X}_f^{(i)} - \tilde{X}_f^{(i)} \tag{18}$$

satisfying :

$$\mathring{e}_{i_f} = \left[A_{22}^{(i)} - G_{i_f} C_2^{(i)} \right] e_{i_f} \tag{19}$$

If the pairs :

$$\left[A_{22}^{(i)} , C_2^{(i)} \right] \text{ and } \left[\tilde{A}_s^{(i)} , \tilde{C}_s^{(i)} \right]$$

are completely observable, then G_{is} and G_{if} exist and the estimation gain matrix G_i can be obtained under a composite form as :

$$\left| \begin{array}{l} G_i = \left[G_{i1}^T , G_{i2}^T \right]^T \\ \text{with :} \\ \left| \begin{array}{l} G_{i1} = \begin{bmatrix} A_{12}^{(i)} \\ H_{12}^{(i+1,i)} \end{bmatrix} \left[A_{22}^{(i)} \right]^{-1} G_{if} + \\ \qquad + G_{is} \left[I - C_2^{(i)} \left[A_{22}^{(i)} \right]^{-1} G_{if} \right] \\ G_{i2} = G_{if} \end{array} \right. \end{array} \right. \tag{20}$$

(I is an identity matrix of appropriate dimensions) which leads to N composite estimators $\phi_{i,c}$ modelled by equations (7).

The redundancies being made on the slow variables, it is also possible to build local reduced observers which estimate only the slow redundant variables.

They are characterized by gains G_{is} obtained in (17) and are influenced by all the dynamical parts of inputs and outputs.

We define this second type of observers as $\phi_{i,s}$ modelled by :

$$\left| \begin{array}{l} \mathring{\hat{X}}_s^{(i)} = \left[\tilde{A}_s^{(i)} - G_{is} \tilde{C}_s^{(i)} \right] \hat{X}_s^{(i)} + G_{is} y_i + \\ \qquad + \left[\tilde{B}_s^{(i)} - G_{is} \tilde{D}_{is} \right] \begin{bmatrix} u_i \\ u_{i+1} \end{bmatrix} + \sum_{\substack{j=1 \\ j \neq i}}^{N} \tilde{H}_{11}^{(ij)} \hat{X}_s^{(j)} \end{array} \right. \tag{21}$$

3. IMPLEMENTATION ON AN EXAMPLE

The state and output equations are given by :

$$\left| \begin{array}{l} \mathring{X} = A\,X + B\,u \\ y = C\,X \end{array} \right. \tag{22}$$

with :

$$X \in \mathbb{R}^6,\ u \in \mathbb{R}^3,\ y \in \mathbb{R}^3$$

$$A = \left[\begin{array}{cc:cc:cc} -2 & 1 & 0 & 0 & 0 & 0 \\ 3 & -10 & 1.6 & 2 & 0 & 0 \\ \hdashline 0 & 1.6 & -1.5 & 1.2 & 0 & 0 \\ 0 & 1 & 2 & -8 & 1.2 & 1.5 \\ \hdashline 0 & 0 & 0 & 0.8 & -1 & 0.7 \\ 0 & 0 & 0 & 0.5 & 1.4 & -5 \end{array}\right]$$

$$B = \begin{bmatrix} B^{(1)} & & 0 \\ & B^{(2)} & \\ 0 & & B^{(3)} \end{bmatrix}, \quad B^{(i)} = \begin{bmatrix} 1 \\ 1 \end{bmatrix} \quad i=1,2,3 \tag{23}$$

$$C = \begin{bmatrix} C^{(1)} & & 0 \\ & C^{(2)} & \\ 0 & & C^{(3)} \end{bmatrix}, \quad C^{(i)} = \begin{bmatrix} 1 & 1 \end{bmatrix} \quad i=1,2,3$$

The augmented state vector is $\tilde{X}$ as :

$$\tilde{X} = \left[\tilde{X}_1^T , \tilde{X}_2^T , \tilde{X}_3^T \right]^T$$

with :

$$\tilde{X}_1 = \begin{bmatrix} x_1 \\ z_1 \\ x_2 \end{bmatrix} \qquad \tilde{X}_2 = \begin{bmatrix} x_2 \\ z_2 \\ x_3 \end{bmatrix} \qquad \tilde{X}_3 = \begin{bmatrix} x_3 \\ \\ z_3 \end{bmatrix} = X_3$$

We will notice the differences between two estimations of the same state vector as :

$$\left| \begin{array}{l} d_1 = (x_2)_{\phi_{1,c}} - (x_2)_{\phi_{2,c}} \\ \\ d_2 = (x_3)_{\phi_{2,c}} - (x_3)_{\phi_{3,c}} \end{array} \right. \qquad (24)$$

and :

$$\left| \begin{array}{l} d_1^* = (x_2)_{\phi_{1,s}} - (x_2)_{\phi_{2,s}} \\ \\ d_2^* = (x_3)_{\phi_{2,s}} - (x_3)_{\phi_{3,s}} \end{array} \right. \qquad (25)$$

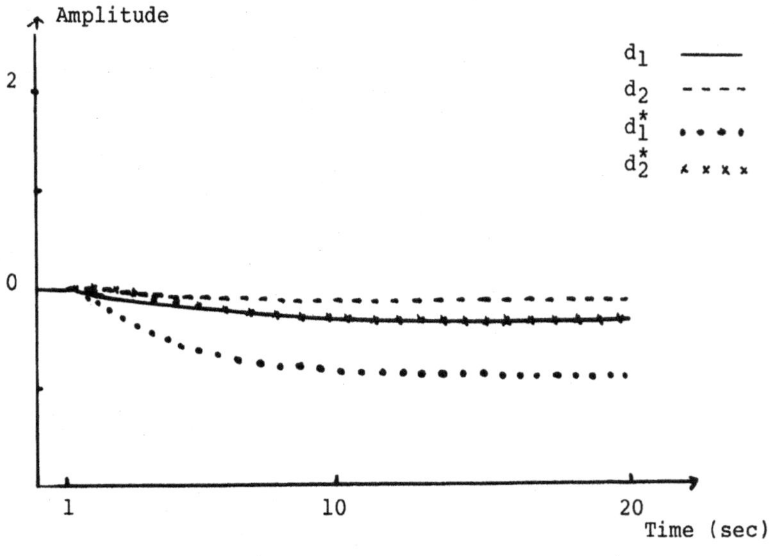

Figure 1. Normal operation

The Figure 1 shows the four values in normal operation. When no failure occurs, we can determine then steady-state values as :

$$| d_1 | \leqq 0.18 \qquad | d_2 | \leqq 0.08$$

$$| d_1^* | \leqq 0.45 \qquad | d_2^* | \leqq 0.18$$

Figures 2, 3, 4 show the cases of output sensor failure.

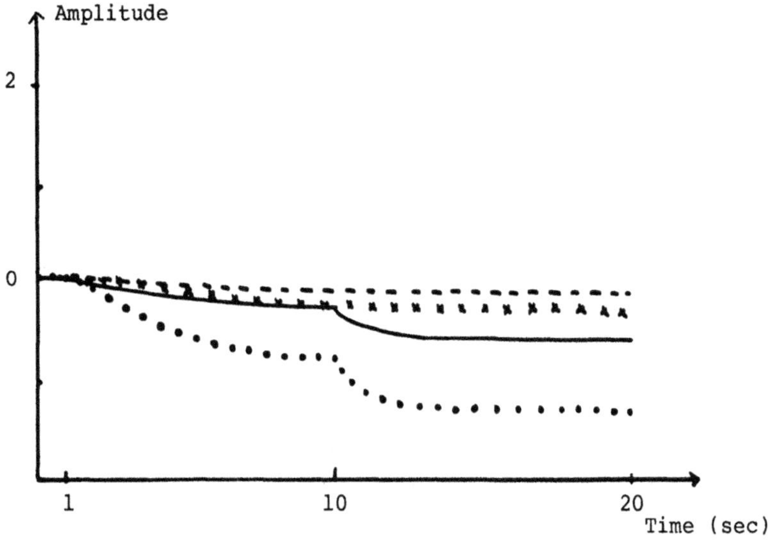

Figure 2. Failure on 1st output sensor

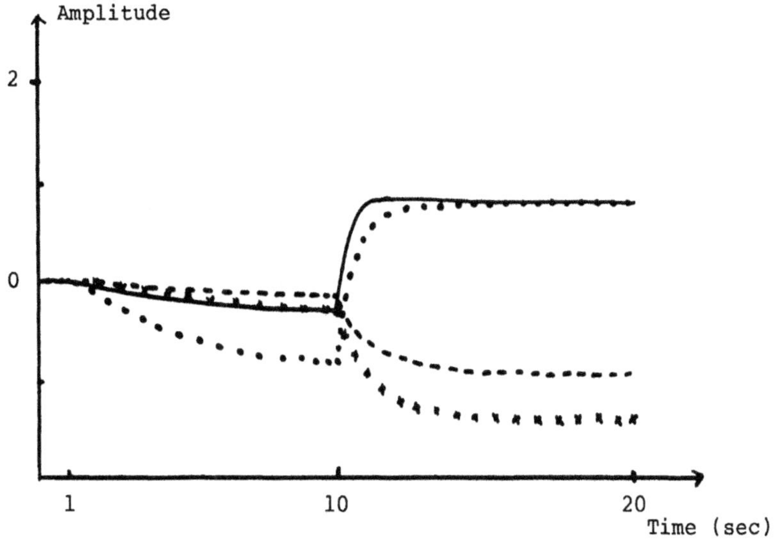

Figure 3. Failure on 2nd output sensor

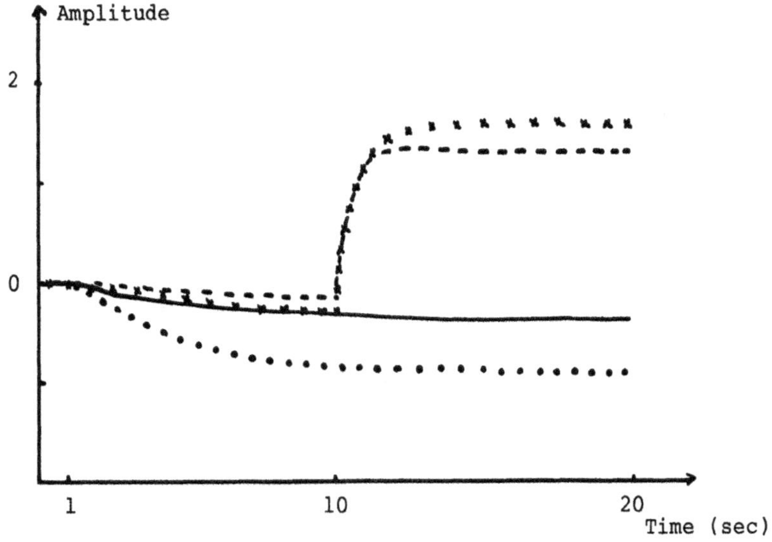

Figure 4. Failure on 3rd output sensor

INTERPRETATION

In normal operation, we see that the local slow observers are less accurate, especially $\phi_{1,s}$ (in d_1^*) than the composite ones, and the sensitivity parameters ε_i, ε_{i+1} must be choosen in that sense.

When a failure occurs in one ouput sensor, its detection is very easy in the two cases and it appears in the example that the amplitude of variations of the differences is greater for the slow local observers.

It means that local reduced observers are sufficiently efficient for failure detection. The monitoring system has only :

$$\sum_{i=2}^{N} n_{i1}$$

differences values to test in real time, and especially the derivatives of the variations to be able not only to detect the fault but also to identify the malfunctioning sensor.

CONCLUSION

The three tasks of a failure detection system are performed by the proposed method. The first one, "alarm", detects if something has gone wrong or everything is fine. The second one "isolation" determines the source of the failure and identifies the faulty sensor. The third one "estimation" decides if a sensor is completely non-operational or simply suffers degradation. It is done here by defining a sensitivity threshold according to errors intervals ε_i. For control purpose it is generally necessary to build full-rank observers, but we saw here that for failure detection purpose only local reduced observers are useful, of minimum--rank which gives efficiency in quickness of detection. The test being made on the derivatives of variations in differences values, the imprecisions introduced by the method do not matter. It is only necessary to know their steady-state values.

REFERENCES

Dauphin-Tanguy G. (1986)
'Failure detection by decentralised observers for complex multi-time scale systems'
Proc. of the 7th International Conference on Analysis and Optimization of Systems, Antibes (France), Lectures Notes in Control and Information Sciences, Springer Verlag, pp. 668-679.

Ikeda M., Siljak D.D. (1980)
'Overlapping decompositions, expansions and contractions of dynamic systems'
Large Scale Systems, n° 1, pp. 29-38.

O'Reilly J., (1979)
'Full-order observers for a class of singularly perturbed linear time--varying systems'
Int. J. Control, **30**, n° 5, pp. 745-756.

Singh M.G., Hassan M.F., Chen Y.L., Pan Q.R., (1983)
'New approach to failure detection in large-scale systems'
IEE Proceedings, **130**, Part D, n° 5, pp. 243-249.

Willsky A.S., (1976)
'A survey of design methods for failure detection in dynamic systems'
Automatica, **12**, pp. 601-611.

SENSOR FAULT DETECTION VIA ROBUST OBSERVERS

J.Wünnenberg, P.M.Frank
University of Duisburg
Department of Electrical Engineering
Measurement and Control Group
Bismarckstraße 81
4100 Duisburg 1
Federal Republic of Germany

ABSTRACT. This paper describes the design and the application of so called "Unknown Input Observers" or "Robust Observers" for failure detection. The fist part of the paper shows a design procedure for a discrete time observer which considers the concept of the "invariant subspace" and of the "almost invariant subspace". Design tool is the Kronecker canonical form which also gives a complete answer to the question of state estimation of discrete time systems acting under unknown inputs. The resulting observer is described by equations that allow the estimation of a linear combination of the state with a delay of a finite number of samplings. This robust state estimation though hardly usable for control purposes can successfully be considered for failure detection.

The second part of the paper shows as an example the application to instrument failure detection (IFD). A new IFD scheme is presented, and the applicability is demonstrated with the aid of simulation examples.

KEYWORDS. Robust observers; fault detection; linear systems; instrument fault detection.

INTRODUCTION

Various methods for the detection of instrument failures in automatic systems using signal processing techniques have been introduced in recent years. (Frank 1986). Basic idea of these IFD schemes is the use of Luenberger Observers that are acting like stabilized system models. By substracting the estimated output signals from the measured ones residuals are generated. These residuals are fed into a logical

S. Tzafestas et al. (eds.), System Fault Diagnostics, Reliability and Related Knowledge-Based Approaches, Vol. 1, 147–160.

device and serve as a criterion to decide whether an instrument failure has occured.

The known IFD schemes work satisfactorily as long as the model of the physical system is well known and no disturbances are present (Wünnenberg et.al., 1985). If this is not true, false alarms may occur or instrument failures may be undetected. Different approaches to reduce these effects have been attempted with varying degrees of success.

One approach to improve the robustness of the state estimation is the use of "Robust Observers" or "Unknown Input Observers". (Engell and Konik, 1986). The first part of the paper presents a design procedure for robust observers. From the existency conditions of a robust observer the consequences for the resulting IFD scheme are deduced. An IFD-scheme considering these design objectives is presented. With the aid of simulation examples it is illustrated to be sensitive against instrument failures and robust against modelled uncertainties. Furtheron it is shown that disturbances that were not or could not be considered during the observer design procedure can be handled by an appropriate choice of the decision thresholds inside the detection logic, as in conventional IFD schemes.

STATE OBSERVATION FOR SYSTEMS ACTING UNDER UNKNOWN INPUTS

In this part of the paper a brief description is given of how to obtain a "Robust Observer". Therefore we assume a discrete time description of a linear dynamical system of the form

$$x_{k+1} = A_0\, x_k + B_0\, u_k + E\, d_k \tag{1}$$

$$y_k = C\, x_k\ . \tag{2}$$

x_k represents the n-dimensional state vektor, u_k the known r-dimensional input vektor, d_k the s-dimensional unknown input vektor and y_k the l-dimensional output vektor. The unknown input distribution matrix E is of the form

$$E = \left[\Delta A,\ \Delta B,\ G\right] \tag{3}$$

where ΔA represents the difference between the nominal system matrix A_0 and the actual system matrix A, similar ΔB. G represents a matrix that distributes external disturbances. A_0, B_0, E and C are of appropriate dimensions.

The problem is now to design an observer such that the estimation error is decoupled from any unknown input signal d_k.

Various attempts have been made to achieve this goal. One of the possible ways for continuous time systems is described in (Engell and

Konik, 1986). Design tool is the "Kronecker Canonical Form" which is easy to handle without geometric theorie that is normally used. In addition numerically stable algorithms for the necessary transformations are available (Konik and Engell, 1986).

In the following the approach is derived for the case of discrete time systems.

OBSERVABILITY UNDER UNKNOWN INPUTS

Purpose of this chapter is to define observability under unknown inputs and to describe a numerical method to prove whether a system is observable under unknown inputs.

Starting from the system description (1) and (2) we perform the regular transformation of the state vector

$$x_k = \left[M,\ C^R \right] \begin{bmatrix} x_k^* \\ y_k \end{bmatrix} \tag{4}$$

where

$$C \left[M,\ C^R \right] = \left[0,\ I_l \right] \tag{5}$$

and I_l is equal to the lth-order identity matrix. This means that the state vector x_k is separated into the measurable part y_k and the unmeasurable x_k^*. x_k^* has to be estimated later by the observer.

From eq.(1), (2) and (4) we obtain the single equation

$$Mx_{k+1}^* - A_0 M x^*_k = B_0 u_k + Ed_k - C^R y_{k+1} + A_0 C^R y_k \tag{6}$$

Multiplying (6) from the left with the regular matrix

$\begin{bmatrix} N \\ E^L \end{bmatrix}$ where

$$\begin{bmatrix} N \\ E^L \end{bmatrix} E = \begin{bmatrix} 0 \\ I_q \end{bmatrix} \tag{7}$$

we obtain

$$NMx_{k+1}^* - NA_0 M x_k^* = NB_0 u_k - NC^R y_{k+1} + NA_0 C^R y_k \ . \tag{8}$$

On the left hand side of eq.(8) we have an expression with the unknown x_k^* . All elements on the right hand side are known or measurable.

By substituting

$$u_k^* = NB_0u_k - NC^Ry_{k+1} + NA_0C^Ry_k \tag{9}$$

eq.(8) reduces to

$$NMx_{k+1}^* - NA_0Mx_k^* = u_k^* \tag{10}$$

which is a system of difference equations that has to be solved. With help of the shift operator z eq.(10) can be rewritten as

$$(zNM - NA_0M)\ x_k^* = u_k^*(z) \tag{11}$$

or

$$H(z)\ x_k^* = u_k^*(z) \tag{12}$$

Kronecker showed (see e.g. Gantmacher, 1974) that by proper choice of the N- and M-Matrix the so called Matrix pencil H(z) can be transformed into the block diagonal Kronecker canonical form.

$$\begin{aligned} H(z) = \text{diag}\Big\{ & 0_{\mu_0,\epsilon_0}\ ;\ L_{\epsilon_1}(z),\ldots,\ L_{\epsilon_q}(z)\ ; \\ & zI_{\beta_I} - J_e^+\ ;\ zI_{\beta_s} - J_e^-\ ;\ zJ_0 - I_\zeta\ ; \\ & L_{\mu_1}^T(z),\ldots,\ L_{\mu_p}^T(z) \Big\} \end{aligned} \tag{13}$$

The $\epsilon_i (i = 0\ldots q)$ are the column indices and the μ_i $(i = 0\ldots p)$ are the row indices. The expression $0_{\mu o,\epsilon o}$ corresponds to zero rows or columns. The matrix L_{ϵ_i} is of dimension $\epsilon_i x(\epsilon_i + 1)$ and has the form

$$L_{\epsilon_i} = \begin{bmatrix} z & -1 & 0 & \cdots & 0 \\ 0 & z & -1 & \ddots & \vdots \\ \vdots & \ddots & \ddots & \ddots & 0 \\ 0 & \cdots & 0 & z & -1 \end{bmatrix} \tag{14}$$

and the corresponding matrix for the row indices is

$$L_{\mu_i}^T = \begin{bmatrix} z & 0 & \cdots & 0 \\ -1 & z & \ddots & \vdots \\ 0 & \ddots & \ddots & 0 \\ \vdots & \ddots & \ddots & z \\ 0 & \cdots & 0 & -1 \end{bmatrix} \tag{15}$$

The matrix J_e^+ is a β_I-dimensional matrix that has only unstable eigenvalues ($|z| \geq 1$). J_e^- represents a β_s-dimensional Jordan matrix with stable eigenvalues only.
J_0 is a Jordan matrix with all eigenvalues identical to zero. Consider now that part x_{0k}^* of the state vector x_k^* that corresponds to the J_0 block x_{0k}^* is determined by the difference equations

$$\begin{aligned}
-x_{02_{k+1}}^* + x_{01_k}^* &= u_{01_k}^* \\
-x_{03_{k+1}}^* + x_{02_K}^* &= u_{02_k}^* \\
&\vdots \\
-x_{0\zeta_{k+1}}^* + x_{0\zeta-1_k}^* &= u_{0\zeta-1_k}^* \\
x_{0\zeta_k}^* &= u_{0\zeta_k}^*
\end{aligned} \tag{16}$$

which are directly derived with the aid of of eq. (12) and (13). It is easily seen, that all components x_{0i}^* are completely determined by the known signals u_{0i}^*. x_{0i}^* is then calculated with a maximum delay of ζ time shifts where ζ is the dimension of the J_0-matrix.

We define the abbrevations

$$\epsilon = \epsilon_0 + \sum_{i=1}^{q} \epsilon_i + q \tag{17}$$

$$\mu = \sum_{i=1}^{p} \mu_i \tag{18}$$

and partition the appropriately chosen matrix M from eq.(4) into the matrices

$$M = \left[M_\epsilon, M_e^+, M_e^-, M_\infty, M_\mu\right] \; . \tag{19}$$

M_ϵ contains the first ϵ columns, M_e^+ the following β_I columns, M_e^- the next β_s, M_∞ the next ζ and M_μ the last μ columns of the M-matrix.

Analogously to the results given in (Engell and Konik, 1986) we state for discrete time systems:

A linear combination of the state variables

$$\omega_k = Rx_k \tag{20}$$

can be reconstructed

i) without delay but with free choice in the eigenvalues of the estimations error dynamic matrix if

$$R\left[M_\epsilon,\ M_e^+,\ M_e^-,\ M_\infty\right] = 0 \tag{21}$$

ii) without delay and without free choice of all eigenvalues of the estimation error dynamic matrix if

$$R\left[M_\epsilon,\ M_e^+,\ M_\infty\right] = 0 \tag{22}$$

iii) with delay of finite number of samplings and with free choise of the dynamics of the estimation error dynamics matrix if

$$R\left[M_\epsilon,\ M_e^+,\ M_e^-\right] = 0\ . \tag{23}$$

In terms of the geometrical theory we can say that the maximal (A,E) invariant subspace in the kernel of C is spannend by $\left[M_\epsilon,\ M_e^+,\ M_e^-\right]$.

Equation (16) gives a solution to the problem of "Disturbance decoupled estimation with anticipation" which is connected to the problem of "almost disturbance decoupling" and is outlined in (Willems, 1982).

We stated conditions for existency, structure and eigenvalues of the resulting observers as well as a basis for all possible matrices R. Numerically stable algorithms for the computation of an upper triangular form that contains all information of the Kronecker canonical form are available and documented (Konik and Engell, 1986).

ROBUST OBSERVERS

Let us assume a structure of the observer that can be expressed by

$$z_{k+1} = F\ z_k + G\ y_k + H\ u_k \tag{24}$$

$$\omega_k = J\ z_k + \sum_{i=0}^{\zeta} L_{1i}\ y_{k+i} + L_{2i}\ u_{k+i}\ . \tag{25}$$

The order of the system is determined by the matrix F. Equation (25) needs a delay of ζ samplings in order to be solvable.

An observer that estimates a linear combination of the state vector according to eq.(20) has to fulfil the following requirements

i) $\lim\limits_{k\to\infty} (\omega_k - R\ x_k) = 0$ for all x_0, z_0 (26)

and

ii) a matrix T must exist such that the equality

$$T\,x_0 = z_0 \qquad (27)$$

implies the equality

$$T\,x_k = z_k \qquad (28)$$

for all k.

From this we obtain the well known observer equations

$$TA_0 - FT = GC \qquad (29)$$

$$TE = 0 \qquad (30)$$

$$H = TB \qquad (31)$$

and

$$\left[J,\ L_{10}^{PI}\right]\begin{bmatrix} T \\ C \end{bmatrix} = R^{PI} \ . \qquad (32)$$

The expressions L_{10}^{PI} and R^{PI} express that only the part of the observer is calculated which satisfies eq.(22). The missing expressions for L_{1i} and L_{2i} can be easily derived from eq.16. The solution of eq.(29) to (32) can be found in an elegant way again by employing the matrix pencil (13). Here we have to refer to (Engell and Konik, 1986).

From the structure of the pencil (13) it becomes evident that an observation of the complete state vector (e.g. rank R=n) is impossible if rank E > rank C. If rank E = rank C it is impossible to observe the complete vector with a free choice in the eigenvalues of the estimation error dynamics matrix. From this we conclude that an IFD-scheme has to use as many output signals as possible in order to be as robust as possible in a geometric sense which means that it makes the rank of the E matrix as high as possible. The structure of the E matrix that is actually chosen depends on the given situation.

A ROBUST IFD SCHEME

In the standard IFD schemes based on observers (Clark, 1979) each of the observers is commonly driven by a single system output only. Considering the results of the previous section one realizes that this is not optimal from the view point of robustness. The observer should be driven by as many outputs as possible. No success was achieved so far to use this theory for the design of a single observer that is driven by all system outputs and generates a residual that allows to detect and localize all possible instrument failures. The next step would be to design a scheme using observers that are driven by all but one outputs of the system. Figure 1 shows as an example the structure of the proposed IFD-scheme for a system with three outputs.

In the following a procedure is given for the design of observers to generate residuals which are sensitive to instrument failures only. For the design of the i-th (i=1,...,l) observer the system state space description (1), (2) has to be extended

$$\begin{bmatrix} x_{k+1} \\ \hline x^R_{i_{k+1}} \end{bmatrix} = \left[\begin{array}{c|c} A_0 & 0 \\ \hline 0 & A^R_i \end{array}\right] \begin{bmatrix} x_k \\ \hline x^R_{i_k} \end{bmatrix} + \begin{bmatrix} B_0 \\ \hline 0 \end{bmatrix} u_k + \begin{bmatrix} E \\ \hline 0 \end{bmatrix} d_k \tag{33}$$

$$y^R_k = \left[\begin{array}{c|c} C & C^R_i \end{array}\right] \begin{bmatrix} x_k \\ \hline x^R_{i_k} \end{bmatrix} + y_F \tag{34}$$

where A^R_i and C^R_i represent addition of parts that have to be selected by the designer. A suitable choise will be demonstrated for the example of application. Y_F represents the signal that is generated by the instrument fault.

The matrix R (eq.(20)) has to be chosen as

$$R^* = \left[0_{t,n},\ R^R_i\right] \tag{35}$$

with $0_{t,n}$ a txn zero matrix and R^R_i again to the freedom of the designer.

In other words: We add uncontrollable and undisturbable states to the system description.

The observer will now estimate the product

$$\omega_{i_k} = R^R_i\, x^R_{i_k} \tag{36}$$

which acts as the residual. The instrument fault should now effect the residual what will always be the case if the eigenvalues of the observer differ from those of the A^R_i matrix. In other words: the residual is affected by the instrument faults if the observer is not simply a model of the (A^R_i, C^R_i) system. The mathematical conditions for existence of the IFD observer is derived from eq.(23)

$$R^* \left[M_\epsilon,\ M_e^+,\ M_e^-\right] = 0 \quad . \tag{37}$$

With this we achieve an observer that generates a residual ω_i which is sensitive against instrument faults in all but one instruments and which is robust against unknown inputs characterized by the product $E \cdot d_k$ in eq.(1).

For the example described in fig.1 we have a functional dependence that can be expressed by

$$\omega_1 = f(y_{F1},\ y_{F2}) \tag{38}$$

$$\omega_2 = f(y_{F2},\ y_{F3}) \tag{39}$$

$$\omega_3 = f(y_{F1},\ y_{F3}) \quad . \tag{40}$$

The y_{Fi} represent the i-th component of the fault vector y_F.

We now have to compare the residual with thresholds that have to be chosen by the designer. The thresholds have to be high enough to avoid false alarms if unmodelled disturbances are present. It is obvious that by introducing thresholds unlike zero one loses the ability to detect small instrument failures.

If we denote the i-th threshold by s_i the detection logic is as follows

if $|\omega_1| > s_1$ and $|\omega_3| > s_3$ $\rightarrow$ failure in instrument 1

if $|\omega_1| > s_1$ and $|\omega_2| > s_2$ $\rightarrow$ failure in instrument 2

if $|\omega_2| > s_2$ and $|\omega_3| > s_3$ $\rightarrow$ failure in instrument 3

This describes the overall robust IFD procedure that can be implemented on a digital computer and has to be carried out on-line.

APPLICATION EXAMPLE

The described IFD scheme was applied to the inverted pendulum control problem. For further reference see (Wünnenberg et.al., 1985).

The state vector of the controlled plant is:

$$x = \begin{bmatrix} r \\ \Phi \\ \dot{r} \\ \dot{\Phi} \end{bmatrix} - \begin{bmatrix} \text{position of base, meters} \\ \text{angular position of pendulum, rad.} \\ \text{velocity of base, m/s} \\ \text{angular velocity of pendulum, r/s} \end{bmatrix} \quad . \tag{41}$$

The components r, Φ and $\dot{r}$ of the system are measured with the aid of three sensors. The three measured variables are

Instrument 1: y_{1_k} = 11.71 volts/(meter of base displacement)

Instrument 2: y_{2_k} = 48 volt/(radian of pendelum displacement)

Instrument 3: y_{3_k} = 8.4 volts/(meter/sec of base velocity)

The most significant disturbance in the physical system is the friction of the base. This acts like an additional input signal u_k if the sample time is sufficiently small. Therefore the unknown input distribution matrix E (eq.1) is chosen to be equal to the control input distribution matrix B_0.

The extended models (eq.(33) and (34)) for the three observers of the IFD scheme having the same structure as described in Fig.1 are chosen to be:

$$\text{for observer 1: } x^R_{1_{k+1}} = x^R_{1_k} \; ; \; y^R_{2_k} = y_{2_k} + x^R_{1_k} \tag{42}$$

$$\text{for observer 2: } x^R_{2_{k+1}} = x^R_{2_k} \; ; \; y^R_{2_k} = y_{2_k} + x^R_{2_k} \tag{43}$$

$$\text{for observer 3: } x^R_{3_{k+1}} = x^R_{2k} \; ; \; y^R_{3_k} = y_{3_k} + x^R_{3_k} \; . \tag{44}$$

It can be seen that the matrix A^R_1 reduces to a scalar value and the systems described by eq.(42) to (44) are models for bias type errors in instrument 2 and instrument 3. The observers are designed according to the previously given procedure where the matrix R*(eq.35) is chosen to be

$$R^* = \left[00001\right].$$

The resulting observer poles are chosen to be 0.5, 0.6, 0.7, 0.8 so that the residuals are affected by the instrument faults

SIMULATION RESULTS

Figures 2, 3 and 4 show the response of the residual functions ω_1, ω_2 and ω_3 to different instrument faults and a parameter variation. A step command of 20 cm base displacement is given at t=0 to the closed loop system initially at rest.

Figure 2 shows that the bias fault of 0.1 volts in instrument 1 excites the residual functions ω_1 and ω_3. The observer acts as expected and the failure can be detected.

Figure 3 shows the response of the residue function to an abrupt bias of 0.5 volts in instrument 2. The residual functions ω_1 and ω_2 are effected and reach the same steady state value as the amplitude of the bias. This was intended by choosing the fault models as done in eq.(37) and (38).

Figure 4 shows the correct response of the residual functions to a scale factor fault in instrument 3. The residual function ω_1 is un-effected and the fault can be detected correctly.

Figure 5 shows the response of the residual function to a parameter variation. Eventhough the parameter variation is quite significant the actual values of the residual functions remain much smaller than the values demonstrated in fig. 2 to 4. It seems that the thresholds that have to be chosen in a real physical environment can be quite low especially if one takes into account that the whole IFD scheme is robust against the most significant disturbance namely the friction of the base.
Considering the equations (30) and (31) and the choice of the E matrix the IFD scheme is also robust against all disturbances in the input signal as well as in the actuator.

CONCLUSIONS

This work shows how to design discrete robust observers. The resulting observer structure may be realizable only with a limited time delay.

The proposed IFD-scheme is shown to be robust with respect to modelled disturbances. If unmodelled disturbances are present the generated residuals are affected and decision thresholds have to be placed. This effects the ability to detect 'small' sensor failures.

Further work has to be done to optimize the extended system model and to use the remaining freedom in the observer design to achieve an IFD scheme that gives a maximum of robustness with respect to the modelled and unmodelled disturbances and a high sensitivity with respect to instrument failures.

The next step will be a theoretical investigation and a practical implementation comparable to the work that is reported in (Wünnenberg et.al., 1985).

ACKNOWLEDGEMENT

Dipl.-Ing. D.Konik of the University of Duisburg provided invaluable advice during the work reported here. Thanks is expressed to Dipl.-Ing. C.Miller of the University of Duisburg for helpful discussions.

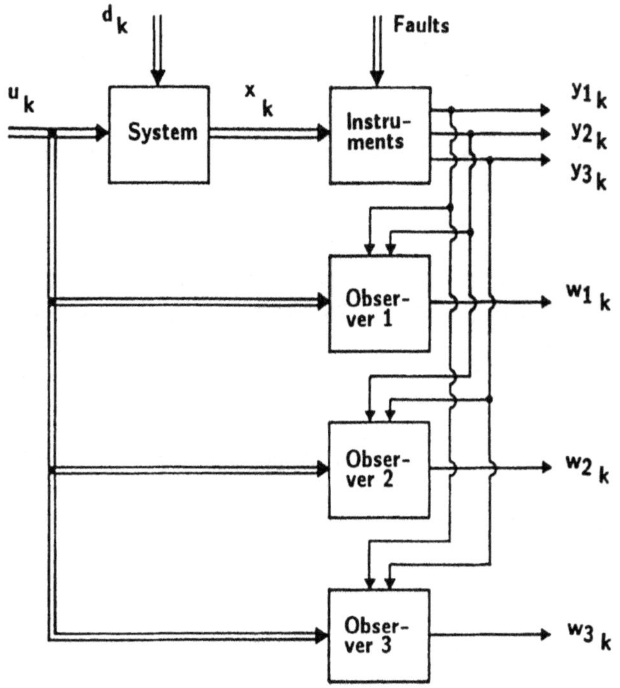

Fig.1: A robust IFD-scheme

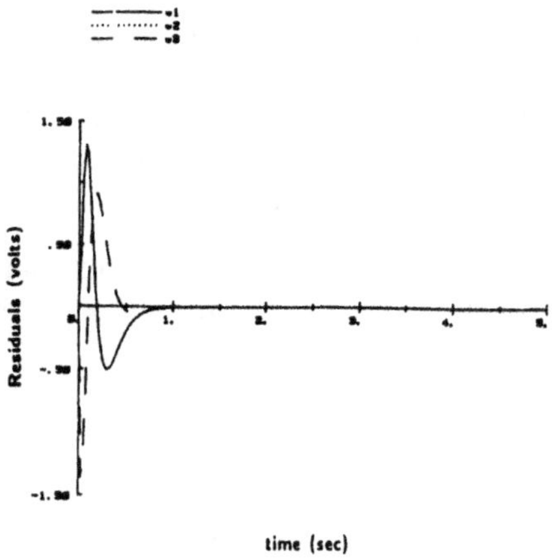

Fig.2: Response of the residual functions to an abrupt bias of 0.1 V in instrument 1

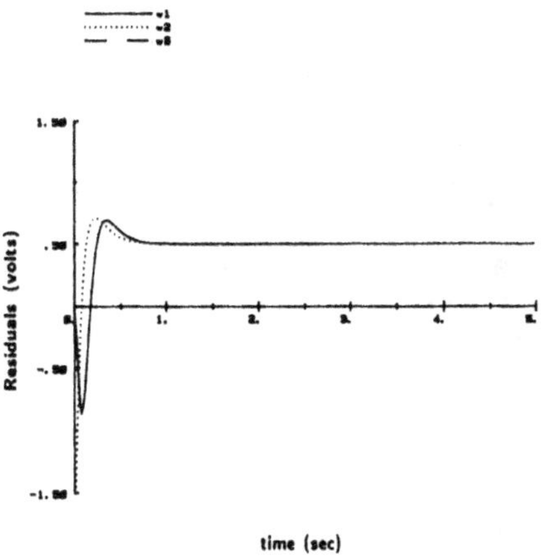

Fig.3: Response of the residual functions to an abrupt bias of 0.5 V in instrument 2

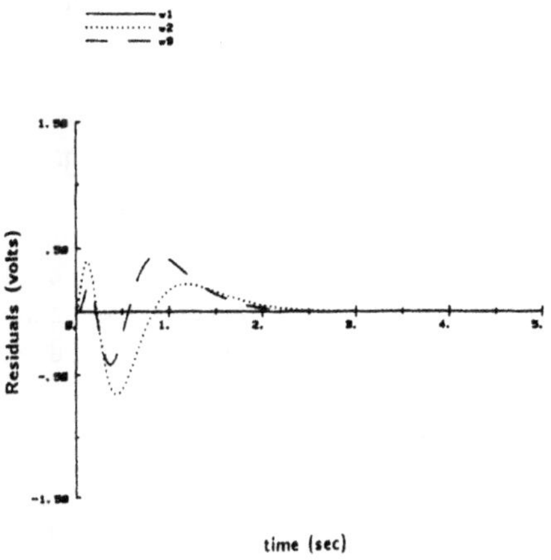

Fig.4: Response of the residual functions to a scale factor fault (0.8 of its nominal value) in instrument 3

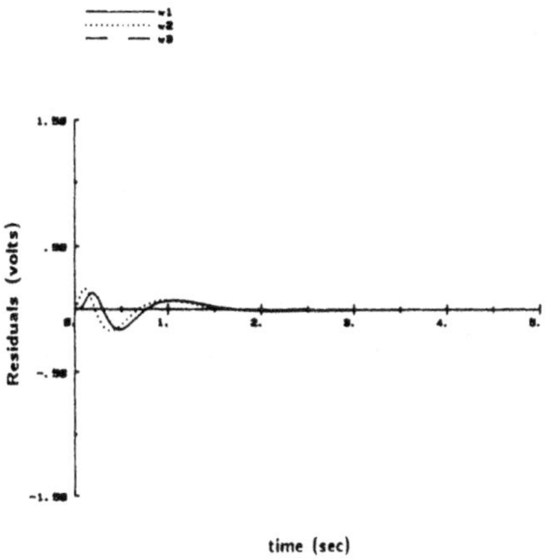

Fig.5: Response of the residual functions to a parameter variation (decrise of 80%) in the mass on top of the pendulum

REFERENCES

Clark, R.N. (1979). 'The Dedicated Observer Approach to Instrument Failure Detection'. *Proc. 18th IEEE CDC Fort Lauderdale*, Florida *pp. 237-241*

Engell, S. and Konik, D. (1986). 'Zustandsermittlung bei unbekanntem Eingangssignal'. *Automatisierungstechnik* at. 34.Jahrgang, vol.1/1986, pp. 38-42, and vol.6/1986, pp. 247-251

Frank, P.M. (1986). 'Fault Diagnosis in Dynamic Systems via State Estimation Methods'. *Survey paper, First European Workshop on Fault Diagnostics, Reliabitlity and Related Knowledge-Based Approaches.* Rhodes, Greece.

Gantmacher (1974). 'The Theory of Matrices' Volume II, *Chelsea Publishing Company*, New York, N.Y.

Konik, D. and Engell, S. (1986). 'Sequential Design of Decentralized Controllers using Decoupling Techniques. Part 2: Numerical Aspects and Application. *Proc. of the 4th IFAC/IFORS Symposium on Large Scale Systems: Theory and Applications.* Zurich, Switzerland

Willems, Jan, C. (1982). 'Almost Invariant Subspaces: An Approach to High Gain Feedback Design - Part II: Almost Conditionally Invariant Subspaces'. *IEEE Transactions on Automatic Control*, vol.AC-27, no.5, pp.1071-1085

Wünnenberg, J., Clark, R.N., and Frank P.M. (1985). 'An Application of Instrument Fault Detection'. *Proc 7th IFAC/IFORS Symposium on Identification and System Parameter Estimation*, York, UK, pp.699-704

DIAGNOSABILITY AND FAILURE DETECTION IN LINEAR DISCRETE DYNAMICAL SYSTEMS

S. Tanaka[†] and P. C. Müller[††]
† Department of Electronics Engineering, Faculty of Engineering, Yamaguchi University, Ube, 755 , Japan
†† Sicherheitstechnische Regelungs- und Meßtechnik, Bergische Universität-GH-Wuppertal, 56oo Wuppertal 1, West Germany

ABSTRACT. This paper analizes first the step-hypothesized generalized-likelihood-ratio(GLR) method for a general failure in linear discrete dynamical systems and discusses the detectionability. We find that a weakly-diagnosable-space can exist for dynamics failure rather than for sensor one. To overcome the problem, two extended GLR methods are next developed, that is, the one making use of another reduced-order GLR method which has effect for the anomaly appearing from near the weakly-diagnosable-space and the another taking into account the estimation of dominant low-frequencies' modes other than a bias. Some comments are maid on the detection methods and the detectionability, and several criteria for optimal sensor location are mentioned about.

1. INTRODUCTION

Failure diagnosis is important from two aspects. The first aspect is improvement of system availability, and the second, which is much more important, the protection to disasters. From this point of view, many failure detection techniques have been developed until now(see, for example,/1/). Especially, a generalized-likelihood-ratio(GLR) method is well-known as its rapid detection speed. But, in its use, it has a drawback such that we must assume a priori several hypotheses which are likely to happen. Step and jump hypotheses have been often used for their convenience. However, practically the anomaly function which adds to the system has in general a very complicated form (time-varying and multi-dimensional) which cannot be necessarily modeled by a simple hypothesis like a step. Despite it, we cannot easily throw the step hypothesis away because of the several fascinating points such that it does not need many parameters to be estimated and that there are often the cases where it can be applied well/5/. But, as far as the authors are aware of, the details has not been strictly discussed.

So, in this paper, we first discuss the detectionability by the method by applying a discriminating measure "divergence" to the distance between the two innovations sequences of different conditions, i.e., normal and abnormal states/3,4,5/, and show that if the system is

S. Tzafestas et al. (eds.), System Fault Diagnostics, Reliability and Related Knowledge-Based Approaches, Vol. 1, 161–175.

observable and the anomaly function has a bias, not necessarily a bias in the sense of arithmetic mean, the anomaly/failure can be in principle detected. It is also shown that if the bias is included in a weakly-diagnosable-space, it is difficult for the anomaly to be detected in the sense that the probability of the miss alarm cannot be made smaller than a specified value. Such a space can exist for dynamics failure rather than for sensor failure. So, we next consider two refined methods to overcome the problem, i.e., the one which makes use of another reduced-order GLR method which has effect for the anomaly appearing from near the weakly-diagnosable-space, and the another which takes into account the estimation of dominant low-frequencies' modes other than a bias.

We furthermore investigate the detection methods and the detectionability somewhat and show that if the input sequence is at least non-trivial, the failure has no weakly-diagnosable-space. Finally, we mention about several criteria for optimal sensor location based on the discussions of detectionability.

2. PROBLEM FORMULATION

We consider here the linear discrete dynamical system defined by

$$x(k+1) = Ax(k) + Bu(k) + \Gamma w(k) \tag{1}$$

$$y(k) = Hx(k) + v(k) \tag{2}$$

where $x(k) \in R^n$:the state, $u(k) \in R^r$:the input, $y(k) \in R^m$: the observation, and $w(k) \in R^p$ and $v(k) \in R^m$ are mutually independent white Gaussian noises with zero-means and covariances Q and R (positive definite). The initial state $x(o)$ is assumed to be Gaussian with mean $\hat{x}(o/-1)$ and covariance $P(o/-1)$ independent of $w(k)$ and $v(k)$. The matrices A and H are also assumed to be of full rank.

When an anomaly/failure occurs to the plant or the sensors, it generally causes changes ΔA, ΔB and ΔH in the system matrices A, B and H. This means that an unexpected time-varying function such as $\Delta Ax(k)$, $\Delta B u(k)$ or $\Delta Hx(k)$ adds to the dynamics or the sensor equations from anomaly occurrence time θ. Therefore, the anomaly can be modeled by the following equations.

$$x(k+1) = Ax(k) + Bu(k) + \Gamma w(k) + \sigma(k+1,\theta_1)f_1(k) \tag{3}$$

$$y(k) = Hx(k) + v(k) + \sigma(k,\theta_2)f_2(k) \tag{4}$$

where $f_1(k)$ and $f_2(k)$ are unknown anomaly vector functions of appropriate dimensions and $\sigma(k,\theta)$ is the unit step function defined by $\sigma(k,\theta) = o$ for $k < \theta$ and $\sigma(k,\theta) = 1$ for $k \geq \theta$. It is clear that the model is valid for a bias type anomaly.

We consider here first step-hypothesized GLR method to detect the failure and discuss the detectionability. Based on the discussions, we next consider some more refined alternatives for the failure detection and mention about optimal sensor location.

3. INNOVATIONS SEQUENCE

3.1. Behavior of the Innovations Sequence

The optimal state estimation of the system (1) and (2) is well-known to be done by the Kalman filter

$$\hat{x}(k/k-1) = A\hat{x}(k-1/k-1) + Bu(k-1) \tag{5}$$

$$\hat{x}(k/k) = \hat{x}(k/k-1) + K(k)\gamma(k) \tag{6}$$

$$\gamma(k) = y(k) - H\hat{x}(k/k-1) \tag{7}$$

$$P(k/k-1) = AP(k-1/k-1)A^T + \Gamma Q \Gamma^T \tag{8}$$

$$P(k/k) = P(k/k-1) - K(k)HP(k/k-1) \tag{9}$$

where

$$K(k) = P(k/k-1)H^T V^{-1}(k) \tag{10}$$

$$V(k) = HP(k/k-1)H^T + R \tag{11}$$

$\gamma(k)$ in (7) is called the innovations and the sequence becomes white Gaussian with mean zero and covariance $V(k)$ when the system is operating normally (we denote it by $\gamma^o(k)$). However, when an anomaly has occurred to the system at time θ, the effect $\Delta\gamma(k;\theta)$ by the anomaly appears on the innovations sequence $\gamma(k)$ from the time θ. Let us concern here only with the dynamics failure which causes $f_1(k)\sigma(k,\theta)$, since the quite similar discussion is possible for other cases.

Then, we have

$$\gamma(k) = \begin{cases} \gamma^o(k) & ; \text{ no-failure} \\ \Delta\gamma(k;\theta) + \gamma^o(k) & ; \text{ failure} \end{cases} \tag{12}$$

where

$$\begin{aligned} \Delta\gamma(k;\theta) &= G_1^o(k;\theta) * f_1(k) \\ &\triangleq \sum_{j=\theta}^{k} G_1^o(k;j) f_1(j) \quad \text{(convolutional sum)} \end{aligned} \tag{13}$$

Here, $G_1^o(k;\theta)$ is a failure signature matrix for dynamics jump/2/.

3.2. A Discriminating Measure "Divergence"

When an anomaly has occurred to the system, the innovations sequence $\gamma(k)$ behaves differently from it does under no failure (see (12)). Hence, whether the anomaly detection can be easily maid or not depends quitely on the difference between the two innovations sequences of the two conditions, i.e., no-failure and the failure. We adopt here divergence to measure the distance between the two sequences and discuss the detectionability by the step-hypothesized GLR method.

Definition.
For the two probability density functions $p_1(y)$ and $p_2(y)$ $(y \in Y)$, the divergence is defined by /6/

$$D(1;2) = \int_Y p_1(y)\ell n\left(\frac{p_1(y)}{p_2(y)}\right)dy + \int_Y p_2(y)\ell n\left(\frac{p_2(y)}{p_1(y)}\right)dy \tag{14}$$

Properties.
(i) $D(1;2) \geq 0$, and $D(1;2) = 0$ if and only if $p_1(y) = p_2(y)$.
(ii) invariant under non-singular transformations.

The divergence is calculated for $\gamma(k)$'s using (12) and it becomes

$$D(H_o;H_1) = \sum_{j=\theta}^{k} \Delta\gamma(j;\theta)^T V^{-1}(j)\Delta\gamma(j;\theta) \tag{15}$$

where H_o and H_1 mean respectively the condition of no-failure and a failure.

4. DETECTIONABILITY BY USUAL STEP-HYPOTHESIZED GLR METHOD

We first consider the detectionability when the usual step-hypothesized GLR method is applied to the failure detection. For simplicity of discussions, we introduce now the inner product defined by

$$(a(k),b(k)) \triangleq \sum_{j=\theta}^{k} a^T(j)V^{-1}(j)b(j) \tag{16}$$

(valid throughout the paper) where $a(k)$ and $b(k)$ are vector functions of dimension m. Using the notation, (15) can be rewritten as

$$\begin{aligned} D(H_o;H_1) &= (\Delta\gamma(k;\theta),\Delta\gamma(k;\theta)) \\ &= \| \Delta\gamma(k;\theta)\|^2 \end{aligned} \tag{17}$$

We decompose now $f_1(k)$ into a constant $\nu_1 \in R^n$ and a time-varying component $\varepsilon_1(k)$ which satisfy

$$(G_1^o(k;\theta)*e_i, G_1^o(k;\theta)*\varepsilon_1(k)) = 0 \ , \quad 1 \leq i \leq n \tag{18}$$

where e_i is the i-th natural basis vector in R^n. Note that ν_1 is not necessarily the arithmetic mean $\bar{\nu}$ of $f_1(k)$, but also a vector which minimizes $\|G_1^o(k;\theta)*(f_1(k)-\nu)\|^2$ with respect to ν. Explicitly writing,

$$\nu_1 = \bar{\nu} + \bar{\Delta}^{-1}\bar{\varepsilon} = \bar{\Delta}^{-1}\bar{f} \tag{19}$$

where $\bar{\Delta}$ is the nxn matrix whose (i,j) element is $(G_1^o(k;\theta)*e_i, G_1^o(k;\theta)*e_j)$, and $\bar{\varepsilon}$ and $\bar{f}$ are the n-dimensional vectors whose i-th components are respectively given by $(G_1^o(k;\theta)*e_i, G_1^o(k;\theta)*(f_1(k)-\bar{\nu}))$ and $(G_1^o(k;\theta)*e_i, G_1^o(k;\theta)*f_1(k))$. From (13), (17) and (18), we obtain

$$
\begin{aligned}
D(H_o;H_1) &= \| G_1^o(k;\theta)*\nu_1 + G_1^o(k;\theta)*\varepsilon_1(k) \|^2 \\
&= \| G_1^o(k;\theta)*\nu_1 \|^2 + \| G_1^o(k;\theta)*\varepsilon_1(k) \|^2 \\
&= \| G_1(k;\theta)\nu_1 \|^2 + \| G_1^o(k;\theta)*\varepsilon_1(k) \|^2 \\
&= \nu_1^T C_1(k;\theta)\nu_1 + D(H_1(\nu_1);H_1) \qquad (20)
\end{aligned}
$$

where

$$G_1(k;\theta) \triangleq \sum_{j=\theta}^{k} G_1^o(k;j) \qquad (21)$$

is a failure signature matrix for a dynamics step and $C_1(k;\theta)$ defined by

$$C_1(k;\theta) \triangleq \sum_{j=\theta}^{k} G_1^T(j;\theta)\, V^{-1}(j)G_1(j;\theta) \qquad (22)$$

becomes the information matrix concerning the constant vector ν_1. $D(H_1(\nu_1);H_1)$ represents the divergence between the two innovations sequences of the true failure H_1 and the dynamics step failure $\nu_1\sigma(k,\theta)$ (we denote it by $H_1(\nu_1)$).

Taking the difference between the two divergences $D(H_o;H_1)$ and $D(H_1(\nu_1);H_1)$, we obtain

$$D(H_o;H_1) - D(H_1(\nu_1);H_1) = \nu_1^T C_1(k;\theta)\nu_1 \qquad (23)$$

If the system is observable, $C_1(k;\theta)$ becomes a positive definite matrix and so the quantity (23) takes the positive value for any anomaly/failure such that $\nu_1 \neq 0$. Furthermore, the information matrix satisfies $C_1(k;\theta) \leq C_1(k+1;\theta)$ for any $k \geq \theta$. This means that we can detect the anomaly by a dynamics step hypothesis, only if ν_1 takes a non-zero value. If we wait to judge till (23) takes an appropriate large value, we can expect to detect the anomaly with a sufficient accuracy.
In fact, calculating the maximum likelihood estimate (MLE) of ν_1 and the generalized likelihood ratio (GLR), we obtain

$$\hat{\nu}_1(k;\theta) = C_1^{-1}(k;\theta) \sum_{j=\theta}^{k} G_1^T(j;\theta)V^{-1}(j)\gamma(j) \qquad (24)$$

$$
\begin{aligned}
\ell(k;\theta) &= \ell n \left[\frac{p(\gamma^k/H_1(\nu_1),\theta,\hat{\nu}_1(k;\theta))}{p(\gamma^k/H_o)} \right] \\
&= (1/2)\hat{\nu}_1^T(k;\theta)C_1(k;\theta)\hat{\nu}_1(k;\theta) \qquad (25)
\end{aligned}
$$

Substituting $\gamma(j) = G_1(j;\theta)\nu_1 + G_1^o(j;\theta)*\varepsilon_1(j) + \gamma^o(j)$ into (24), we find that the MLE $\hat{\nu}_1(k;\theta)$ obeys the Gaussian distribution with mean ν_1 and covariance $C_1^{-1}(k;\theta)$ and thus the probability density of the GLR $\ell(k;\theta)$ becomes a non-central Chi-squared one with non-centrality parameter

$$\delta^2 = \nu_1^T C_1(k;\theta)\nu_1 \qquad (26)$$

This quantity is nothing but (23) and this fact shows us that the foregoing statement is right. The quantity (23) is also equivalent to the divergence $D(H_o;H_1(\nu_1))$ between the no-failure H_o and the dynamics step failure $\nu_1\sigma(k,\theta)$.

From these considerations, we can define the detectionability of the failure $f_1(k)$ which has a bias ν_1 by the divergence $D(H_o;H_1(\nu_1))$, when the step-hypothesized GLR method is applied. Since $C_1(k;\theta) \leqq C_1(k+1;\theta)$ holds for any $k \geqq \theta$, the divergence generally increases with k. However, we see from the following theorems that it depends on the system and the anomaly.

Theorem 1.
We assume the stationary state. Then, we let $\Pi \overset{\Delta}{=} A(I_n - K^*H)$, where K^* is a Kalman gain in the stationary state and I_n is the $n \times n$ identity matrix. Then, if the condition

$$H(I_n - \Pi)^{-1}\nu_1 = 0 \tag{27}$$

is satisfied, the divergence $D(H_o;H_1(\nu_1))$ is saturated by increasing k, i.e., the divergence is bounded.
(proof)
Using the recursive equations on the failure signature matrix $G_1^o(k;\theta)$ (see /2/, for example), we can obtain the effect by the dynamics step ν_1 on the innovations by

$$\begin{aligned}\Delta\gamma(k;\theta) &= \sum_{j=\theta}^{k} G_1^o(k;j)\nu_1 \\ &= H\,[\,I_n + \sum_{j=\theta}^{k-1}(\,\Pi_{k-1}\Pi_{k-2}\cdots\Pi_j)]\nu_1\end{aligned} \tag{28}$$

where

$$\Pi_j \overset{\Delta}{=} A(I_n - K(j)H)$$

Thus, in the stationary state, we have

$$\Delta\gamma(k;\theta) = H\sum_{j=o}^{k-\theta}\Pi^j\nu_1 \tag{29}$$

Since we are assuming the stationary state, Π is a stable matrix. This is easily verified by considering that the recursive equation of the estimation error $e(k/k) = x(k) - \hat{x}(k/k)$ is given by

$$e(k/k) = (I_n - K(k)H)Ae(k-1/k-1) + (I_n - K(k)H)\Gamma w(k-1) - K(k)v(k) \tag{30}$$

and that the matrices $(I_n - K(k)H)A$ and $A(I_n - K(k)H)$ are similar to each other. Combining the fact with (29), we can conclude that under the condition (27) the divergence $D(H_o;H_1(\nu_1))$ is saturated by k (see (17)). (Q.E.D.)

This theorem means that for the anomaly satisfying (27) we can not make the probability of miss alarm smaller than a certain value, even if we used many observation data. We call the space constructed by such a ν_1 "a weakly-diagnosable-space". The space is the null space of the matrix $H(I_n - \Pi)^{-1}$ and the dimension of the space becomes (n-m). The detectionability in the space depends also on the direction of ν_1 (see

(23) or (26)). As an illustration, we consider the case where A is a stable matrix. Then, we easily have

$$H(I_n - \Pi)^{-1} = [I_m + HA(I_n - A)^{-1}K^*]^{-1}H(I_n - A)^{-1} \tag{31}$$

So, the space becomes the null space of $H(I_n - A)^{-1}$. This space is equivalent to one such that the effect of the anomaly on observation y(k) vanishes meanwhile for the element ν_1. This fact agrees with the intuitive consideration.

The similar discussions are possible for sensor failure and the composite one causing both sensor and dynamics biases.

Theorem 2.
If the system is observable and the matrix $H(I_n - \Pi)^{-1}(I_n - A)$ is not of full rank, a weakly-diagnosable-space exists, and the space is given by the range space of Ω by H, i.e., $H\Omega$, where Ω is the null space of $H(I_n - \Pi)^{-1}(I_n - A)$.
(Proof)
As similarly as in Theorem 1, we can obtain the effect by the sensor step ν_2 as follows :

$$\Delta\gamma(k;\theta) = [I_m - H(AK(k-1) + \sum_{j=\theta}^{k-2} \Pi_{k-1}\Pi_{k-2}\cdots\Pi_{j+1}AK(j))]\nu_2 \tag{32}$$

So, in the stationary state, we have

$$\Delta\gamma(k;\theta) = [I_m - H(\sum_{j=o}^{k-\theta-1} \Pi^j)AK^*]\nu_2 \tag{33}$$

and furthermore by letting $(k-\theta)$ sufficiently large we obtain

$$\Delta\gamma(k;\theta) = [I_m - H(I_n - \Pi)^{-1}AK^*]\nu_2 \tag{34}$$

We now investigate the following matrix identity.

$$\begin{aligned} [I_m - H(I_n - \Pi)^{-1}AK^*]H &= H - H(I_n - \Pi)^{-1}AK^*H \\ &= H(I_n - \Pi)^{-1}[(I_n - \Pi) - AK^*H] \\ &= H(I_n - \Pi)^{-1}(I_n - A) \end{aligned} \tag{35}$$

Here, Π is a stable matrix and H is a matrix of full rank. So, if (I_n-A) is non-singular (although this is the usual case), the right-hand side of (35) is of full rank and thus we have a non-singular matrix $[I_m - H(I_n - \Pi)^{-1}AK^*]$. This means that $\Delta\gamma(k;\theta)$ takes a non-zero value for any non-zero vector ν_2. However, when $(I_n - A)$ is singular, it is possible for the right-hand side matrix $H(I_n - \Pi)^{-1}(I_n - A)$ to become a matrix of non-full-rank. If it has not a full rank m (we let the rank be m′ , m′ < m), the matrix $[I_m - H(I_n - \Pi)^{-1}AK^*]$ becomes no longer a non-singular one and thus $\Delta\gamma(k;\theta)$ in (34) can take the value zero even for a non-zero bias ν_2. Such a vector ν_2 is given by $H\omega$, where ω is any vector in the null space of $H(I_n - \Pi)^{-1}(I_n - A)$. As the null space of H is included in that of $H(I_n - \Pi)^{-1}(I_n - A)$, the dimension of weakly-diagnosable-

space, i.e., that of the range space $\{H\omega\}$ is clearly $(m-m')$. (Q.E.D.)
From Theorem 2, we obtain the following corollary immediately.

Corollary 1.
If the system is observable and the matrix $(I_n - A)$ is non-singular, sensor failure has no weakly-diagnosable-space for the bias ν_2 (in the similar sense as in (19)).

Theorem 3.
If the system is observable, any cancellation of effects by dynamics and sensor biases on the innovations sequence cannot happen (and thus any failure causing biases on both the dynamics and the sensors can be detected in principle by the step-hypothesized GLR method).
(the proof is omitted)

Theorem 4.
A weakly-diagnosable-space can exist for the composite failure case. Let the biases (in the sense of (19)) in the dynamics and the sensors be $\nu_1 \in R^n$ and $\nu_2 \in R^m$. Then, the condition under which such a space exists is that there exists non-zero vectors ν_1 and ν_2 such that

$$H(I_n - \Pi)^{-1}\nu_1 + [I_m - H(I_n - \Pi)^{-1}AK^*]\nu_2 = 0 \tag{36}$$

(proof)
The condition can be derived by combining Theorem 1 with Theorem 2. (Q.E.D)
It is obvious that the weakly-diagnosable-space can be constructed by such a ν_1 and ν_2. Especially when A is a stable matrix, the space is obtained as the null space of $[H(I_n - A)^{-1} \mid I_m]$.

5. EXTENDED METHODS

5.1. Parallel Step-Hypothesized Methods

In the previous chapter, we have seen that there can exist a weakly-diagnosable-space especially for dynamics failure. So, it is desirable to implement another method together with the usual GLR one which is particularly sensitive to the vector ν_1 in the weakly-diagnosable-space and near the space.

Assuming that the bias ν_1 has appeared from this space and calculating the GLR, we obtain

$$\ell(k;\theta) = (1/2)\hat{\beta}^T(k;\theta)C_1(k;\theta)\hat{\beta}(k;\theta) \tag{37}$$

where

$$\hat{\beta}(k;\theta) = S[S^T C_1(k;\theta)S]^{-1}(S^T \sum_{j=\theta}^{k} G_1^T(j;\theta)V^{-1}(j)\ \gamma(j)) \tag{38}$$

Here, S represents the matrix of basis vectors of the weakly-diagnosable-space. If ν_1 has actually occurred from the space, the GLR (37) obeys a non-central Chi-squared density with non-centrality parameter

$\delta^2 = \nu_1^T C_1(k;\theta)\nu_1$ $(=D(H_o;H_1(\nu_1))$, that is,

$$\ell(k;\theta) = (1/2)\{(\delta+\omega_1)^2 + \omega_2^2 + \ldots + \omega_\alpha^2\} \tag{39}$$

where α is the dimension of the space and $(\omega_1, \omega_2, \ldots, \omega_\alpha)^T$ is a Gaussian random vector with mean zero and unit covariance. We call the GLR method a reduced-order one.

On the other hand, when we used the usual GLR method, the GLR obeys

$$\ell(k;\theta) = (1/2)\{(\delta+\omega_1)^2 + \omega_2^2 + \ldots + \omega_n^2\} \tag{40}$$

Only one difference between (39) and (40) is the number in the random variables $\{\omega_i\}$. For reference, we show in TABLE I the probability of miss alarm P_M for various δ when a threshold which allows false alarm of 5% is selected (the case of n=5 and $\alpha = 2, 3$ is illustrated). From the table, we see that the probability can be considerably decreased by using the reduced-order one for the anomaly whose ν_1 appears from the space.

TABLE I Probability of miss alarm for various anomaly level δ (a threshold allowing false alarm of 5% is adopted).

Anomaly level δ	Probability of miss alarm P_M Usual-order n=5	Reduced-order α=3	Reduced-order α=2
0.5ε	0.7301	0.6675	0.6133
0.6ε	0.6152	0.5347	0.4703
0.7ε	0.4843	0.3949	0.3294
0.8ε	0.3518	0.2657	0.2084
0.9ε	0.2333	0.1613	0.1181
1.0ε	0.1399	0.0877	0.0595

where $\varepsilon = 3.841$

The method has also effect for the bias ν_1 appearing from near the weakly-diagnosable-space, because the divergence $D(H_o;H_1(\nu_1))$ does not make much difference between one calculated for anomaly projected to the space. Practically, we judge that a failure has occurred when one of the two GLR's computed by the parallel step-hypothesized method exceeds a threshold.

5.2. Another Method

The anomaly function $f_1(k) = \Delta Ax(k) + \Delta Bu(k)$ has generally a component of low frequency other than a bias. So, we can expect a much more accurate and rapid failure detection by taking into account the estimation of the dominant low-frequency modes.

The usual dynamics step hypothesis is nothing but approximating the anomaly function $f_1(k)$ by a linear combination of $e_i\sigma(k,\theta)(1 \leq i \leq n)$, where e_i denotes the i-th natural basis vector of R^n. Here, we further consider appropriate basis functions $\phi_{ij} = e_i\psi_{ij}(k,\theta)\sigma(k,\theta)(1 \leq j \leq p_i$, $1 \leq i \leq n)$ other than the step functions $\phi_{io} = e_i\sigma(k,\theta)$ $(1 \leq i \leq n)$ in order to be able to express the anomaly function $f_1(k)$ much more precisely. Although the selection of the functions $\{\psi_{ij}\}$ is mentioned about later, the anomaly function $f_1(k)$ is approximated by a linear combination of these basis functions ϕ_{ij} $(o \leq j \leq p_i,\ 1 \leq i \leq n)$. The direct computation of the MLE of the coefficients $\{c_{ij}\}$ is not always necessary (it becomes first necessary when the failure has been detected and then the isolation and estimation of the failure are required). Because by applying the Gram-Schmidt's orthogonalization procedures to the functions $\varepsilon_{ij} = G_1^o(k;\theta)*\phi_{ij}(k)$ $(o \leq j \leq p_i,\ 1 \leq i \leq n)$ and calculating the orthogonal projection of $\gamma(k)$ on the resultant orthonormal basis functions ε_{ij}^*, we can obtain

$$\zeta = \min_{\{c_{ij}\}} \| \gamma(k) - G_1^o(k;\theta)*(\sum_{i,j} c_{ij}\phi_{ij}(k))\|^2$$
$$= \min_{\{c_{ij}\}} \| \gamma(k) - \sum_{i,j} c_{ij}\varepsilon_{ij}(k) \|^2 \tag{41}$$

and thus we can easily obtain the GLR

$$\ell(k;\theta) = (1/2)\{ \| \gamma(k) \|^2 - \zeta \} \tag{42}$$

The merit of the procedure is that we can remove the orthonormal bases ε_{ij}^* which do not play important role in expressing $\gamma(k)$. Then, bases having a large component of high frequency are the first to be removed. This is particularly important from the viewpoint of decreasing the probability of false (and miss) alarms. When q orthonormal bases are used to approximate $\gamma(k)$, the probability density of the GLR under no failure obeys the Chi-squared one with q degreesof freedom. So, we must take into account the number of using functions in determining the threshold for the GLR. The number of functions $\{\phi_{ij}\}$, i.e., that of $\{\varepsilon_{ij}^*\}$ can be also a priori decreased by considering the linear independence of $\{\varepsilon_{ij}\}$. This method has clearly effect for the anomaly which has not necessarily a bias.

We next consider the selection of the basis functions $\{\phi_{ij}\}$. Although we can consider any smooth functions like descretized triangular or spline functions as the basis functions to express the total anomaly function $f_1(k) = \Delta Ax(k) + \Delta Bu(k)$, we can also use (and it is better to use) the information about the state $x(k)$ and the input $u(k)$ such that $f_1(k)$ can be effectively extracted out from the innovations sequence. One idea is to use the functions $e_iu_j(k)\sigma(k,\theta)$ $(1 \leq j \leq r,\ 1 \leq i \leq n)$ and $e_im_j(k)\sigma(k,\theta)$ $(1 \leq j \leq n,\ 1 \leq i \leq n)$ as $\{\phi_{ij}\}$, where $u_j(k)$ is the j-th component of the input $u(k)$ and $m_j(k)$ is the j-th mode function of the closed loop system : $x(k+1) = (A-BF)x(k)$, where F is a feedback coefficient matrix. As an alternative approach to approximate $\Delta Ax(k)$, we can also make use of the observation $y(k)$. This approach is effective, especially when some state variables can be resolved by the observations. Since the observations are corrupted by the noise, we must first filter them by a low-pass filter and obtain approximate curves of some state components. Letting these functions be denoted by $t_j(k)$ $(1 \leq j \leq \alpha)$, then we have the basis functions $e_it_j(k)$ $(1 \leq j \leq \alpha,\ 1 \leq i \leq n)$ to express $\Delta Ax(k)$.

We can also consider another convenient method which makes a preliminary processing on the innovations sequence $\gamma(k)$ and constructs a set of desirable basis functions in the innovations-space. That is, we filter the innovations by a low-pass filter and obtain the m fundamental basis functions $e_i^*\bar{\varepsilon}_i(k)$ $(1 \leq i \leq m)$, where e_i^* is the i-th natural basis vector of R^m and $\bar{\varepsilon}_i(k)$ is the output of the filter for the i-th innovation. When there exists time regions where any function of these takes large value over some extent, we can further introduce some orthonormal basis functions (of low-frequencies) weighted by the corresponding function and the region. By considering all the functions obtained like this as the functions $\{\varepsilon_{ij}(k)\}$ in (41), we can calculate the GLR by (42).

6. SOME COMMENTS ON THE DETECTION AND DETECTIONABILITY

6.1. Actuator Failures

When a bias adds to the input of i-th actuator, i.e., the case where $u(k)$ takes the form of $u(k) + \alpha e_i \sigma(k,\theta)$, the step-hypothesized approach is directly applicable, because the failure can be exactly modeled by the dynamics step $\alpha b_i \sigma(k,\theta)$, where b_i represents the i-th column of B. So, if b_i happens to be located near the weakly-diagnosable-space (if exists), the failure is difficult to be detected. In such a case, the method of 5.1 should be used.

Other failures of i-th actuator can be modeled by the anomaly function $\Delta b_i u_i(k)\sigma(k,\theta)$ adding to the dynamics equation, where Δb_i is a change in b_i and $u_i(k)$ is the i-th component of $u(k)$. Then, we have

$$\begin{aligned} \Delta\gamma(k;\theta) &= [G_1^o(k;\theta) * u_i(k)]\ \Delta b_i \\ &= G_1^*(k;\theta)\ \Delta b_i \end{aligned} \qquad (43)$$

This means that we can detect the anomaly by a dynamics step hypothesis for the new failure signature matrix $[G_1^o(k;\theta)*u_i(k)]$ instead of using the method of 5.2 which assumes $\{\phi_{ij}\} = e_j u_i(k)$ $(1 \leq j \leq n)$ and that the diagnosability of the failure can be governed by the information matrix

$$C_1^*(k;\theta) \triangleq \sum_{j=\theta}^{k} G_1^{*T}(j;\theta) V^{-1}(j) G_1^*(j;\theta) \qquad (44)$$

We obtain the following theorem.

Theorem 5.
If $u_i(k)$ is non-trivial, i.e., not constantly zero and time-varying, then the actuator failure has no weakly-diagnosable-space for the change Δb_i.

The extension to the multi-actuator's failure is easy. We can also suggest the dual control which aims not only the state control, but also a rapid and more accurate failure detection.

6.2. Plant Failures

In section 5.2 we presented a general method. We show here a convenient method which makes only use of the input sequence. Observations and mode functions of the closed loop system are available as the information to express $\Delta Ax(k)$. But, the quantity which can be truly trusted is only the input sequence $u(k)$. So, we can try to detect the anomaly only by using the input $u(k)$, i.e., by extracting the term $\Delta Bu(k)$ from the anomaly function $f_1(k) = \Delta Ax(k) + \Delta Bu(k)$. Its effectiveness depends of course on the failure. That is, when $\Delta Bu(k)$ occupies a dominant part of the function $f_1(k)$, it works well. But, otherwise, it will need much time to detect the anomaly. At any rate, by selecting an appropriate matrix $\Delta B^{\ddagger} \varepsilon R^{nxr}$ such that

$$(\overset{o}{G}_1(k;\theta)*(\Delta Ax(k) - \Delta B^{\ddagger}u(k)),\ \overset{o}{G}_1(k;\theta)*(\Delta B+\Delta B^{\ddagger})u(k)\ ) = 0 \qquad (45)$$

we can obtain

$$D(H_o;H_1) - D(H_1(\Delta B');H_1) = \|\ \overset{o}{G}_1(k;\theta)*(\Delta B'u(k))\ \|^2$$

$$= \|\ [\overset{o}{G}_1(k;\theta)*u(k)]\mathrm{Column}(\Delta B')\|^2 \qquad (46)$$

where $\Delta B' \overset{\Delta}{=} \Delta B + \Delta B^{\ddagger}$ and $\mathrm{Column}(\Delta B')$ is the vector composed of the elements of $\Delta B'$. (46) is just the divergence $D(H_o;H_1(\Delta B'))$. The detectionability is governed by the information matrix computed for the new failure signature matrix $[\overset{o}{G}_1(k;\theta)*u(k)]$ and we see that there exists no weakly-diagnosable-space for the change $\Delta B'$, if $u(k)$ is non-trivial.

If the observation noise is small and some state variables can be resolved by the observations, they can be also used as the known information.

7. OPTIMAL SENSOR LOCATION

7.1. Detectionability and Separability Indices

We have seen that an information matrix plays an important role in detectionability. We define here some detectionability and separability indices to describe the ability of the system in detecting the bias which appears in anomaly function (extension to a general case is also possible as described later). For examples, for the bias ν_1 appearing from R^n, we can define

$$J_1(k;\theta) \overset{\Delta}{=} \min_{\|\nu_1\|=1} D(H_o;H_1(\nu_1))$$

$$= \lambda_{min}[C_1(k;\theta)] \qquad (47)$$

and for the case where ν_1 appears from a subspace $\mathrm{Span}\{s_1, s_2, \ldots, s_q\}$, we can similarly define

$$J_1'(k;\theta) \overset{\Delta}{=} \lambda_{min}[S^T C_1(k;\theta)S] \qquad (48)$$

where $S \overset{\Delta}{=} [s_1, s_2, \ldots, s_q]$ and $\lambda_{min}[\cdot]$ denotes the minimum eigenvalue of the matrix. We can also consider the case where two types of anomalies can occur from two subspaces S_1 and S_2 (we let $S_1 \cap S_2 = \{0\}$). Then, we need to define a quantity to describe the easiness of the separation. We can define the separability index by a function of the minimum angle φ between the two functional subspaces $\Delta\gamma(k;\theta)$ $(\nu_1 \varepsilon S_1)$ and $\Delta\gamma(k;\theta)$ $(\nu_2 \varepsilon S_2)$, i.e.,

$$\tilde{S} \triangleq (\sin\varphi)^2$$
$$= \lambda_{min}[(S_1^T C_1(k;\theta)S_1)^{-1/2}\tilde{G}(S_1^T C_1(k;\theta)S_1)^{-1/2}] \qquad (49)$$

where $S_1 \triangleq [s_{1j}]$ and $S_2 \triangleq [s_{2j}]$ are the matrices composed of the basis vectors $\{s_{1j}\}$, $\{s_{2j}\}$ of S_1 and S_2, respectively, and $\tilde{G}$ is the Gram matrix for orthogonal components of $\{G_1(k;\theta)s_{1j}\}$ to $\{G_1(k;\theta)s_{2j}\}$ (in the sense of (16)).

The separability between the dynamics and the sensor failures can be also defined in the same way.

By the way, as the detectionability index characterizing the ability of detecting the specified anomaly functions $g_j(k)$ $(1 \leq j \leq \alpha)$(n-dimension) , we can define

$$J_1''(k;\theta) \triangleq \min_{1 \leq j \leq \alpha} \| \overset{o}{G}_1(k;\theta) * g_j(k) \|^2 \qquad (50)$$

where it is assumed that $g_j(k)$ are all normalized in the sense of Euclidean norm. This index can be made use of to obtain a sensor location which is especially sensitive to the anomaly represented by the anomaly functions.

7.2. Optimal Sensor Location

An effectiveness of the observation system can be evaluated by the detectionability and separability indices. Adding to them, observability index must be also taken into account. So, desirable criteria for sensor location are such that which maximizes the three indices. For example, for $\nu_1 \varepsilon S_1$ or S_2(where $S_1 \cap S_2 = \{0\}$), we can consider

$$c_1(\min_{S_1, S_2} J_1'(k;\theta)) + c_2\tilde{S} + c_3\lambda_{min}[P^{-1}(k/k)] \longrightarrow \max_H \qquad (51)$$

where $c_1, c_2, c_3 \geq 0$. We can also consider other various multi-objectives -optimization methods. These criteria have effect for the systems which have flexibility in the location of sensors such as that in distributed parameter systems. Addition of other sensors to obtain a desired property concerning the detectionability/separability and/or the weakly-diagnosable-space is also possible.

8. NUMERICAL EXAMPLES

We first apply the parallel step-hypothesized method to the simple stable system defined by

$$x(k+1) = \begin{bmatrix} 0.98 & 0 \\ 0 & 0.94 \end{bmatrix} x(k) + w(k) \tag{52}$$

$$y(k) = \begin{bmatrix} 1 & 1 \end{bmatrix} x(k) + v(k) \tag{53}$$

where $Q = 0.001I_2$, $R = 0.001$, $x(0) = (0.663, 1.19)^T$, $\hat{x}(0/-1) = (0.,0.)^T$, and $P(0/-1) = I_2$. We assume that the anomaly $\Delta A = \mathrm{diag}(-0.007, 0.021)$ occurs at the time $\theta = 50$. The detection method of 5.1 is applied with window 15 and threshold which allows a false alarm of 5%. Simulations are done for 100 different noise sequences to obtain the detection rate. Then, we get the detection rate 82% (only rate for correct detection) by the proposed method and only 51% by the usual GLR method. We see that the proposed one is much superior to the usual one. This example is such that the jump $\Delta Ax(k)$ occurs from near the weakly-diagnosable-space Span $\{(-1, 3)^T\}$. Other types of jump are now being considered to see how much the addition of the reduced-order GLR method can effect the increase of the probability of the false alarm.

Next, we apply the extended method of 5.2 to the magnetic suspension control system of a linear motor car. We consider here a simple model defined by /7/

$$\dot{x}(t) = A(L)x(t) + b(L)u(t) \tag{54}$$

$$u(t) = -Fx(t) = -(f^1, f^2, f^3)x(t) \tag{55}$$

$$y(t) = \begin{bmatrix} 1 & 0 & 0 \\ 0 & 0 & 1 \end{bmatrix} x(t) \tag{56}$$

where

$$A(L) = \begin{bmatrix} 0 & 1 & 0 \\ 0 & 0 & 1 \\ \frac{K_sR}{mL} & \frac{K_s}{m} - \frac{K_i\tilde{K}_{\dot{s}}}{mL} & -\frac{R}{L} \end{bmatrix}, \quad b(L) = \begin{bmatrix} 0 \\ 0 \\ -\frac{K_i}{mL} \end{bmatrix}$$

and $m = 16Kg$, $R = 8\Omega$, $K_s = 57000N/m$, $K_i = \tilde{K}_{\dot{s}} = 114N/A$, $L = 0.5Vs/A$, $f^1 = -9123.5$, $f^2 = -470.2$, $f^3 = -7.1$ are assumed. By assuming sampling time = 0.015s and $x(0) = (0.01, 0.05, 0.0)^T$, $\hat{x}(0/-1) = x(0)$, $P(0/-1) = 10^{-4}I_3$, $Q = 9x10^{-8} I_3$, $R = 9x10^{-8}I_2$, we obtain the discrete system (1) and (2). Let now the anomaly correspond to the variation of the inductance value L from 0.5 to 0.54 (+8% variation) at the discrete time 15. Then, the application of the methods of 5.2 (with the information about x(k) by the closed -loop system and the input u(k)) gives us the results, that is, 76% by the GLR method based on the estimation of $\Delta Ax(k)$, and 82% by that based on the estimation of $\Delta Bu(k)$. These detection rates are much higher than that (64%) of the step-hypothesized GLR method. In the example, window 10 and threshold allowing a false alarm of 5% are assumed and the number of used basis functions is fixed as $n = 3$ in each of proposed methods. So, if we try to select the number of basis functions adaptively corresponding to the situation, we can expect much higher detection rate.

9. CONCLUSIONS

The paper analized first the step-hypothesized GLR method for a general failure in linear discrete dynamical systems and discussed the detectionability. By the analysis, we found that a weakly-diagnosable-space can exist for dynamics failure rather than for a sensor one. To overcome the difficulty, two extended methods were developed, that is, the method using a step- hypothesized reduced-order GLR one with the usual one, and the another one taking into account the estimation of the anomaly function by the available informations on the state and the input.

Next, based on the discussions on the detectionability and separability, desirable criteria for optimal sensor location were mentioned about.

Although the effectiveness of the proposed detection methods and the criteria was verified by a simulation/3/, it still remains to be further investigated.

ACKNOWLEDGEMENT

We would like to thank Alexander von Humboldt Foundation in West Germany for the support for this research.

REFERENCES

1) A.S.Willsky, 'A Survey of Design Methods for Failure Detection in Dynamic Systems', Automatica, 12, 6o1/611 (1976)
2) A.S.Willsky and H.L.Jones, 'A Generalized Likelihood Ratio Approach to the Detection and Estimation of Jumps in Linear Systems', IEEE Trans., AC-21, 1o8/112 (1976)
3) S.Tanaka, M.Sagara and T.Okita, 'On Optimal Sensor Locations for Anomaly Diagnosis', SICE Trans., 20 , 60/67 (in Japanese) (1984)
4) S.Tanaka, 'Diagnosability of Systems and Optimal Sensor Location', GAMM-Tagung, Dortmund, April (1986)
5) A.S.Willsky, 'Detection of Abrupt Changes in Dynamic Systems', Lecture Notes in Control and Information Sciences, 77 , 27/49 (1986)
6) S.Kullback, Information Theory and Statistics, John-Wiley & Sons (1959)
7) W.Breinl and P.C.Müller, 'State-Observer with Low Sensitivity and Its Application to Maglev Vehicle Suspension Control', Regelungstechnik, 30 , 403/411 (1982)

Fault Diagnosis in Dynamic Systems Using a Robust Output Zeroing Design Method

R.J Patton and S.W. Willcox
Department of Electronics,
University of York,
Heslington,
York YO1 5DD, England

ABSTRACT
A new method of analysing faults in the m measurements of an nth order dynamic system is presented. The estimation error space of an observer is used for fault diagnosis. The design uses a robust observer in which the weighted linear combination of m estimation errors is equivalent to the output zeroing problem. In order to design a fault monitor which is insensitive to r (multivariable) input and parameter-induced disturbances, n-1 eigenvectors are assigned to the null space $\mathbf{N}(\underline{\mathbf{W}}\mathbf{C}_0)$ where $\underline{\mathbf{W}}$ is a weighting matrix and $\mathbf{C}_0$ is the nominal (no-fault) measurement matrix. Motions away from the null space regime are due to parameter variations, disturbances or sensor faults. The problem is to distinguish the variations due to sensor faults from all other uncertain signals. By assigning the eigenstructure of a dual problem and exploiting an orthogonality relation between the eigenstructures of the dual systems a robust design can be achieved. The effects of parameter variations are nulled and the fault monitor is rendered sensitive to sensor faults defined in a given spectral bandwidth.

1. INTRODUCTION

In an earlier paper[1] a comparison of two techniques of Instrument Fault Diagnosis (I.F.D.) was given. The work has been formalised using eigenstructure assignment[2] and the present paper describes a further extension to the basic concept with a consideration of the generalised multivariable problem. Use is made of the concept of **analytical redundancy**[3] to design a match between the m measurable components of the estimation error vector. The estimation error space has been chosen in preference to the well known approaches using the state-space[4,5,6,7,8].

The aim has been to develop a way of using observers as robust fault monitors. In this context a robust fault

S. Tzafestas et al. (eds.), System Fault Diagnostics, Reliability and Related Knowledge-Based Approaches, Vol. 1, 177–192.

monitor provides an invariant tracking response in the presence of widely changing system parameters. For this application it is not necessary for the state estimates to actually track all the variations of the non-linear system states. In order to maintain or guarantee robustness in this sense certain stuctural properties need to be satisfied whilst minimising a sensitivity measure $\underline{S}$.

The degree of robustness is achieved in each fault monitor by assigning the eigenstructure of the estimation error space of the observer such that an 'output-zeroing' condition is satisfied. By invoking the output-zeroing problem in this way a suitable modal match in the components of the estimation error can be generated. This approach is used together with the assignment of a narrow band robustness filter in the eigenspace to reduce the local parameter sensitivity properties of the monitor. The robust monitor must be insensitive to disturbances and uncertain variations due to parameter changes within the plant.

A recent paper[2] has discussed the application of this new idea to the problem of detecting the measurement faults in the longitudinal motion of an aircraft flying through turbulence. An overlapping state-space decomposition of the equations of motion was described such that, with 4 measurements and 4 observers (each based on a different sub-system) faults on any sensor can be uniquely located. The idea of using a bank of observers using an overlapping decomposition has not been reproduced in this paper in order to concentrate on the robust observer design aspects. The main aim of the present work is to show the extension of the method to the general multivariable case with more than one uncertain input signal.

One sub-system of the longitudinal motion example given in the earlier paper[2] is used to illustrate the concepts and the paper gives a second example of the design of a fault monitor for the lateral motions of the same aircraft. The complete designs are given together with simulation results for the lateral motions using the non-linear aircraft model.

Before entering into the design method and describing the examples it is first necessary to outline the nature of the most significant problems associated with the use of observers in the application. In order to do this a brief description of a well known 'dedicated observer' approach[5] is outlined.

2. ANALYTICAL REDUNDANCY USING BANK OF SINGLE-INPUT OBSERVERS

This is a well known approach in which m observers are each dedicated to one measurement. In general, the state-space structure of each observer is based on the same model, however, as the measurement systems are different, the eigenstructures are not the same. It is assumed that the state-space modelling approach is valid and that for small parameter variations the following notation is applicable:

$$A = A_0 + \Delta A \; ; \; B = B_0 + \Delta B \; ; \; C = C_0 + \Delta C \qquad (2.1)$$

The terms ΔA , ΔB and ΔC denote uncertainty in the system and its measurements. It should be noted that whilst ΔC can be used to indicate a measurement (sensor) fault this is only a symbolic representation of the fault. The simulations used in this work use more detailed and realistic non-linear failure models.

The observer (estimator) dynamics are based on the nominal system parameters using the controllable and observable triple $[A_0, B_0, C_0]$. The m observer equations in the bank are given by:

$$\begin{aligned}
\dot{\underline{x}}_1 &= A_0\underline{x}_1 + B_0\underline{u} + K_1(C_{01}\underline{x} - C_{01}\underline{x}_1) + \Delta C_{01}\underline{x} \\
\dot{\underline{x}}_2 &= A_0\underline{x}_2 + B_0\underline{u} + K_2(C_{02}\underline{x} - C_{02}\underline{x}_2) + \Delta C_{02}\underline{x} \\
&\cdot \quad \cdot \quad \cdot \quad \cdot \quad \cdot \\
\dot{\underline{x}}_m &= A_0\underline{x}_m + B_0\underline{u} + K_m(C_{0m}\underline{x} - C_{0m}\underline{x}_m) + \Delta C_{0m}\underline{x}
\end{aligned} \qquad (2.2)$$

The estimated measurements in each observer can be used to infer a sensor fault, however, a more powerful approach follows through taking the difference between state-estimate vectors as follows:

Define: $\underline{e}_1 = \underline{x} - \tilde{\underline{x}}_1 \; ; \quad \underline{e}_2 = \underline{x} - \tilde{\underline{x}}_2 \; ; \; e_i = \underline{x} - \tilde{\underline{x}}_i$

Also: $\underline{e}_{ik} = \underline{e}_i - \underline{e}_k = \tilde{\underline{x}}_k - \tilde{\underline{x}}_i$, for $i,k = 1,2,..,m \;\; i \neq k$

For a non-linear process with small parameter deviations and **no measurement faults:**

$$\begin{aligned}
\dot{\underline{e}}_i &= (A_0 - K_iC_{0i})\underline{e}_i + \Delta A\underline{x} + \Delta B\underline{u} \\
\dot{\underline{e}}_k &= (A_0 - K_kC_{0k})\underline{e}_k + \Delta A\underline{x} + \Delta B\underline{u}
\end{aligned} \qquad (2.3)$$

$$\dot{\underline{e}}_{ik} = A_0\underline{e}_{ik} + K_kC_{0k}\underline{e}_k - K_iC_{0i}\underline{e}_i \; ; \; i,k = 1,2,\ldots,m \;\; i \neq k \qquad (2.4)$$

Now $\underline{e}_i$ and $\underline{e}_k$ depend on the inputs $\Delta A\underline{x}$ and $\Delta B\underline{u}$ to different extents as $K_iC_{0i} \neq K_kC_{0k}$. Furthermore, it is not simple , in general to arrange for these terms to be equal. This approach based on differencing the state estimates of dissimilar observers has been discussed by Frank et al.[8,9] and these authors have provided some interesting design schemes based on linear system data. Up to the present time the design problem to achieve the required robustness has proved to be difficult and not well suited to failure analysis in uncertain systems. Apart from the problem of parameter sensitivity, the main shortcoming is the lack of definition in eigenstructure robustness. Results arising from the application of this scheme to a non-linear system are presented in earlier papers[1,2]. This background has stimulated the development of the new more robust approach.

3. ANALYTICAL REDUNDANCY IN OBSERVATION ERROR SPACE

By using the state-space in an I.F.D. scheme redundant state estimate information can be used to detect and locate faults. The above analysis shows that the state-space approach has no generalised design tactic to ensure robustness without the addition of considerable complexity. These problems have been discussed by Watanabe et al.[10].

Instead of designing directly in the state-space it seems natural to use the state-estimation error space of the nominal system at the design stage. By using eigenstructure assignment together with an output-zeroing approach powerful insight can be gained into the effects of parameter variations and disturbances. The approach is thus made via the s and frequency domains using, in addition, the modal structure of the estimation error. The output to be zeroed is essentially a weighted linear combination of the estimation error signals i.e. $\underline{\mathbf{W}}C_0\underline{e}(t)$ for a given fault monitor observer.

We define the following to be a sensitivity signal and, additionally, a fault detection signal as follows:

$$\underline{\mathbf{S}} = \underline{\mathbf{W}}C_0\underline{E}(s) \quad \text{(s-domain)}$$

$$\text{or } \underline{f}(t) = \underline{\mathbf{W}}C_0\underline{e}(t) \quad \text{(time domain)}$$

If $\underline{\mathbf{S}}$ remains close to zero in the presence of disturbance and parameter variations in the monitored system then the measure $\underline{S}$ is insensitive to uncertain effects. It is, however, of crucial importance that $\underline{S}$ is designed to be sensitive to sensor faults. This can be achieved by eigenstructure assignment in the estimation error space.

The malfunction of a sensor can be modelled (for analysis purposes) as a simple additional term $\Delta C\underline{x}$ in the estimation error state equation as follows:

$$\dot{\underline{e}} = (A_0 - KC_0)\,\underline{e} + \Delta A\underline{x} + \Delta B\underline{u} + K\,\Delta C\underline{x} \tag{3.2}$$

The objective of the design is to make the time response $\underline{e}(t)$ insensitive to terms $\Delta A\underline{x}$ and $\Delta B\underline{u}$ whilst sensitive to sensor malfunctions denoted by the term $K\,\Delta C\underline{x}$. If we consider, for generality an n state multivariable non-linear problem with m measurements and r inputs then one can attempt to make the time response insensitive to terms like $\Delta A\underline{x}$ and $\Delta B\underline{u}$ by using the s-domain as follows:

$$\underline{E}(s) = (sI_n - D)^{-1} B_0 \underline{u}^*(s) \qquad (3.3)$$

where $\underline{E}(s)$ denotes the Laplace Transform of $\underline{e}(t)$,

and: $$B \approx dB_0/dt\ (\Delta t) \quad \text{(to first order)} \qquad (3.4)$$

Now $B\underline{u}(t) = s\, B_0\ \Delta t * \underline{u}(s)$, where $*$ denotes the convolution operation and, for convenience of notation, $B_0\underline{u}^*(s)$ is used to replace $sB_0\,\Delta t*\underline{u}(s)$. D is the closed-loop observer matrix given by:

$$D = (A_0 - KC_0) \qquad (3.5)$$

In most problems the nominal input direction matrix B_0 may not be well known. It is therefore important to be able to exclude information concerning B_0 from the robust observer design. The more general way of 'nulling' the effects of input uncertainties in the fault monitor is to assign an eigenstructure to D such that:

$$\underline{\mathbf{W}}C_0\ \underline{m}_i = 0 \quad , \text{ for } i = 1, n-m \qquad (3.6)$$

with $\underline{m}_i$ distinct **right** eigenvectors of D. For this problem the rank $r(\underline{\mathbf{W}}C_0)$ is m. The $\underline{m}_i$ eigenvectors thus $\varepsilon\ N(\underline{\mathbf{W}}C_0)$ and motions in this null space become insensitive to parameter variations of A lying in the range space $R(C_0)$. This is the dual result of the well known 'output zeroing' in multivariable control[11,12]. The physical interpretation is that, for the nominal system $[A_0\ ,\ B_0\ ,\ C_0]$, n-m modes are forced to become unobservable in the estimation error space.

A more useful result follows if the rank $r(\underline{\mathbf{W}}C_0)$ is designed to be unity instead of m. A match in one modal component of the estimation error can then be made such that, for motions in the null space $N(\underline{\mathbf{W}}C_0)$, $\underline{\mathbf{W}}C_0$ becomes a left anhilator of $\underline{E}(s)$. In this case n-1 modes in the estimation error space are made unobservable and hence parameters associated with these modes have little effect on the weighted estimation error signal. If motions of the estimations error system are maintained close to or within the null space, then:

$$\underline{\mathbf{W}}C_0\underline{E}(s) = 0 \ , \ \forall s \qquad (3.7)$$

for some weighting matrix $\underline{\mathbf{W}}$. The geometry of the null-space is <u>defined</u> by the assignment of the eigenstructure of the observer. This null space is not, except in rare cases, coincident with the null-space arising from a fault condition, i.e.:

$$\underline{\mathbf{W}}(C_0 + \Delta C)\,\underline{X}(s) \neq 0 \tag{3.8}$$

A rapid response in the estimation error will thus enable the fault to be detected. The following section describes the design procedure based on eigenstructure assignment.

4. ROBUST OBSERVER (FAULT MONITOR) DESIGN

Consider $\underline{v}_i$ to be a left eigenvector of the **dual** (control) problem given by:

$$\dot{\underline{z}} = (A_0^T - C_0^T K^T)\underline{z} = D^T \underline{z} \quad , \quad \underline{z} \in R^n \tag{4.1}$$

It follows, by transposition, that $\underline{v}_i^T$ is a left eigenvector of the observer problem, i.e.:

$$\underline{v}_i^T D = \underline{v}_i^T \lambda_i \tag{4.2}$$

The above property leads to a well known result that the dual sets of right eigenvectors of D and D^T corresponding to **distinct** eigenvalues are orthogonal.

i.e. $\quad \langle \underline{m}_i , \underline{v}_k \rangle = 0 \quad , \quad i \neq k$

$$\text{for } D^T \underline{v}_i = \underline{v}_i \lambda_i \quad \text{and} \quad D\underline{m}_i = \underline{m}_i \lambda_i \tag{4.3}$$

Use is made of this result here and it is also extended to include simple degeneracy of multiplicity m or conjugacy. In order to achieve $r(\underline{\mathbf{W}}C_0) = 1$ there are **two** possible approaches depending on the magnitudes of n and m as follows:

(i) **For n-1 odd** we proceed using **generalised** eigenvectors of D. For this case the nullity of $N(\underline{\mathbf{W}}C_0)$ is odd. To guarantee that the **one** range space eigenvector and the n-1 null-space eigenvectors span disjoint eigenspaces m eigenvectors must be linearly dependent.

However, by using the Jordan canonical form J for D:

$$\underline{E}(s) = T(sI_n - J)^{-1} T^{-1} B_0\, \underline{u}^*(s) \tag{4.4}$$

A transformation matrix T can be found such that $\underline{\mathbf{W}}C_0$ becomes a **left** anhilator of $\underline{E}(s)$. This is achieved when T contains the **generalised** eigenvectors corresponding to the degenerate modes[13].

Thus an n-1 ordered set of generalised eigenvectors of D form n-1 linearly-independent columns of T. In order to assign repeated eigenvalues the QR algorithm can be used along with an iterative eigenstructure assignment algorithm.

(ii) **For n-1** even the null-space eigenstructure can be partitioned into complex-conjugate sets. In this case the design can be achieved using conventional methods of determining the allowable subspaces for the complex-conjugate eigenvectors of D.

In either case only one eigenvector is assigned. However, a more rigorous analysis can be performed by computing the appropriate unobservable sub-space using a well known decomposition procedure[14]. If $\underline{v}_i$ is an assignable eigenvector of the dual system matrix D^T, then **at least one** right eigenvector can be assigned such that $\underline{v}_i \underline{m}_k = 0$, for $i \neq k$. The eigenvectors of this null-space are thus orthogonal to the equivalent range space eigenvector in the dual problem. Thus the null-space is assigned by assigning one right eigenvector to the dual problem. In the generalised case n-1 pole-zero cancellations occur in $\underline{W}C_0F(s)$ such that the dynamic order of F(s) is reduced from n to 1. The eigenvector $\underline{v}_i$ should be assigned such that when transposed it is equal to $\underline{W}C_0$.

5. DESIGN EXAMPLES

<u>Example 1</u> (n = 3, m = 2, r = 1), [A_0, C_0] observable)

$$A_0 = \begin{bmatrix} -2.930 & 32.770 & 0 \\ -0.416 & -0.645 & 0 \\ 0.000 & 1.000 & 0 \end{bmatrix}$$

$$B_0 = \begin{bmatrix} 5.318 \\ -13.580 \\ 0.000 \end{bmatrix} \qquad C_0 = \begin{bmatrix} 1 & 0 & 0 \\ 0 & 0 & 1 \end{bmatrix} \tag{5.1}$$

For s=-2 ; $\underline{v}_3^T = [-0.086 \quad 1 \quad 1.404]$ (5.2)

This gives a strong weighting for the s=-2 mode, as required. In practice the vector $\underline{v}_3^T = [p_1 \quad 1 \quad p_2]$ is assigned. The elements p_1, p_2 must be non-zero otherwise an arbitrary choice can be made. The ratio p_1 to p_2 gives the weighting factor of the two innovation signals. The dual closed-loop system matrix is:

$$D^T = \begin{bmatrix} -29.758 & -2.858 & 0.329 \\ 32.770 & -0.645 & 1.000 \\ -71.943 & 7.167 & -11.609 \end{bmatrix} \tag{5.3}$$

The **Modal** matrix is:

$$V = \begin{bmatrix} 0.095 & 0.095 & -0.061 \\ -0.213 & -0.213 & 0.712 \\ 1.000 & 1.000 & 1.000 \end{bmatrix} \quad \begin{matrix} (\underline{v}_1) & (\underline{v}_2) & (\underline{v}_3) \\ (s=-20) & (s=-20) & (s=-2) \end{matrix} \tag{5.4}$$

The (transformed) observer problem has:

$$D = \begin{bmatrix} -29.758 & 32.770 & -71.943 \\ -2.858 & -0.645 & 7.167 \\ 0.329 & 1.000 & -11.609 \end{bmatrix} \tag{5.5}$$

The generalised **Modal matrix** is:

$$T = \begin{bmatrix} 1.000 & 1.000 & 1.000 \\ 0.169 & 0.181 & 1.196 \\ -0.059 & -0.070 & 0.160 \end{bmatrix} \quad \begin{matrix} (\underline{m}_1) & (\underline{m}_2) & (\underline{m}_3) \\ (s=-20) & & (s=-2) \end{matrix} \tag{5.6}$$

$\underline{m}_2$ is a generalised eigenvector based on the mode s = -20 which, along with $\underline{m}_1$, satisfies $(D - \lambda I)^2\underline{m}_i = 0$, for i=1,2 and $\lambda = -20$.[13] The vectors $\underline{m}_1$, $\underline{m}_2$ and $\underline{m}_3$ form a basis for the Jordan canonical form for D, ie:

$$J = T^{-1}DT \tag{5.7}$$

The orthogonality condition (4.3) above is satisfied with:

$$\underline{v}_3^T\underline{m}_1 = \underline{v}_3^T\underline{m}_2 = 0 \tag{5.8}$$

$$\text{i.e.} \quad \underline{m}_1 \text{ and } \underline{m}_2 \in N(\underline{v}_3^T)$$

Furthermore, $\underline{W}C_0\underline{m}_1 = 0$ is satisfied, with:

$$\underline{W} = [1 \quad 16.94] \tag{5.9}$$

The **parity** signals are thus given by:

$$C_0\underline{E}(s) = \begin{bmatrix} e_1(s) \\ e_3(s) \end{bmatrix} = \begin{bmatrix} \dfrac{-16.94u^*(s)}{(s + 2)} \\ \dfrac{u^*(s)}{(s + 2)} \end{bmatrix} \tag{5.10}$$

Note that, $F(s) = \underline{W}C_0\underline{E}(s) = 0$, $\forall s$ corresponding to the nominal system. First order **analytical redundancy** has been achieved using the range space mode of s = -2.

Example 2 (n = 7 , m = 2 , r = 2) **Aircraft Lateral Motion**

The A_0 and B_0 matrices for this example are given in a paper in this workshop proceedings by Mudge & Patton[15]. The measurement matrix for this particular problem is given by:

$$C_0 = \begin{bmatrix} 0 & 1 & 0 & 0 & 0 & 0 & 0 & 0 \\ 0 & 0 & 0 & 0 & 1 & 0 & 0 & 0 \end{bmatrix} \qquad (5.11)$$

Now n-1 = 6 and the null-space can be assigned with an entirely complex eigenstructure. Once again the range space mode must be reasonably slow to obtain the best filtering action and insensitivity to parameter variations away from the null-space regime. This results in the potential cancellation of 3 complex-conjugate mode pairs leaving one slower real mode. A reasonable choice is as follows:

$$s_1 = -1.0,\ s_{2,3} = -5 \pm j5\ ,\ s_{4,5} = -12 \pm j12\ ;\ s_{6,7} = -25 \pm j25\ ;$$

These modes (eigenvalues) are assigned to the dual problem matrix D^T and $\underline{v}_1$ is assigned as the range-space eigenvector:

$$\underline{v}_1^T = [0.010\ \ -0.096\ \ -0.047\ \ -0.100\ \ 0.711\ \ 0.044\ \ 0.686]$$

Thus: $\underline{\mathbf{W}} = [-0.096\ \ 0.711]$ (5.12)

The **right** eigenvectors of D are:
(real parts)

$$\underline{\mathbf{M}}_R = \begin{bmatrix} (\underline{m}_1) & (\underline{m}_2) & (\underline{m}_3) & (\underline{m}_4) & (\underline{m}_5) & (\underline{m}_6) & (\underline{m}_7) \\ 0.860 & 0.997 & 0.997 & 0.995 & 0.995 & 0.866 & 0.866 \\ 0.004 & 0.004 & 0.004 & 0.010 & 0.010 & 0.041 & 0.041 \\ 0.031 & 0.030 & 0.030 & 0.074 & 0.074 & 0.044 & 0.044 \\ 0.510 & 0.072 & 0.072 & 0.066 & 0.003 & 0.008 & 0.295 \\ 0.001 & 0.001 & 0.001 & 0.003 & 0.003 & 0.008 & 0.008 \\ -0.009 & 0.002 & 0.002 & 0.025 & 0.025 & -0.032 & -0.032 \\ -0.002 & -0.003 & -0.003 & -0.003 & -0.003 & -0.033 & -0.033 \end{bmatrix} \qquad (5.13)$$

(imaginary parts)

$$\underline{\mathbf{M}}_I = \begin{bmatrix} 0.000 & 0.000 & 0.000 & 0.000 \\ 0.000 & \pm 0.005 & \pm 0.012 & \pm 0.039 \\ 0.000 & \pm 0.017 & \pm 0.007 & \pm 0.032 \\ 0.000 & +0.013 & \pm 0.001 & \pm 0.385 \\ 0.000 & \pm 0.001 & \pm 0.003 & \pm 0.004 \\ 0.000 & \pm 0.019 & \pm 0.013 & \pm 0.065 \\ 0.000 & 0.000 & 0.000 & \pm 0.049 \end{bmatrix} \qquad (5.14)$$

Note that perfect cancellation of all three complex modes has not occurred in the weighted signal as evidenced by the presence of the 0.51 real entry in $\underline{m}_1$. The small effect of the remaining modal terms in the eigenstructure suggests that the weighting matrix $\underline{\mathbf{W}}$ can be taken with good approximation as:

$$\underline{W} = [-0.096 \quad 0.711] \tag{5.15}$$

This results in a fault-detection signal as follows:

$$\underline{W}C_0\underline{E}(s) = [-0.096 \quad 0.711] \begin{bmatrix} e_2(s) \\ e_5(s) \end{bmatrix} \underline{u}^*(s) \tag{5.16}$$

$$= [-0.096 \; 0.711] \begin{bmatrix} \dfrac{7.406}{(s+1)} \\ \\ \dfrac{1.0}{(s+1)} \end{bmatrix} \underline{u}^*(s) = \underline{0} \tag{5.17}$$

This is valid as the parasitic mode $12 \pm j12$ is fast enough to not have a strong effect on the matched signal. The result holds for the nominal system $[A_0, B_0, C_0]$. During parameter variations the range-space mode at $s=-1$ acts as a robust filter; the resulting sensitivity to parameter changes is thus low. A sensor failure will cause the time response of $\underline{W}C_0\underline{E}(s)$ to deviate substantially.

6. RESULTS

Figures 1 to 3 show the results obtained using the new scheme described above. This I.F.D. scheme was applied to a non-linear model of the lateral motion of an unmanned aircraft. This aircraft model uses a VSS control system which, in this case, corrects an initial offset on p (roll rate). The response which occurs due to this initial condition is used to test the sensitivity of the I.F.D. scheme in the presence of parameter variations and pilotic inputs. Each diagram shows the response of the system to no-fault and fault situations. For each case the two measurements, p and ψ , are shown together with the pair of weighted elements of the measurement error vector. The final diagram shows the fault detect signal, $f^2(t)$, which is the square of the weighted estimation errors. Note that, f(t) is defined by equation (3.1). When one observer monitor is considered (as in the examples given), f(t) is a scalar signal. In the multi-observer case described in an earlier paper[2], f(t) becomes a vector of signals, the square of each component of which should be used for fault detection. Figure 1 shows the no-fault case. The response of the system to the aircraft performing the manoeuvre can be seen. The most significant feature being the very small response of $f^2(t)$. The slight imperfection of the design, in this case, is clearly seen as an oscillatory mode **close to** $-12 \pm j12$ is repeatedly excited by disturbances. The lateral motion system is non-minimum phase for this aircraft and a robust VSS sliding mode control law design has been used[15]. This switching is sufficient to excite the fast mode. By fine tuning of the eigenstructure assignment it is possible to remove all traces of this mode in the response. However, as long as the weighted estimation error signals match closely, the presence of a small parasitic mode does not seriously affect the fault detection capability of the monitor; this is a question of robustness. Figure 2 shows the response to a step offset change on the ψ measurement. The effect of this fault can be seen on the two components of the measurement error vector, the ψ state estimate and is also clearly shown on the weighted estimation error signal. Figure 3 shows another common sensor fault, that of a noisy output. Again the response of the I.F.D. scheme can be clearly seen, allowing the onset of the fault to be easily detected by placing a low threshold on $f^2(t)$.

7. CONCLUDING DISCUSSION

It has been shown that, by using the estimation error in the observation space, a powerful approach to fault diagnosis can be achieved. The new method constitutes a robust observer fault detection filter in which the effects of plant parameter variations and process disturbances are minimised using eigenstructure assignment. The robustness aspects have made it possible to place low thresholds for rapid detection of faults - even in an unstable loop integrity situation. The use of the observation error sub-space has removed the need to compute the state estimation vector and the designs can thus be implemented easily in real time computer systems. The designs do not involve the nominal directions of the input matrix B_0 and the effects of uncertain signals are removed by weighting m signals in the estimation error. The second example given has shown that, even for a 7th order multivariable system, a very effective fault detection design can be achieved. The order of the problem has, however demonstrated the difficulty in making the precisely correct eigenstructure assignment. This problem is associated with limitations in non-iterative methods of assigning the eigenstructure; the chief limitation being that a preferred eigenvector may not be assignable to the closed-loop system as it may not lie in an allowable sub-space. One other significant problem is the need for further information on eigenstructure assignment of observers , particularly for the degenerate case. There is a wealth of important literature available on eigenstructure assignment as applied to the control problem[16,17,18] and this has been invaluable in this work. However, the subject of eigenstructure assignment in observers is a hitherto lacking topic with much scope for research.

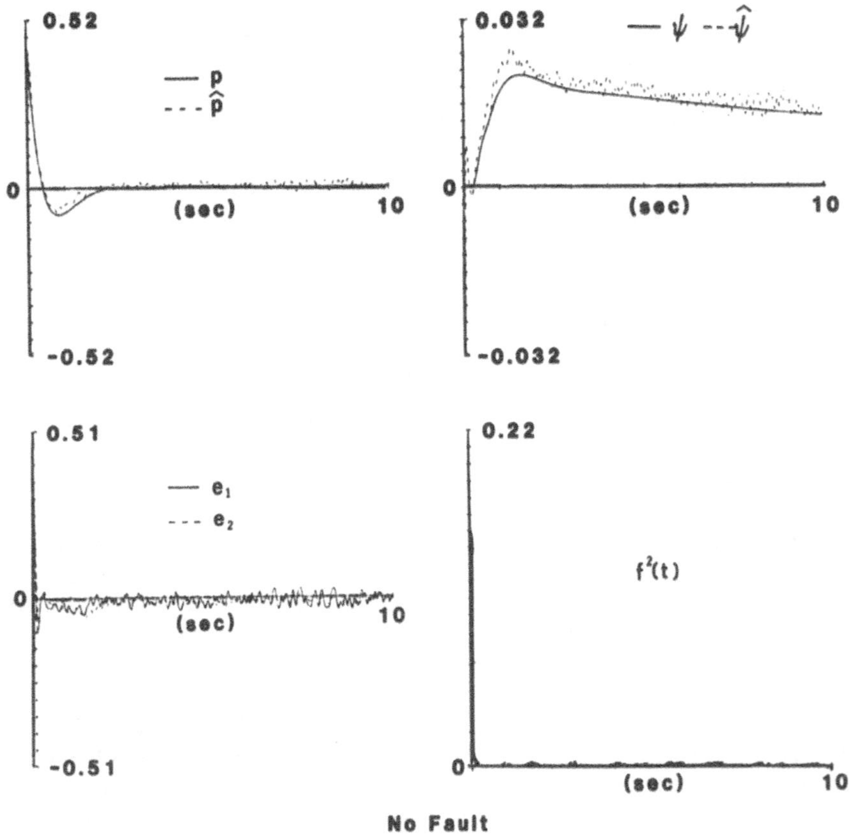
0.52
— p
- - - $\hat{p}$
0
(sec)
10
-0.52
0.032
— ψ - - $\hat{\psi}$
0
(sec)
10
-0.032
0.51
— e1
- - - e2
0
(sec)
10
-0.51
0.22
f²(t)
0
(sec)
10
No Fault

Figure 1.

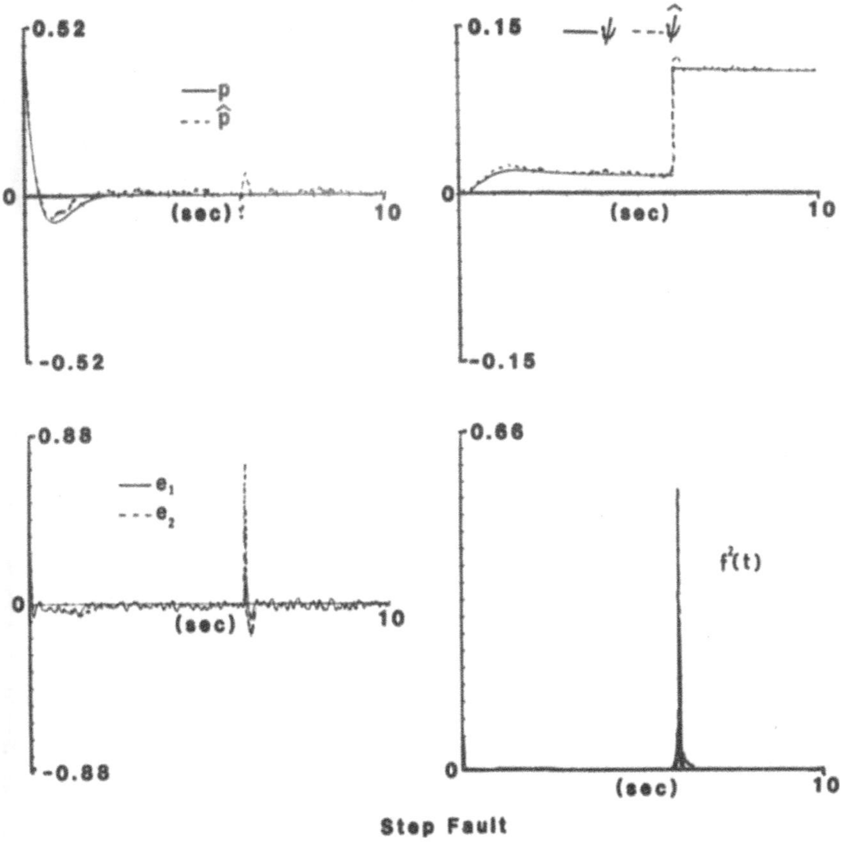
0.52
—p
--- $\hat{p}$
0
(sec)
10
-0.52
0.15
—ψ ---$\hat{\psi}$
0
(sec)
10
-0.15
0.88
—e_1
---e_2
0
(sec)
10
-0.88
0.66
$f^2(t)$
0
(sec)
10
Step Fault

Figure 2.

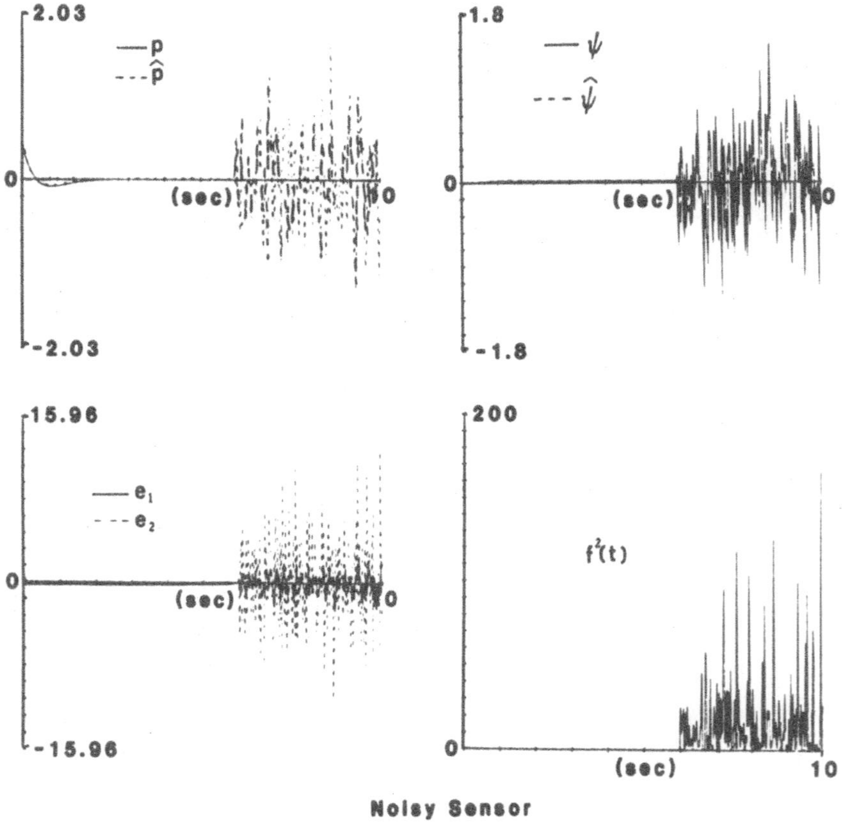
2.03
p
p̂
0
(sec)
10
-2.03
1.8
ψ
ψ̂
0
(sec)
10
-1.8
15.96
e1
e2
0
(sec)
10
-15.96
200
f²(t)
0
(sec)
10
Noisy Sensor

Figure 3.

8. ACKNOWLEDGEMENTS

This work has been carried out with the support of the Procurement executive UK MOD. The authors would also like to thank Professors Frank and Wunnenburg (Duisburg, FRG) for helpful discussions. Thanks also to Miss Sarah Mudge at York who assisted in the complex eigenstructure computations. The model of the Machan umanned aircraft has been provided by GEC Avionics.

9. REFERENCES

1. **Patton R.J. & Willcox S.W. 'A Comparison of two Techniques of IFD based on a Non-Linear Stochastic Model of an Aircraft'** Proceedings of the 7th IFAC/IFORS Symposium on Identification & System Parameter Estimation, York, England, July 3-7 1985.
2. **Patton R.J. & Willcox S.W. 'A Parameter Insensitive Technique for Aircraft Sensor Fault Analysis Using Eigenstructure Assignment and Analytical Redundancy'**, Paper No. 86-2029-CP, Proceedings of the AIAA Guidance, Navigation and Control Conference , Williamsburg , Virginia , USA August 1986.
3. **Willsky A.S. 'A Survey of Design Methods for Failure Detection in Dynamic Systems'**, Automatica 12, pp601-611, 1976.
4. **Clark R.N. 'A Simplified Instrument Failure Detection Scheme'**, IEEE Trans AES-14, pp558-563, 1978.
5. **Clark R.N. 'The Dedicated Observer Approach to Instrument Failure Detection'** Proceedings 18th IEEE CDC Conference, pp237-241, 1979.
6. **Clark R.N. & Setzer W 'Sensor Fault Detection in a System with Random Disturbances'** IEEE Trans AES-16, pp468-473, 1980.
7. **Frank P.M. , Elmadbouly E. & Keller L. 'The Application of Observers to Instrument Failure Detection'** Proceedings of the IFAC Symposium on Theory & Application of Digital Control, New Delhi, 5-7 January 1982.
8. **Frank P.M. & Keller L 'Sensitivity Discriminating Observer Design for Instrument Failure Detection'** IEEE Trans AES-16, pp460-467, 1980.
9. **Frank P.M. 'Robust Instrument Failure Detection via Luenberger Observers in Nuclear Power Plants'**IFAC-CIGRE Symposium on Control Applications for Power Systems Security, Florence, 26-28 September, 1983.
10. **Watanabe K & Himmelblau D. M. 'Instrument Fault Detection in a System with Uncertainties'** Int. J. Systems Science, 13 (2), pp137-158, 1982.
11. **MacFarlane A.G.J. & Karcanias N. 'Poles & Zeros of Linear Multivariable Systems: A Survey of Algebraic, Geometric and Complex Variable Theory'**,Int. J. Control 24 ,

pp33 - 74 , 1976.

12.**El-Ghezawi O.M.E. , Billings S.A. & Zinober A.S.I. 'Variable Structure Systems and System Zeros'**, Proc IEE Part D, 130, (1), Jan, 1983.

13.**Brogan William L. 'Modern Control Theory'** Quantum Publishers Inc. 1974

14.**Kudva P & Gourishankar V.'On the Stability Problem of Multivariable Model Following Systems'**,Int. J. Control 24 1976, pp33-74.

15. **Mudge S.K. & Patton R.J. 'The Design and Assessment of a Sliding Mode Controller for the Lateral Motion model of an Aircraft'** Proceedings of the 1st European Workshop on Fault Diagnostics, Reliability and Related Knowledge-Based Approaches, Rhodes Sept. 1986, Reidel Press.

16. **D'Azzo J.J.'Synthesis of Digital Flight Control Systems by the Method of Entire Eigenstructure Assignment'** Proc IEEE NAECON Conference, Dayton, Ohio, USA, 1979, pp343-347.

17.**Porter B. & D'Azzo J.J. 'Algorithms for Closed-Loop Eigenstructure Assignment by State Feedback in Multivariable Linear Systems'**, Int. J. Control 27, pp943-947, 1978.

18.**Mielke R.R. , Tung L.I. & Marefat M. 'Design of Multivariable Feedback Controllers via Spectral Assignment'** NASA (Unclassified) Report No. CR-169889, 1983.

VALIDATION OF MEASUREMENTS IN CONTROL SYSTEM BY MEANS OF ANALYTICAL REDUNDANCY

M. HAMAD, M. STAROSWIECKI
Centre d'Automatique de Lille
Université des Sciences et Techniques de Lille Flandres Artois
59655 Villeneuve d'Ascq Cédex

ABSTRACT. At the physical level of control systems, faulty instruments lead to generate faulse informations which will be used in upper levels of the control hierarchy. Accordingly, the necessity to validate the measurements leads to supervision structures which use physical and/or analytical redundancy of the sensors. Analytical redundancy uses the system's model in order to generate statical or dynamical relationships linking the values of the measured variables. These relationships generate residuals whose supervision allows the validation of the informations issued from the physical level through the sensors.
In this communication, we present an approach which generates residuals with minimal variance. The coherence degree of a variable is defined and used to localize failures by means of a signature which is independant of the kind of the failure.

1. INTRODUCTION

In the hierarchical structure of a control system, orders are elaborated at each level from informations which are provided from lower ones. At the physical level (i.e the controled process) informations are issued from measurement devices including sensors and processing systems.

Faulty instruments lead to generate false informations which are used in upper levels. Thus these levels elaborate orders and statements which do not correspond to the real state of the process. Accordingly, it is necessary to validate these informations before their treatement at different hierarchical levels |1|.

The validation procedure consists to check the coherence between measurements themselves |2|, |3|, and between measurements and real values corresponding to a given operation point. Such a procedure is based upon the existence of either hardware or analytical redundancy. Analytical redundancy uses the system's model in order to generate statical or dynamical relationships linking the values of the measured variables. These relationships generate residuals whose supervision allows the validation of the informations issued from the physical

S. Tzafestas et al. (eds.), System Fault Diagnostics, Reliability and Related Knowledge-Based Approaches, Vol. 1, 193–201.

level through the sensors. The general scheme of the decision procedure appears on fig. 1

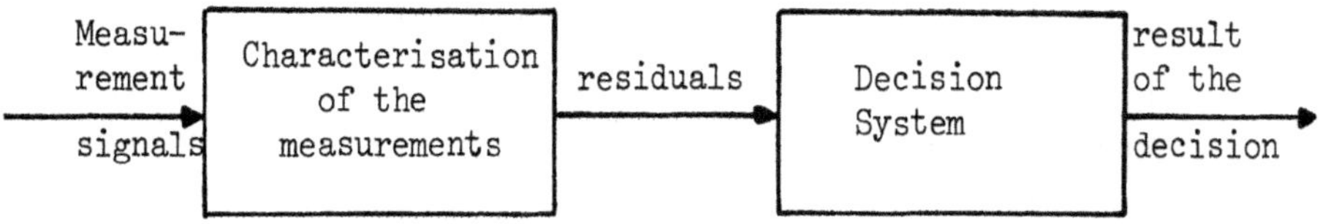

Figure 1. The decision procedure.

The decision problem can be formulated in different ways :
- the detection problem consists to decide if there is a faulty sensor or not within the set of sensors
- the isolation problem cnsists to recognize the faulty instrument
- the diagnosis problem consists to recognize the nature of the failure.

In all cases, the validation of the data which are transmitted to the higher levels of the control hierarchy supposes that the two first problems have been solved.

2. RESIDUAL GENERATION

Let us consider a control system described by the following invariant model :

$$x(k+1) = A\,x(k) + B\,u(k) + \eta(k) \tag{1}$$

$$y(k) = C\,x(k) + \varepsilon(k) \tag{2}$$

(1) is the state equation ; (2) is the measurement equation with the following notations : x, y and u are state, measurement and control vectors. n, m and q are their dimensions. A, B, C are matrices with corresponding dimensions. η and ε are additive noises applying respectively on the system and on the sensors.

The equations (1) and (2) describe the normal operation of the process. Any occuring failure will lead to have these equations not verified.

2.1. Parity space approach

The parity space approach |4| is based only on the model of the sensors provided by the measurement equation (2). It exploites the excess information due to the physical redundancy (m > n) to elaborate relationships among the outputs of the sensors. The vector defined by the residuals of these relationships is called the parity vector.

$$p(k) = \Omega\,y(k) = \Omega\,\varepsilon(k)$$

$$(\Omega \text{ such that } \Omega C = 0) \tag{3}$$

This method cannot be applied in the absence of physical redundancy ($m \leqslant n$).

The generalized parity space approach |5| creates excess information via the use of the analytical redundancy relationships which exists in the system model (1) and (2). Considering an observation window of length l, one obtains :

$$\begin{vmatrix} y(k) \\ y(k+1) \\ \vdots \\ \vdots \\ y(k+l) \end{vmatrix} = \begin{vmatrix} C \\ CA \\ \vdots \\ \vdots \\ CA^{l} \end{vmatrix} x(k) + \begin{vmatrix} 0 & & 0 \\ CB & & \\ \vdots & & \\ \vdots & & \\ CA^{l-1}B & ..CB & 0 \end{vmatrix} \begin{vmatrix} u(k) \\ u(k+1) \\ \vdots \\ \vdots \\ u(k+l) \end{vmatrix}$$

$$+ \begin{vmatrix} 0 & & 0 \\ C & & \\ \vdots & & \\ \vdots & & \\ CA^{l-1} & ..C & 0 \end{vmatrix} \begin{vmatrix} \eta(k) \\ \eta(k+1) \\ \vdots \\ \vdots \\ \eta(k+l) \end{vmatrix} + \begin{vmatrix} \varepsilon(k) \\ \varepsilon(k+1) \\ \vdots \\ \vdots \\ \varepsilon(k+l) \end{vmatrix} \tag{4}$$

which is written under condensed form :

$$Y(k,l) = O_l x(k) + C_l U(k,l) + D_l \eta(k,l) + \varepsilon(k,l) \tag{5}$$

The elimination of the state vector leads to the generalized parity vector, defined by :

$$p(k,l) = \Omega Y(k,l) - \Omega C_l U(k,l) = \Omega D_l \eta(k,l) + \varepsilon(k,l) \tag{6}$$

(Ω is such that $\Omega O_l = 0$).

The failure detection problem consists to check whether the parity vector, or the generalized one, results only from noises as indicated by equation (3) or (6) or does not. For that, each of its components is compared to a given threshold whose value results from the solution of a decision problem in which H_o is the no failure hypothesis while H_1 is the failure hypothesis |6|. It is clear that the performance of the optimal decision is decreasing with the variance of the parity vector in the H_o case.

2.2. Minivar residual approach |7|

If we assume the system is stable at known operating point it is possible to generate a set of redundancy relationships whose residuals have minimal variance. We call this method Minivar Residual Approach MVRA. The equation (1) and (2) considered in the neighbourhood of the operating point (x^*,u) become :

$$X(k+1) = AX(k) + \eta(k) \tag{7}$$

$$Y(k) = CX(k) + \varepsilon(k) \tag{8}$$

$X(k) = x(k) - x^*(k)$, $\eta(k)$, $\varepsilon(k)$ are independant, Zero mean, random vectors, with covariance matrices Σ, Q and R.
$Y(k) = y(k) - Cx^*(k)$ is then a random vector Zero mean and covariance matrix $V = C\Sigma C'+R$.

The task of the MVRA is to transform the deviation vector (difference between measurement and theoretical vectors) into a new vector which is called minival residual vector. This vector is obtained by solving a constrained quadratic minimisation problem.

2.2.1. Statical relationships. Consider the following parity vector :

$$P(k) = Y(k) = CX(k) + \varepsilon(k) \tag{9}$$

P(k) is a random vector with Zero mean and covariance matrix V. Any linear combination of the components of P(k) can be used as a residual for the decision problem. We shall look for these which have minimal variance.

2.2.2. Dynamical relationships. The observation window of length l, applied to equations (7), (8) gives :

$$Y(k,l) = O_l X(k) + D_l \eta(k,l) + \varepsilon(k,l) \tag{10}$$

The parity vector $P(k,l) = Y(k,l)$ has Zero mean and covariance matrix $V_l = O_l \Sigma O'_l + D_l \bar{Q} D'_l + \bar{R}$

with $\bar{Q} = \text{diag}\{Q,\dots,Q\}$, $\bar{R} = \text{diag}\{R,\dots,R\}$

We shall in this case apply the same approach as in the statical one.

2.2.3. Determination of the residuals. Let $r_i(k,l) = \omega'_i P(k,l)$ be the ith generated residual. In both cases (statical or dynamical) the optimal residual generation problem is formulated as follows :

Find w_i $\quad i=1,\dots,\sigma$ $\quad$ such that :

$w'_i V_l w_i$ $\quad$ minimum

under the constraints $\quad w'_i w_i = 1$; $w'_i w_j = 0 \quad j \neq i$

This is a principal components analysis problem |8| whose solution leads to retain the σ eigenvectors of the matrix V_l wihch correspond to the σ smallest eigenvalues. The variance of the ith residual is then equal to λ_i (the ith eigenvalue) and the number σ of the retained residuals is such that λ_σ has still an acceptable value. Let $R(k,l) = WY(k,l)$ be the vector of the r_i's.

It is possible to show that when x^* is the least square estimate of x, the parity space (or generalized parity space) is a subspace of the minivar residual space (or generalized minivar residual space).

In the absence of failure, the residual vector is a random vector with known mean and covariance. When a failure occurs, these characteristics of the residual vector change.

The failure detection problem consists to decide whether the residual vector lies in the noise area or not. The isolation problem leads then to the recognition of the failed sensor.

3. FAILURE DETECTION AND ISOLATION OF THE FAULTY INSTRUMENT

As in the parity space approach, the columns of the matrix W define m failure directions in the statical case (or m failure subspaces in the dynamical case) corresponding to the m sensors in the residual space |4|, |5|.

In fact, in the dynamical case, the apparition of a failure causes the residual vector to describe a transient trajectory in the residual subspace, which depends on the kind of the failure. This trajectory is the failure signature, which has to be recognized for the detection and the isolation of the failed instrument. The difficulty of such an approach lies in the infinity of possible signatures and in the fact that the failure subspaces of differents sensor may overlap.

This is why we shall use a voting scheme, in order to define failure signatures which are independant on the nature of the failure.

The use of a voting scheme supposes the 2 following conditions : detection condition : each sensor is included in at least one relationship ; isolation condition : $\forall\ (y_i, y_j)$ the set of the relationships in which y_i intervenes is not identical to the set of the relationships in which y_j intervenes.

3.1. Degree of coherence of a measurement

Let us consider a given sensor "y_j" and let "σ_j" be the number of relationships in which "y_j" intervenes. As we perform tests concerning all the relations, the set of the σ relationships including "y_j" will be split into two parts :

- The set of those whose residual is less than the decision threshold. Let σ_{jc} be their number.

- The set of those whose residual is greater than the decision threshold : their number is $\sigma_j - \sigma_{jc}$.

The *degree of coherence* of the measurement y_j is then defined by :

$$dj = \frac{\sigma_{jc}}{\sigma_j} \tag{11}$$

3.2. Failure signature of the sensor

Let us suppose that a failure occurs on the ith sensor. The degree of coherence of the measurements is given by :

$$j = 1,2,\ldots,m \quad d_j(k) = \begin{cases} 0 & \text{if } j = i \\ 1 - \dfrac{\sigma_{ji}}{\sigma_j} & \text{if } j \neq i \end{cases} \tag{12}$$

where σ_{ji} is the number of relationships in which both y_j and y_i intervene.

The vector $D_i = (d_j(i),\ j = 1,\ldots m)$ is called *failure signature of the sensor* y_i.

It appears clearly that when no sensor is failed, all the degrees of coherence should be equal to "1". This is the no-failure signature, noted $D_0 = (1,1,\ldots,1)$.

In order to give an idea of the detection and isolation algorithm and of the failure signatures, let us consider the following example fig. 2 where four statical relationships (13) link the outputs of six sensors.

$$\begin{aligned} ay_1 - ay_2 - by_3 - by_4 &= 0 \\ by_1 - ay_3 + by_4 - ay_5 &= 0 \\ -ay_3 + ay_4 - by_5 - by_6 &= 0 \\ ay_1 - by_4 - by_5 + ay_6 &= 0 \end{aligned} \tag{13}$$

where $a \neq 0$ and $b \neq 0$.

We can note that four residuals are sufficient to detect and isolate six different single failures. In fact, it is possible to show that, in the ideal case, the maximal number of sensors which can be supervised by the use of σ residuals is equal to $2^{\sigma}-1$, |9|. The supervision algorithm simply consists in the comparison, at each instant, of the vector of the degrees of coherence to its different possible profiles.

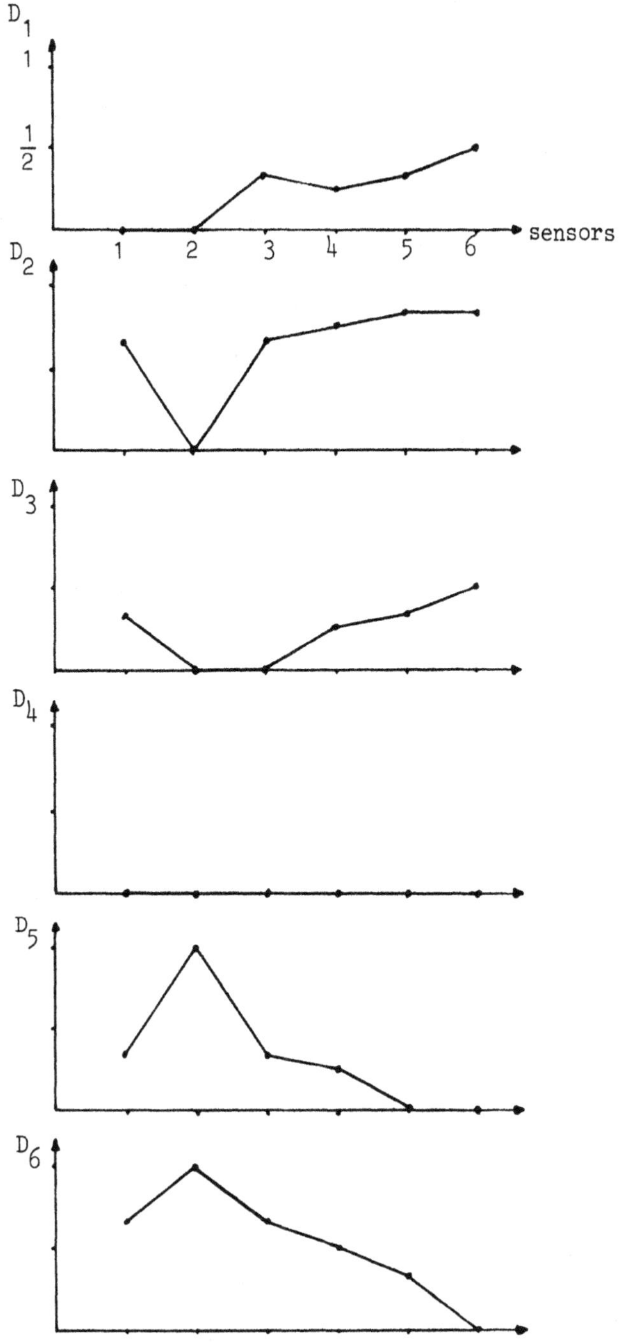

Figure 2. Single failure signatures.

3.3. Multiple failures

The case in which multiple failures occur simultaneously rises two kinds of problems :

. multiple failures may cancel each other so that we obtain the no failure signature (invisibility case)

. multiple failures can be detected, but exhibit a signature that is the same than an existing single failure signature.

The first case leads to a non detection of the failure, while the second case leads to a false isolation of the failed instrument.

The conditions under which no invisibility case can exist are linked with the independance of the columns of the matrix W.

The following result can be shown : Let $I \subset \{1,2,\ldots,m\}$ be a subset of indexes corresponding to a given subset of sensors. A necessary and sufficient condition under which any simultaneous failure of this subset of sensors will be detectable is that the submatrix W_I constructed with the corresponding columns of the matrix W has rank card I |9|.

In the detectability case, one can define, for each subset of sensors, it's corresponding failure signature.

Fig. 3 shows the signature of double failures sensor 2 and 6 in the preceding example.

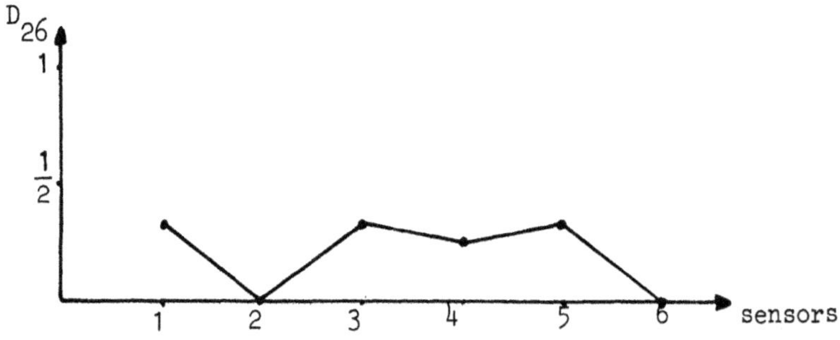

Figure 3. Double failure signature.

4. CONCLUSION

The MVRA is used to generate statical and dynamical relationships applying to the measured variables of a control system. The degree of coherence of each measurement is computed on the basis of the result of a decision procedure applied to each of the relationships. The choice of the best relations is not a trivial one. In fact they should have :

- a small variance, in order to obtain a good decision threshold
- a short observation window, in order to accelerate the failure detection
- a weak effect of the failure of one sensor on the degrees of coherence of the others (see (11)).

RERERENCES

|1| Tylee, J.L., Hon, A.L. "New concepts in Nuclear Power Plant Instrumentation and Control" 22th IEEE CDC, San Antonio, Texas, December 1983

|2| Desai, M.N., Ray, A., "A Fault Detection and Isolation Metholology" 20th IEEE CDC, San Diego, Calofornia, December 1981, pp 1363-1369

|3| Deckert, J.C., and all, "A signal Validation Methodologie for Nuclear power plants" presented at American Control Conference, Arlington, Virginia, June, 1982

|4| Potter, J.E., Suman, M.C., "Thresholdless Redundancy Menagement with Array of Skewed Instruments" Integrity in Electronic Flight Control Systems, AGARDOGRAPH - 224, pp 15-1 to 15-25, 1977

|5| Chow, E.Y., Willsky, A.S., "Analytical Redundancy and the Design of Robust Failure Detection Systems" IEEE Trans. on AD, Vol. AC-29, N° 7, July 1984

|6| Arques, P.Y., "Décisions en Traitement du Signal" Masson, 1979

|7| Staroswiecki, M., Hamad, M., "Validation of measurements and Detection of Sensor's Failures in Control Systems" EUSIPCO-86, The Hague, The Netherland, Sept 2-5, 1986

|8| Seber, G.A.F., "Multivariate Observations" J. WILY and Sons, 1984

|9| HAMAD, M., "Validation des mesures et détection des capteurs défaillants dans un système de controle commande" Thèse de Doctorat en Automatique, N° 41, Juin 1986, Université des Sciences et Techniques de Lille Flandres Artois.

FAULT DETECTION AND LOCATION IN MRAC ROBOTIC SYSTEMS

A. Pouliezos
Tech.Univ. of Krete
Chania, Krete
Greece

S.Tzafestas, G.Stavrakakis
Control and Robotics Group
Natl. Tech. Univ., Athens
Greece

ABSTRACT. Fault detection, identification and monitoring plays a primary role in systems engineering. In this paper the problem of fault detection in Model Reference Adaptively Controlled (MRAC) robotic systems is investigated. The MRAC model assures convergence to suitable reference models for a class of processes, one of which is the manipulator. However, sudden changes, modelled as faults, in the process parameters may lead to degradation of performance and even to instability. Using the Discrete Square Root Filter (DSF) it is possible to monitor the performance of the system and make the appropriate decisions in the event of a failure. The whole fault monitoring scheme is easily added to an existing MRAC system. Simulation studies in the case of faults in the manipulator's D.C. motors complete the paper and demostrate the usefulness of the method.

1. INTRODUCTION

The increasing demands on reliability and safety of industrial processes and their elements lead to the development of methods for improving the supervision and monitoring as part of the overall control of the process. This is also true for fine processes with highest demands on reliability and safety, e.g. the robotic manipulators {1-7}.

In the next sections the elementary functions of process supervision are cosidered. The methods for faults detection are described for the DC - motor actuators of a robotic manipulator.

A fault is to be understood here as a nonpermitted deviation of a physical parameter from the nominal value which leads to the inability to fulfil the intended purpose. After the effect of a fault is known, a decison on the action to be taken can be made. If the fault is evaluated to be tolerable, the operation may continue and if it is conditionally tolerable a change of operation has to be applied. However if the fault is not tolerable the operation must be immediately stopped and the fault eliminated.

S. Tzafestas et al. (eds.), System Fault Diagnostics, Reliability and Related Knowledge-Based Approaches, Vol. 1, 203–217.

2. DYNAMIC MODELS FOR THE MRAC ROBOTIC MANIPULATOR

A lot of results on the control problem of robotic manipulators have been derived and widely disseminated. Among the control algorithms developed, the so called **Model Reference Adaptive Control** (MRAC) approach seems to provide a robust method for the control of processes with variable parameters and/or unknown parts.

Nicosia and Tomei {1} use the complete nonlinear time varying model including all second order terms, resulting from a generalized application of Lagrangians dynamics, to derive a robust MRAC algorithm for the manipulator based on Popov's hyperstability theory. In their paper the DC-motor actuators'dynamics are neglected resulting in sudden fractuations for the derived control torques.

Tzafestas and Stavrakakis {2} extended this original approach by introducing the DC-motors'dynamics with some improvement concerning the on line calculation of the controller parameters. The results can be compared with those obtained in {1}. It is easy to see that the DC-motor dynamics are taken into account for the control voltage calculation, the control voltage is smooth and easier to apply in practice than the control torque proposed in {1}. The global dynamic model of a **3-dof** (degrees of freedom) robotic manipulator is derived based on the Lagrangian approach and the dynamic equations for the armature circuit and the mechanics of the actuators DC-motors. These equations have the following form (see Fig. 1):

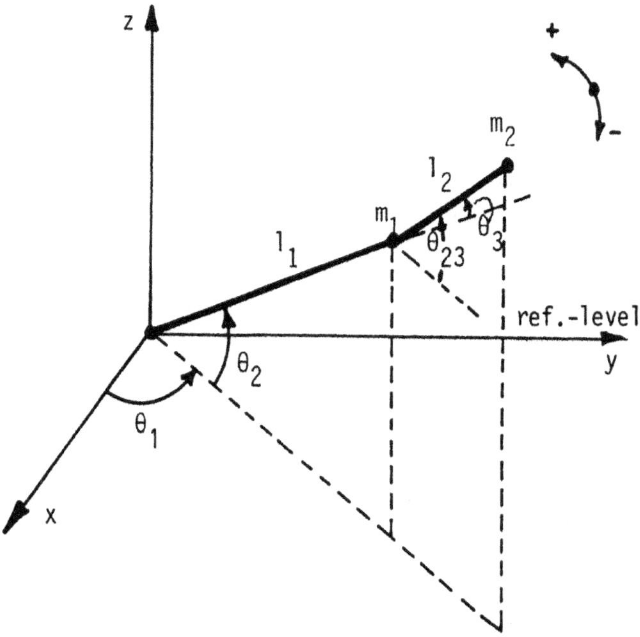

Fig.1. Geometry of the three-link manipulator under study.

$$\begin{bmatrix}\dot{\mathbf{T}}\\ \dot{\boldsymbol{\omega}}\end{bmatrix}=-\mathbf{G}^*(\boldsymbol{\theta})\{\mathbf{C}^*(\boldsymbol{\theta})\mathbf{h}^*(\boldsymbol{\omega},\mathbf{T})-\boldsymbol{\Gamma}^*(\boldsymbol{\theta})\}+\mathbf{G}^*(\boldsymbol{\theta})\,\mathbf{S}\,\mathbf{V} \tag{1}$$

where: $\boldsymbol{\theta}^T(t)=\left[\theta_1(t)\ \ \theta_2(t)\ \theta_3(t)\right]$, $\mathbf{V}^T(t)=\left[V_1(t)\ V_2(t) V_3(t)\right]$

$$\boldsymbol{\omega}^T=\left[\dot{\theta}_1\ \dot{\theta}_2\ \dot{\theta}_3\right]$$

$$\mathbf{G}^*(\boldsymbol{\theta})=\begin{bmatrix}K_{mN} & 0\\ 0 & G(\theta)\end{bmatrix}\varepsilon R^{6x6},\quad \mathbf{C}^*(\boldsymbol{\theta})=\begin{bmatrix}0 & \mathbf{I} & \mathbf{K}_{mN}^{-1} & K_{RL}\\ D(\boldsymbol{\theta}) & RN & & -I\end{bmatrix}\varepsilon R^{6x12}$$

$$\mathbf{h}^*(\boldsymbol{\omega},\mathbf{T})=\begin{bmatrix}\mathbf{h}(\omega)\\ \omega\\ \mathbf{T}\end{bmatrix}\varepsilon R^{12},\quad \boldsymbol{\Gamma}^*(\boldsymbol{\theta})=\begin{bmatrix}0\\ \Gamma(\theta)\end{bmatrix}\varepsilon \mathbf{R}^6,\quad \mathbf{S}=\begin{bmatrix}-K_{mN}^{-1}\,K_{mL}\\ 0\end{bmatrix}\varepsilon \mathbf{R}^{6x3}$$

with

$$\mathbf{K}_{mN}=\text{Diagonal}\left[\frac{K_{mi}^2N_i^2}{L_i}\right],\ K_{RL}=\text{Diagonal}\left[\frac{R_i}{L_i}\right],\ \mathbf{N}=\text{Diagoanl}\left[N_i\right]$$

$$\mathbf{R}=\text{Diagonal}\left[\rho_i\right],\ K_{mL}=\text{Diagonal}\left[\frac{K_{mi}N_i}{L_i}\right],\ \mathbf{J}_m=\text{Diagonal}\left[J_{mi}\right],\ i=1,2,3$$

$$\mathbf{G}(\boldsymbol{\theta})=\left[\mathbf{A}(\boldsymbol{\theta})+J_m\mathbf{N}\right]^{-1}$$

The symbols used have the following meaning:

N_i, K_{mi}, R_i, L_i (i=1,2,3): are the gear ratio, the DC torque constant, the armature resistance, the armature inductance of the DC-motors, respectively.

V_i : is the armmture voltage (control input) of the ith DC-motor

ρ_i : is the viscous friction coefficient of the ith DC-motor

J_{mi} : is the moment of inertia of the rotor of the ith DC-motor

$\omega_{mi}=N_i\dot{\theta}_i=N_i\omega_i$: is the reflected shaft angular velocity

$\mathbf{A}(\boldsymbol{\theta})$: is the (3x3) generalized inertial matrix which is symmetric and positive definite.

$\boldsymbol{\Gamma}(\boldsymbol{\theta})$: is the (3x1) vector of the gravitational torques

$\mathbf{D}(\boldsymbol{\theta})\mathbf{h}(\dot{\boldsymbol{\theta}})$: involves the centrifugal and the Coriolis torques.

(with $\mathbf{h}(\dot{\boldsymbol{\theta}})$=vector of $\dot{\theta}_i\dot{\theta}_j$, i=1,2,3, j=i,...,3

$i_i(t)$:is the armature current of the ith joint actuator

$T_i(t)=K_{mi}N_i i(t)$:is the reflected electromagnetic torque of the ith joint actuator

$T_{Li}(t)=T_i(t)-J_{mi}\dot{\omega}_{mi}-\rho_i\omega_{mi}$:is the torque generated at the ith joint (reflected load of the ith DC motor)

The **MRAC** algorithm which provides a robust trajectory following method of robots using the DC-motor actuators dynamics is fully presented in {2}.

The values of the DC-motors constants play important role for the calculation of the control law at any instant. On the other hand the correct operation of the whole robot depends strongly on the correct operation of the DC-motor actuators i.e. on the conservation of their physical properties (parameters) during operation.

Sudden changes of the physical parameters of the actuators which may must, therefore, be automaticaly monitored in order to achieve a reliable and safe operation of the manipulator.

3. FAULT MONITORING FOR THE DC-MOTOR ACTUATORS OF A ROBOT

The Model Reference Adaptive Controlled robotic system described in the previous section must be supervised automaticaly. The first stage of this supervision consists of the detection of changes (faults) in the dc motor actuators based on theoretically derived motor models and parameter estimation. After the effect of the fault is known, a decision on the action to be taken can be made. In {3} - {7} many methods of fault monitoring in industrial processes are reported. The generalized structure of fault detection methods is given in Fig.2 and involves three distinguished stages.

In many cases, the following variables can be assumed to be given:

— Measurements of the input $\mathbf{y}(t)$ and the output $\mathbf{u}(t)$ of the process.
— More or less exact a-priori information about the static and dynamic behaviour of the process.

3.1. Mathematical model for the DC-motor actuator

The dynamic model for the DC-motor actuator is given analyticaly in Appendix I. Define $\mathbf{u}^T(t)=\left[V_i(t) \quad T_{Li}(t)\right]$, $\mathbf{y}(t)=\left[i_i(t) \ \omega_i(t)\right]^T$. Then the model (I_1) of appendix I can be formulated as a Continuous-Time **Multi-Input Multi-Output** (MIMO) of two (r=2) differential equations, for the actuator of each link. That is:

$$\mathbf{y}^{(1)}(t)+\mathbf{A}_1\mathbf{y}^{(1)}(t)+\mathbf{A}_0\mathbf{y}(t)=\mathbf{b}_0\mathbf{u}(t) \qquad (2)$$

where:

$$\mathbf{A}_1=\mathbf{0}\in\mathbf{R}^{4\times4},\ \mathbf{A}_0=\begin{bmatrix} \dfrac{R_i}{L_i} & \dfrac{K_{mi}N_i}{L_i} \\ -\dfrac{K_{mi}}{J_{mi}} & \dfrac{\rho_i}{J_{mi}} \end{bmatrix},\quad \mathbf{b}_0=\begin{bmatrix} \dfrac{1}{L_i} & 0 \\ 0 & -\dfrac{1}{J_{mi}N_i} \end{bmatrix}$$

Define: $\theta_1=\dfrac{R_i}{L_i},\quad \theta_2=\dfrac{K_{mi}N_i}{L_i},\quad \theta_3=\dfrac{1}{L_i},\quad \theta_4=-\dfrac{K_{mi}}{J_{mi}},\ \theta_5=\dfrac{\rho_i}{J_{mi}},\ \theta_6=-\dfrac{1}{J_{mi}N_i}$

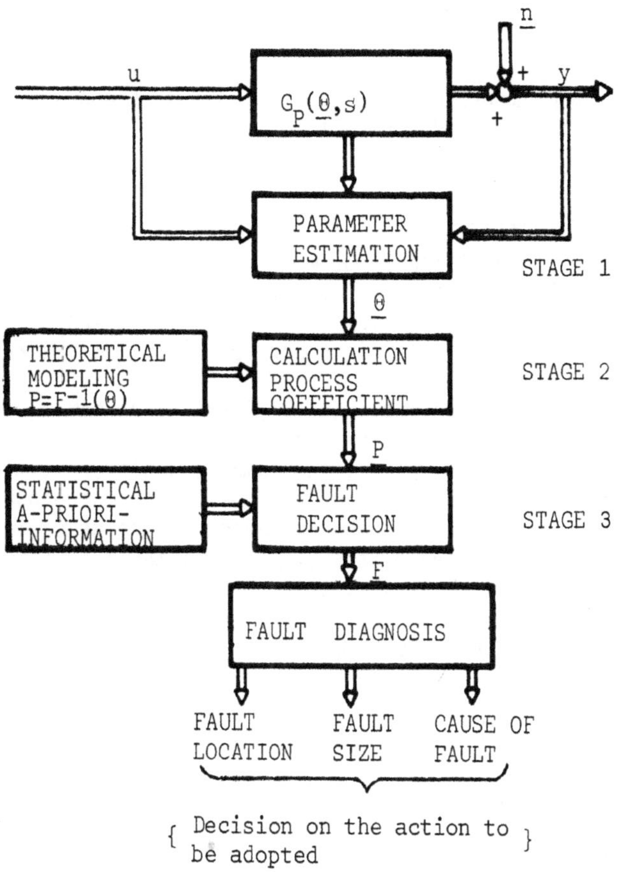

Fig. 2. Generalized structure of fault detection method.

i.e. $\theta^T = \begin{bmatrix}\theta_1 & \theta_2 & \theta_3 & \theta_4 & \theta_5 & \theta_6\end{bmatrix} \varepsilon \mathbf{R}^6$

Then, if measurements of $\mathbf{y}(t)$ and $\mathbf{u}(t)$ are available equation (2) leads to:

$$\mathbf{e}(t)=\mathbf{y}^{(1)}(t)-\mathbf{\Psi}(t)\theta \tag{3}$$

with

$$\boldsymbol{\Psi}(t)=\begin{bmatrix} -\mathbf{y}^T(t) & u_1(t) & 0 & 0 \\ 0 & 0 & -\mathbf{y}^T(t) & u_2(t) \end{bmatrix} \quad \mathbf{R}^{2x6}$$

3.2. Process parameter estimation using a discrete Square-root filter (DSF)

In this stage, measurements of the input and output signals are considered to be available at discrete times $t=kT_0$, $k=1,2,\ldots,N$ with T_0 the sampling time. This means, that measurements of $i_i(k)$, $\omega_i(k)$, $V_i(k)$, $T_{Li}(k)$, $k=0,1,\ldots,N$ are available.

In the present case of the 3-DOF MRAC robot these measurements are obtained through simulation (numerical integration) of the complete robot model (1), where the input voltage $\mathbf{V}(t)$ is deremined via the control algorithm proposed in {2} in order to follow the desired trajectories.

The parameter estimation requires the first order time derivatives of the noisy output signal $\mathbf{y}(t)$. These derivatives can be calculated from the sampled measurements $\mathbf{y}(k)$ using numerical differentation. The simplest way is to replace the derivatives by the corresponding (backward) differences. To reduce the influence of the noise interpolation formulas, interpolation by third-order polynomials or Newton interpolation can be used. For calculating higher-order derivatives the use of state variable filtering is recommended {3}.

The **Discrete Square Root Filtering** (DSF) method combined with a moving window of N samples over the data, provides a better method than the classical **RLS** for the detection of parameters changes. This is due to the numerical properties of the **DSF**. For details refer to Kaminski et.al. {5}.

By observing the **MIMO**-linear continuous-time system over $k=1,2,\ldots,N$ samples the following discrete time set of N equations is obtained from the continuons time system (3):

$$\begin{aligned} \mathbf{e}(1)&=\mathbf{y}^{(1)}(1)-\boldsymbol{\Psi}(1)\theta \\ \mathbf{e}(2)&=\mathbf{y}^{(1)}(2)-\boldsymbol{\Psi}(2)\theta \\ &\cdots\cdots\cdots\cdots \\ \mathbf{e}(N)&=\mathbf{y}^{(1)}(N)-\boldsymbol{\Psi}(N)\theta \end{aligned} \tag{4}$$

or

$$\mathbf{E}(N)=\mathbf{Y}^{(1)}(N)-\underline{\boldsymbol{\Psi}}(N)\theta \tag{5}$$

with

$$\mathbf{E}^T(N)=\left[\mathbf{e}^T(1)\ \ldots\ \mathbf{e}^T(N)\right]\varepsilon\mathbf{R}^{(2Nx1)}$$

$$\mathbf{Y}^{(1)^T}(N)=\left[\mathbf{y}^{(1)^T}(1)\ \ldots\ \mathbf{y}^{(1)^T}(N)\right]\varepsilon\mathbf{R}^{(2Nx1)}$$

$$\underline{\Psi}(N)=\begin{bmatrix}\psi(1)\\ \\ \psi(N)\end{bmatrix} \in R^{2N\times 6}$$

The meaning of the moving window of width N over the data is shown by the following definition

$$E(N+i)=\left[\mathbf{e}^T(i+1) \;\ldots .\; \mathbf{e}^T(N+i)\right]$$

$$\mathbf{Y}^{(1)}(N+1)=\left[\mathbf{y}^{(1)^T}(i+1) \;\ldots .\; \mathbf{y}^{(1)^T}(N+i)\right]$$

$$\underline{\Psi}(N+i)=\begin{bmatrix}\psi(i+1)\\ \vdots \\ \Psi(N+i)\end{bmatrix} \qquad \text{for } i=0,1,2,\ldots$$

The parameters $\boldsymbol{\theta}$ can be estimated by minimizing the cost function $V(i)$:

$$V(i)=||\mathbf{E}(N+i)||^2 = \sum_{k=1+i}^{N+i} e^T(k)\mathbf{e}(k) \qquad \text{for } i=0,1,2,\ldots \quad (6)$$

Let us define

$$S(N+i)=\Big[\underbrace{\Psi(N+i)}_{6} \quad \underbrace{\mathbf{Y}^{(1)}(N+i)}_{1}\Big]\Big\}2N \text{ and } \boldsymbol{\theta}^*=\begin{bmatrix}-\boldsymbol{\theta}\\ \hline I\end{bmatrix} \in R^7$$

Then, the condition (6) becomes:

$$\min_{\boldsymbol{\theta}} V(i)=\min_{\boldsymbol{\theta}}||S(N+i)\boldsymbol{\theta}^*||^2 \quad i=0,1,2,\ldots \qquad (7)$$

The least squares estimator of the parameter vector $\boldsymbol{\theta}$ is then given by the well-known estimation equation:

$$\hat{\theta}_{N+i}=\left[\Psi^T(N+i)\Psi(N+i)\right]^{-1}\Psi(N+i)\mathbf{y}^{(1)}(N+i) \;, \quad i=1,2,\ldots \qquad (8)$$

This equation is not recursive and the parameters are biased for any type of noise $\mathbf{e}(t)$.

The basic idea of the Discrete Square Root Filtering applied here is as loffows:

Given:

$$\mathbf{S}(N+i)\boldsymbol{\theta}^*=\mathbf{E}(N+i)$$

Find an orthogonal transformation $\mathbf{T}\in\mathbf{R}^{2N\times 2N}$ such that

$$\mathbf{T}(N+i)\mathbf{S}(N+i)=\begin{bmatrix}\overbrace{\mathbf{W}(N+i)}^{7}\\ \hline \mathbf{0}\end{bmatrix}\begin{matrix}\}7\\ \}2N-7\end{matrix} \;, \; \mathbf{T}(N+i)^T\mathbf{T}(N+i)=\mathbf{I},\; i=0,1,2,\ldots \qquad (9)$$

where $\mathbf{W}(N+i)$ is upper triangular, or, equivalently, find $\mathbf{W}(N+i)$ and $E'(N+i)=\mathbf{T}(N+i)\mathbf{E}(N+i)$ directly.

This transformation does not change the cost function, i.e.

$$V(i)=||E(N+i)||^2=||\mathbf{T}(N+i).\mathbf{E}(N+i)||^2=||\mathbf{E}'(N+i)||^2 \qquad (10)$$

From (10) it follows directly that:

$$V(i)=\left\|\mathbf{E}'(N+i)\right\|^2=\left\|\begin{array}{c}\mathbf{E}_1'(N+i)\\ \hline \mathbf{E}_2'(N+i)\end{array}\right\|^2=\left\|\begin{bmatrix}\mathbf{W}_{\theta\theta}(N+i) & \mathbf{W}_{\theta E}(N+i)\\ \mathbf{\Omega} & \mathbf{W}_{EE}(N+i)\end{bmatrix}\begin{bmatrix}-\boldsymbol{\theta}\\ \mathbf{1}\end{bmatrix}\right\|^2 \tag{11}$$

where $\mathbf{W}_{\theta\theta}\varepsilon\mathbf{R}^{6x6}$, $\mathbf{W}_{\theta E}\varepsilon\mathbf{R}^{6x1}$ and $\mathbf{W}_{EE}\varepsilon\mathbf{R}$ i.e.

$$\mathbf{W}(N+i)=\begin{bmatrix}\mathbf{W}_{\theta\theta}(N+i) & \mathbf{W}_{\theta E}(N+i)\\ \mathbf{\Omega} & \mathbf{W}_{EE}(N+i)\end{bmatrix}$$

and $\mathbf{W}_{\theta\theta}$ is also upper triangular.

Eq. (11) implies that:

$$\min_{\boldsymbol{\theta}} V(i)=\min_{\boldsymbol{\theta}}||\mathbf{E}_1'(N+i)||^2$$

where

$$\begin{aligned}&-\mathbf{W}_{\theta\theta}(N+i)\boldsymbol{\theta}+\mathbf{W}_{\theta E}(N+i)=\mathbf{0}\\ &\mathbf{E}_2'(N+i)=\mathbf{W}_{EE}(N+i)\end{aligned} \tag{12}$$

Eq. (12) leads to the following estimator of $\boldsymbol{\theta}$:

$$\hat{\boldsymbol{\theta}}(N+i)=\mathbf{W}_{\theta\theta}^{-1}(N+i)\mathbf{W}_{\theta E}(N+i) \tag{13}$$

and the minimum value of $V(i)$ is given by $\mathbf{W}_{EE}(N+i)$, $i=0,1,2,\ldots$. Choosing the appropriate $\mathbf{T}$ may be interpreted as "compressing" the data matrix $\mathbf{S}$ into the upper triangular matrix $\mathbf{W}$. The triangular form of $\mathbf{W}$ facilitates the computation of $\mathbf{W}_{\theta\theta}^{-1}$ and reduces the dimension of the matrix to be inversed compared with eq. (8). Schmidt derived an algorithm for constructing the transformation matrix T which is given in Appendix II of Kaminski et al. {5}.

3.3. Calculation of the process physical coefficients

The physical process coefficients of the simulated robot DC-motor actuators were selected to be:

$p_1=R_i$	$p_4=J_{mi}$	Vmax=12V
$p_2=L_i$	$p_5=\rho_i$	
$p_3=K_{mi}$	$N_i=64$	

These are values in the range of the coefficients for DC-motor drives found in the literature.

The relation ship between process model parameters and physical process coefficients $\mathbf{p}$ is given by:

$$p_1=\frac{\theta_1}{\theta_3} \qquad (14.a)$$

$$p_2=\frac{1}{\theta_3} \qquad (14.b)$$

$$p_3=\frac{\theta_2}{\theta_3 N_i} \qquad (14.c)$$

$$p_4=\frac{-\theta_2}{N_i\theta_4\theta_3} \qquad (14.d) \text{ and also } p_4=-\frac{1}{\theta_6 N_i} \qquad (14.e)$$

$$p_5=-\frac{\theta_2\theta_5}{\theta_4\theta_3 N_i} \qquad (14.f)$$

The case of a fault occurence into the gear box is considered as an event w.p. 0. Under this consideration the relations (14.a) to (14.f) hold for the physical process coefficients. However, the case of a fault in the gear box i.e. modification of the no-fault value of the gear ratio can be detected via the fault occurence or not in the parameters p_3, p_4, p_5 of the 4th stage.

The physical process coefficient vector **p** will be used in stage 3 of the on line fault identification method as input data.

3.4. Fault detection and identification for the robot DC-motor actuator

Fault detection and identification involves the DECISION that a fault has occured (fault diagnosis), the LOCALIZATION of the fault and of his cause and the ESTIMATION of the fault size.

After calculation of the physical process coefficients $\mathbf{p}(k)$ in stage 2, let us consider $\mathbf{p}(k)$ as a Gaussian vector with its components statisticaly independent and its realizations $\mathbf{p}(i)$ and $\mathbf{p}(j)$ in the different sample instants $i\neq j$ statisticaly independent. It can be considered that the mean vector $\boldsymbol{\mu}(k)=E\,\mathbf{p}(k)$ and the covariance matrix $\boldsymbol{\sigma}^2(k)=E\left[(\mathbf{p}(k)-\boldsymbol{\mu}(k))\,(\mathbf{p}(k)-\boldsymbol{\mu}(k))^T\right]$ are invariant for the non-error case, i.e.

$$\boldsymbol{\mu}(k)=\left[\bar{\mu}_1\ldots\bar{\mu}_6\right]^T;\ \text{const.},\quad \sigma^2(k)=\text{diag}\left[\bar{\sigma}_1^2,\ldots,\bar{\sigma}_6^2\right]:\ \text{const.}$$

Under these conditions the joint probability density function over N samples is defined as:

$$f(\mathbf{p}(1),\ldots,\mathbf{p}(N)) = \prod_{i=1}^{N} f(\mathbf{p}(i)) \qquad (15)$$

A fault is defined by a significant deviation of the mean μ_i and/or variance σ_i^2 of $\mathbf{p}_i(k)$ — the ith component of $\mathbf{p}(k)$ — from the non-error value $\bar{\mu}_i$, $\bar{\sigma}_i^2$. Moreover, it is considered that only one fault may occur at a time.

This is a classical hypothesis test problem and can be handled by the formulation of (m+1) hypothesis H_i, $0 \leq i \leq m$, where m=6 in the case of the DC-motor of each robot joint actuator.

$$H_i = \begin{cases} \text{no fault in the mean and/or the variance of } \mathbf{p}\text{: } (i=0) \\ \text{fault of type } i \text{ (significant deviation of mean and/or variance of } p_i),\ 1 \leq i \leq m. \end{cases}$$

Each hypothesis H_i can be assosiated with a Ganssian conditional density function, where $\mu(H_i)$ and $\sigma^2(H_i)$ denote conditional mean and variance for hypothesis H_i. Therefore, the non-error case is described by $\mu(H_0)$, $\sigma^2(H_0)$, whereas $\mu(H_i)$, $\sigma^2(H_i)$, $1 \leq i \leq m$, describe a fault of type i.

When the no-fault mean and variance are not known i.e. the values fo the parameters are not known a priori, an estimation algorithm of $\mu(H_0)$ and $\sigma^2(H_0)$ is needed.

The output of the fault identification stage is a fault vector:

$$F = \left[i, \hat{\mu}_i(H_i),\ \hat{\sigma}_i(H_i) \right]^T, \quad 1 \leq i \leq m \tag{16}$$

where i represent the type of error, characterized by $\hat{\mu}_i(H_i)$ and $\hat{\sigma}_i(H_i)$. The fault detection and localization is possible by computing the **Logarithmic Likelihood Ratios**. The algorithm used here for estimating the non-error statistics and the fault detection and localization is reported with details in 4 .

4. SIMULATION RESULTS

The physical process coefficients of the simulated DC-motor robotic actuator are:

$R=1.1\Omega$ $\quad J_m=0.001\ \text{Kgm}^2$
$L=1\text{mH}$
$K_m=0.0224\ \text{Vsec}$ $\quad p=0.1\dfrac{\text{kgm}^2}{\text{sec}}$

These values give the a-priori knowlege of the actuator parameters used for the calculation of the **MRAC** controller appled on the robot in 2 .

The simulation of the effect of a fault on the resistance R at sample time k=105 msec on the MRAC controller performance is shown in fig. 8. Figures 3÷5 show the performance of the controller and the fault detector in the normal case. Figures 6 and 7 show the performance of the fault detector which operates very well.

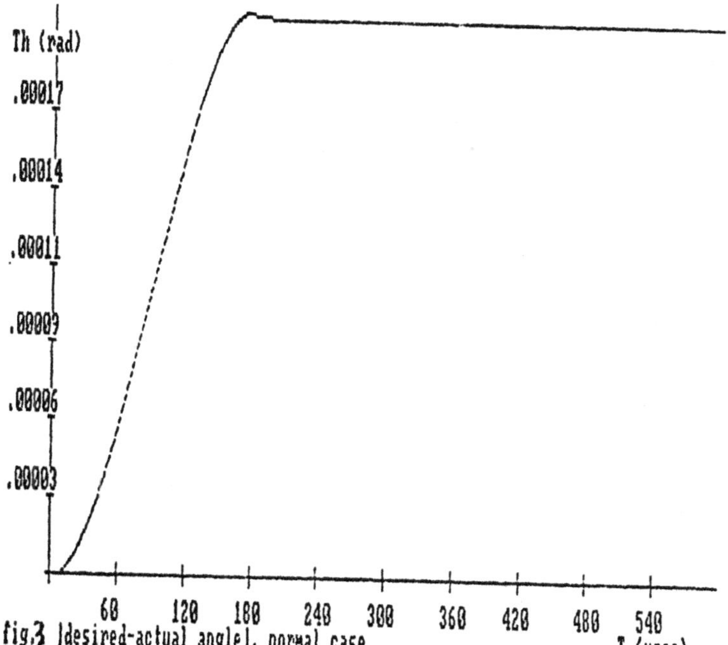

fig.3 |desired-actual angle|, normal case

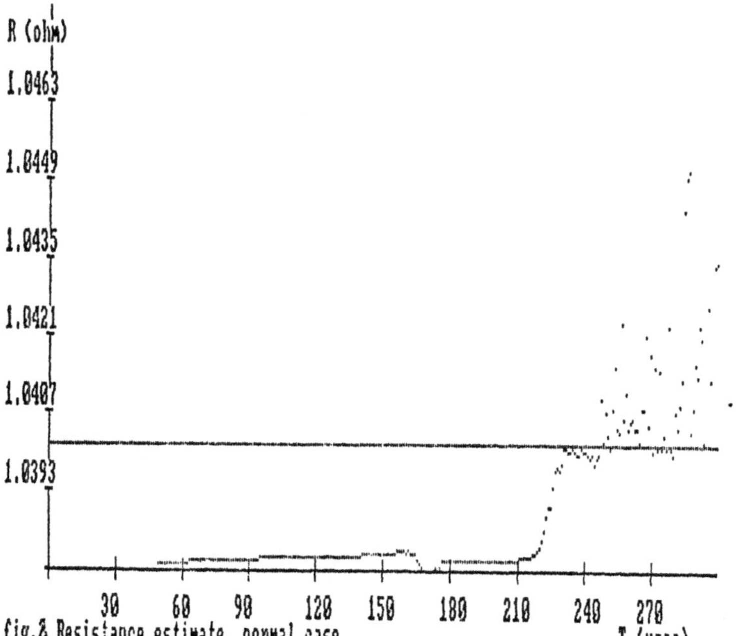

fig.4 Resistance estimate, normal case

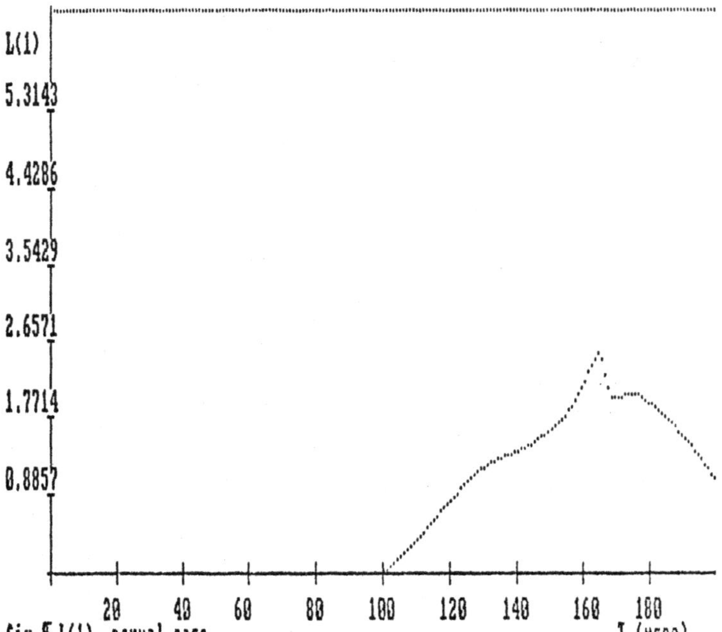

fig.5 L(1), normal case

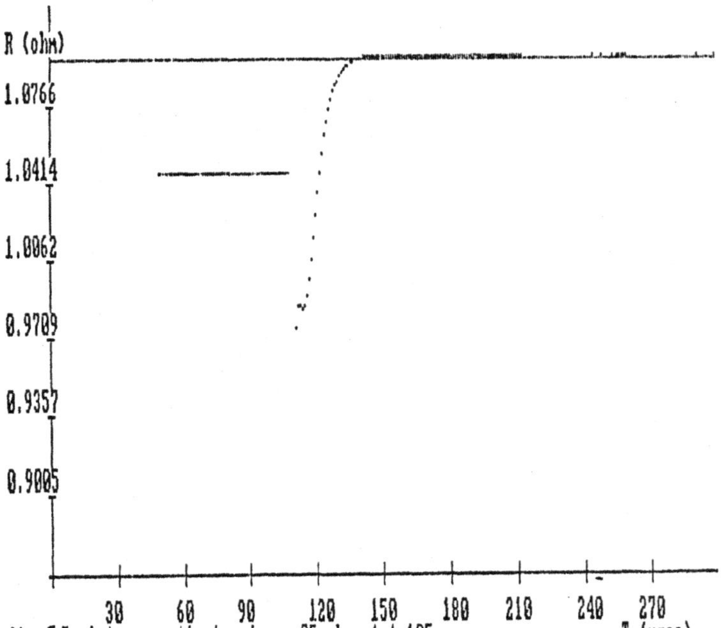

fig.6 Resistance estimate, jump .05 ohm at t=105

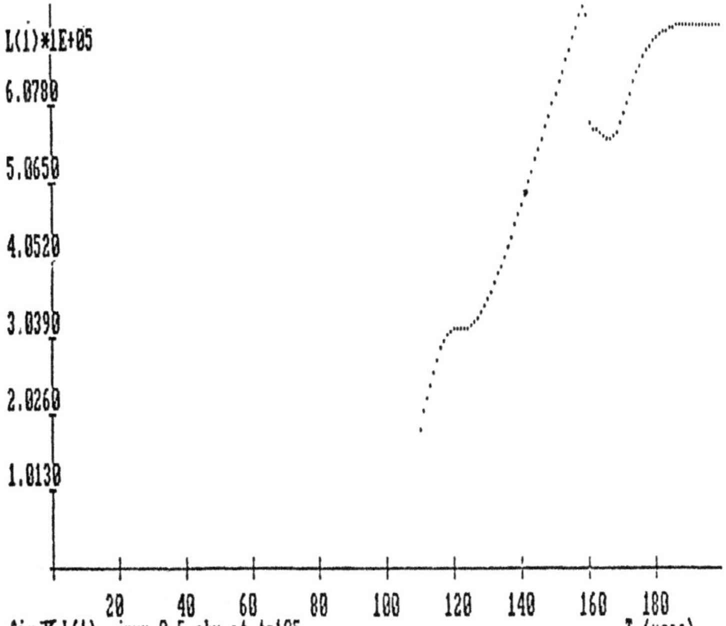

fig.7 L(1), jump 0.5 ohm at t=105

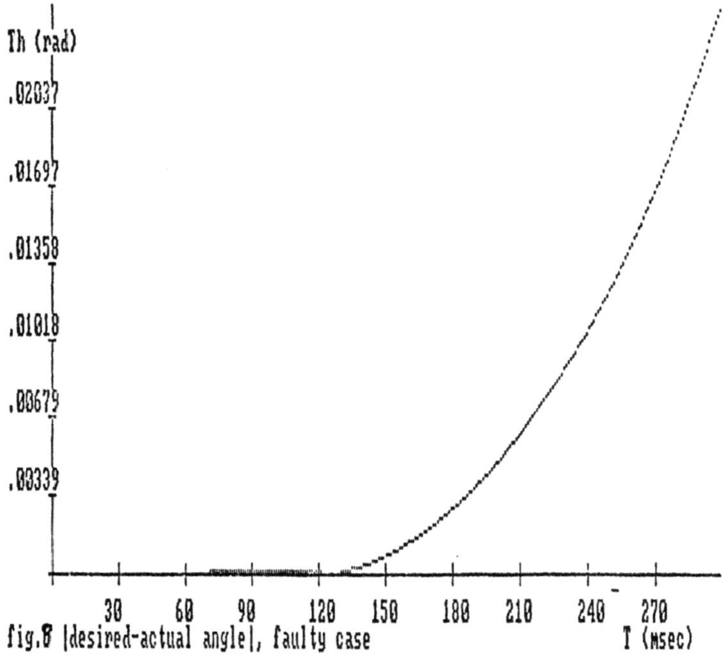

fig.8 |desired-actual angle|, faulty case

5. APPENDIX I:

The dynamic equations for the armature circuit and the mechanics of a d.c. motor leads to the state-space representation for the actuator of the link i of the robot:

$$\begin{bmatrix} \frac{di_i}{dt} \\ \\ \frac{d\omega_i}{dt} \end{bmatrix} = \begin{bmatrix} -\frac{R_i}{L_i} & -\frac{K_{mi}N_i}{L_i} \\ \\ \frac{k_{mi}}{J_{mi}} & -\frac{p_i}{J_{mi}} \end{bmatrix} \begin{bmatrix} i_i \\ \\ \omega_i \end{bmatrix} + \begin{bmatrix} \frac{1}{L_i} & 0 \\ \\ 0 & -\frac{1}{J_{mi}N_i} \end{bmatrix} \begin{bmatrix} V_{di}(t) \\ \\ T_{Ldi}(t) \end{bmatrix} \qquad (I_1)$$

Where V_{di} is the calculated control voltage from the MRAC controller and T_{Ldi} is the corresponding torque generated at the ith joint.

6. CONCLUDING REMARKS

The present paper constitutes a first piece of work towards the end of applying efficient dynamic fault detection and location algorithms to controlled robotic systems. Regarding the technique presented in this paper we note the following:

— The least squares estimator is very sensitive to parameter and other variations

— The window length is critical for false alarms and missing detections

— The window length should be much larger for meaningful system reorganization

— The response time of the detection scheme must be minimized by using custom hardware.

Some aspects which are currently under study are:

— Comparision of various estimation algorithms with reference to the fault detection and location problem.

— Application of the algorithm (s) to real robotic systems.

— Consideration of the case where noise is present.

REFERENCES

1. S.Nicosia and P.Tomei, "Model Reference Adaptive Control Algorithms for Industrial Robots", Automatica Vol. 20, No 5, 1984.

2. S.G.Tzafestas and G.Stavrakakis, "Model Reference Adaptive Control of Industrial Robots with Actuator Dynamics", IFAC/IFIP/IMACS Int. Symp. on Theory of Robots (Vienna, Dec.3-5, 1986).

3. R.Isermann "Process fault detection based on Modelling and Estimation Methods-A Survey", Automatica, Vol.20, No 4 (1984).

4. G.Geiger, "Fault Identification Using a Discrete Square Root Method", Int.J.Modelling and Simulation, Vol.6, No 1, 1986.

5. P.Kaminski et al., "Discrete Square Root Filtering: A Survey of Current Techniques", IEEE Trans. AC-16,6,1971.

6. S.G.Tzafestas, "Optimal Supervision and Fault Detection Techniques of Technological Systems", Proc. 1st National Aerospace Symposium, pp. 455-520 , 12-15 Oct., 1982.

7. S.G.Tzafestas, "Reliability and Fault Detection Techniques of Large Scale Systems", in Real Time Control of Large Scale Systems (G. Schmidt, M.Singh. A.Titli, S.Tzafestas, Editors), Springer Verlag, 1985.

A THREE LEVEL STRATEGY FOR FAULT DETECTION AND DIAGNOSIS IN INDUSTRIAL PROCESSES

T. CECCHIN, J. RAGOT, D. SAUTER
Laboratoire d'Automatique et de Recherche Appliquée - CRAN
Centre de Recherche sur la Valorisation des Minerais
Rue du doyen Marcel Roubault - 54500 Vandoeuvre
France

ABSTRACT. Having proved the importance of detection and diagnosis of faults that may affect an industrial installation, the authors propose a strategy for the diagnosis of working state applicable to all the processes represented by deterministic models. This strategy is applied to supervise a mineralurgical circuit.
KEYWORDS. Failure detection, failure localization, diagnosis of working state, instrumentation failure, process failure, knowledge-based system, grinding circuit.

1. POSITION OF THE PROBLEM

The diagnosis of the working state of an installation is a treatment lying between the acquisition phase of the informations and the control phase of this installation.

The treatment is necessary to have representative information about the proceeding state in order to act on the proceeding. This representativeness implies that informations must be sufficient (observability and governability concepts), that they must be collected from judiciously positioned sensors and that they have to be free from errors being detrimental to the interpretation of the phenomena represented by these informations.

It is also a delicate treatment, because there are various types of informations (logical, analogical, determinist, fuzzy ...) and because they are relative to different subsets (the process itself, the actuators, the regulation systems, the chains of measurement acquisition ...).

Put simply, this treatment comprises two aspects :

The first concerns the detection and localization of failures ("failure" may be defined as an alteration in the functioning of a proceeding element : sensor bias, actuator locking, chocking up ..., the noises of measurement are not considered as failure) that may affect the process and also the instrumentation set-up, except the sensors.

The second aspect forms part of the data validation. It consists in detecting failures in sensors and in correcting the doubtful measurement if the case arises.

Several approaches are possible in order to realize the detection and diagnosis of failures.

S. Tzafestas et al. (eds.), System Fault Diagnostics, Reliability and Related Knowledge-Based Approaches, Vol. 1, 219–233.

Firstly, material redundancy, consists in utilizing several sensors for the same measurement (1). This technique is especially used for "high risk" systems (aerospace and nuclear industry).

Generally the sensors are trebled, then the detection of a sensor error is realized by majority vote. This method is limited to sensor defects and in addition to that, it is inoperative when a fault of common mode occures (2).

The second approach, the analytical redundancy, frees you from this problem of common mode and enables detection of failures existing in all elements of the proceeding. It consists in testing measurement coherence by linking the measurements with the aid of models.

When there is no mathematic model of process, a "data base" model can be used. The detection and diagnosis of failures are then realized by the use of a method analyzing the data : factorial analysis of the correpondences, analysis of the main components, discriminant analysis, pattern recognition (3).

Another recent but apparently very promising technique consists in the use of an knowledge based approaches (4). This system can be used either in an autonomous way for detection and diagnosis (5), or together with another method (6). In the latter case, the role of the knowledge-based system is to confirm the presence of a fault and to determine its cause.

The strategy we are proposing is based on the last approach. It supposes the knowledge of the mathematic model of the process. It comprises three levels :

- A local level where information given by sensors and detectors are independently exploited. This level only enables failure detection.
- A global level where measurement coherence is tested from one or several models. This level enables the detection and the localization of failures and in the case of sensor failures the estimation of its amplitudes is possible.
- A central level where the results obtained in the preceeding levels are synthesized. The synthesis is realized by an "knowledge based system" proposing a diagnosis. Figure 1 represents the general framework of the proposed algorithm.

2. DESCRIPTION OF THE STRATEGY (7)

2.1. Local level (see Figure 2)

2.1.1. Test of the instantaneous or integrated measurement, test of the logical variables. This first test consists in comparing the obtained instantaneous or integrated measurement (for an integrated measurement the duration of the sample-talking is non-zero, the measurement is obtained after division of a quantity by an interval of time) at extreme thresholds and in testing the logical values given by the detectors. This test enables a rapid detection of breakdowns.

2.1.2. Constitution of observation windows. A set of several successive measurements called "window" is collected for every measurement in order to make the following test and treatments possible.

In real time environment the window is sliding ; the last measurement being the standard one.

On this window statistical properties of the measured signal are studied. During normal functioning, the tested parameters develop in an uncertain way around a medium value, according to a partially known law of distribution : the greater the number of samples, the greater the significance of the obtained results.

The adjustment of the window's width is delicate and has to be realized by learning. A narrow window permits the quick detection failures of average importance, a wider window permits the detection of permanent failures of small amplitude, but its utilization is harmful as regards the speed of detection. In order to obtain a satisfactory compromise of detection rapidity-security, several windows of different width can be used together.

Moreover it is judicious, in order to guarantee the significance of the statistical tests, that the minimum number of elements contained in the smallest window be at least equal to 30.

2.1.3. Test of steady state. The aim of this test is to determine whether the examined parameters are in static or in dynamic state, so that appropriate treatments can be realized afterwards.

When the distribution law of measurements is postulated to be normal, the test of the mean square of the successive differences is utilized (8) : suppose the measurements x_i represents a gaussian sequence, then the variable r :

$$r = \frac{1}{2} \frac{\sum_{i=1}^{n-1} (x_{i+1} - x_i)^2}{\sum_{i=1}^{n} (x_i - \bar{x})^2}$$

where $\bar{x}$ is the mean of the sequence and n the window's length

as a probable value equal to 1.

If n is bigger than 25, we assume that :

$$u = (1-r) \sqrt{(n-1)*(n+1)/2}$$

follows a zero-mean standardized gaussian law.

The decision rule used is : $u > \varepsilon$: dynamic state case
$u \leq \varepsilon$: static state case

When the distribution law is anormalous it will be substituted by the test of the series (8).

2.1.4. Drift test. This test is aimed at detecting the slow variations. Although these drifts are not inevitable defects, their existence could reinform a diagnosis on a higher level. It is based on the exploitation of the results obtained by the previous test, utilized with the smallest and the largest window : the small window may now be considered ; if the steady successive test, utilizing the small window detects a steady state for a duration equal to the large window and if for the same duration the test utilizing the large window detects a dynamic state, then a drift is present.

2.1.5. Standard deviation test. This test is realized on all variables declared to be in a steady state. The value of the standard deviation of the measurements is compared to thresholds determined by learning. If the standard deviation exceeds the thresholds, there is a modification of the noise due to bad contacts or to a modification of the process' noise.

2.2. Global level (see Figure 3)

At the global level, the coherence of the measurements is tested by the utilization of the models which link them. These models can be static (balance equations in the case of exact structures ; in the other cases regressions), or dynamic (recurrent equations, discrete impulse response).

If the examined system is complex, it is judicious to split it up into several subsystems, every subsystem forming a homogeneous whole, having minimum connections whith the other subsystems.

2.2.1. Test of the subsystem state. The first stage of the global level consists in identifying the working state (static or dynamic) of the subsystems so that a model and a supervision method appropriate to every subsystem can be used.

The accepted principle is the following one : a subsystem is considered to be in static state if all its inputs outputs variables are in static state. In the opposite case, a dynamic model is used.

2.2.2. Detection, localisation and estimation using a static model.
* Presentation of the model.

There are N balance equations which are independant and have an exact structure, joining n physical parameters x_i^* i = 1, ..., n. We have :

$$F_1\ (x_1^*, \ldots, x_n^*) = 0 \quad n > N$$
$$\vdots$$
$$F_N\ (x_1^*, \ldots, x_n^*) = 0$$

In this system F_j (j = 1, ..., N) represents a linear function or non-linear function whith or whithout a constant term. In fact, the physical variables are only known thanks to their measurements x_i or to their supposed value (afterwards also called measurement), the supposed value being the command given to a control device turned on this variables.

Every measurement is corrupted by an error ε_i :

$$\varepsilon_i = \Delta_i + a_i$$

where : Δ_i represent a systematic error,

$\{a_i\}$ represent a gaussian noise sequences with $E(a_i) = 0$ and $E(a_i^2) = \sigma_i^2$

The balance equations applied to the measured variables are writen as follows :

$$F_1 (x_1, \ldots, x_n) = b_1$$
$$\vdots$$
$$F_N (x_1, \ldots, x_n) = b_N$$

Where b_j "residual error" of the j^{st} equation characterizes the uncertain and systematic errors associated whith this j^{st} balance.

* Failure detection.

The failure detection is realized by comparing the absolute value of the residual error of every balance equation with a threshold determined by learning and depending on the measurement's precision. A failure will be declared present if at least one residue is superior to its threshold c_j ($j = 1, \ldots, N$).

However this test has the inconvenience of using every balance equation separately without considering the information redundancy that can exist between several balance equations.

In order to compensate this lack, a complementary test is used consisting in the comparison -for every measurement- of the estimations for that variable given by the different equations where they intervene. This test is modified as follows in order to assure uniformity of the expressions :

The comparision of the estimations two by two of the same variable generates a new expression, whose proximity to zero constitues a coherence test; threshold used being now c_j ($j = N + 1, \ldots, N'$).

The failure detection test is therefore based on the following system :

$$\left.\begin{array}{l} F_1 (x_1, \ldots, x_n) = b_1 \\ \vdots \\ F_N (x_1, \ldots, x_n) = b_N \end{array}\right\} \text{test equations on the residus of the balances}$$

$$\left.\begin{array}{l} F_{N+1} (x_1, \ldots, x_n) = b_{N+1} \\ \vdots \\ F_{N'} (x_1, \ldots, x_n) = b_{N'} \end{array}\right\} \text{test equations on the redundancy of the equations}$$

* Failure localization.

We propose a classification of the M different physical elements composing the subsystems (sensors, actuators, process elements) studied by decreasing probability to be at the origin of failure.

To every j balance equation ($j = 1, \ldots, N'$) we associate a set S_j containing the index of the physical elements of the part of the process described by this equation.

The principle of the classification is based on the following idea : If there is a faulty element it is very probable that the residus of the balance equations associated to the sets S_j (j=1,...,N'), containing the element in question, will be superior to their thresholds. For every i physical element of the process (i=1,...,M) let us introduce the vector :

$$E_i^T = [e_{i1}, e_{i2}, \ldots, e_{iN'}] \text{ with } e_{ij} = \begin{cases} 1 \text{ if } i \in S_j \\ 0 \text{ if } i \notin S_j \end{cases}$$

Furthermore we define a vector F as follows :

$$F^T = [f_1, f_2, \ldots, f_{N'}] \quad \text{with } f_j = \begin{cases} 1 \text{ if } b_j > c_j \\ 0 \text{ if } b_j \leq c_j \end{cases}$$

The proximity of these two vectors indicates a link between cause and effect.

The index of distance used is :

$$dE_iF = \sum_{j=1}^{N'} |e_j - f_j|$$

When all the distances dE_iF have been calculated, every element is aligned in comparision with the other, according to the obtained distances. In principle, the faulty element must be at the head of the classification.

Supposing there could be several physical elements faulty at the same time, the principle of the localization remains the same. It would suffise to generalize the construction of the vector associated to a combination as follow :

$$E_{i,k}^T = (e_{i1} \oplus e_{k1}, \ldots, e_{ij} \oplus e_{kj} \ldots, e_{iN'} \oplus e_{kN'})$$

where $\oplus$ is the "logical or".

* Estimation of the sensors' magnitudes failures.

When the measurements delivered by one or several sensors are suspect, it is important to correct them, that is the value of the biases by which they are modified must be determined. To resolve this problem a system of equations is formed having the Δ_i (systematic biases) as unknowns.

We will not discuss the practical techniques of the problem solving. Moreover the uniqueness and even the existence of a solution is not assured.

When, due to the circuit structure, the obtained localization is of little precision, it is possible that there will be several combinations of sensors being classified identically. It is necessary to realize a certain number of tests on the incriminated combinations in order to determine which combination is at the origin of the failure :

- physical coherence test on the estimated variables.
- test on the residus of the balance equations recalculated with the

estimated variables : all the residus recalculated with the estimated variables must be inferior to their thresholds.
- test on the value of the estimated biases : every estimated bias must be superior to the precision of the corresponding sensor.

It is not possible to affirm that there exists a unique combination which can satisfy the precedent constraints.

2.2.3. Detection, localisation and estimation using a dynamic model.

* Presentation of the model.

In our study we are limited to the processes that can be represented by linear models described by recurrent equations or by discrete impulse response.

The parameters of these models are determined by the use of an identifying method by minimisation of a structure distance (9, 10).

* Failure detection.

The failure detection is executed by comparing at thresholds, for every output, the mean and the standard deviation of the absolute value of the deviation measured ouput - estimated output.

* Failure localization.

The failure localization is realized according to the same principle as the one employed for the static model.

* Failure characterization.

A failure characterization can be carried out by analysing for every equation, the distribution law of the estimation's residus, all the failure do not have a priori an identic "signature". We are limited to the proceedings applicable to all kinds of processes. 3 classes of failures can be distinguished according their effects on the residus :
- the failures implying at the same time a modification of the standard deviation and of the average of the residus. This kind of defect may be due to a locking of actuator, a non constant deperdition, a rupture of a process-organ, an evolution of the characteristics of the process, a gain error or a sensor breakdown.
- the faults expressing a modification of the standard deviation of the residus without a change of the means' value. This modification is imputable to a variation of the process' noise, to a sensor noise due to the aging, or to a false contact.
- the failures implying a modification of the mean without change of standard deviation's value. This is due to sensors' bias or constant deperdition.

* Estimation of the sensors failures.

When biases can be observed it is possible to plan their determination. In this case, one comes down to the solution of a system of linear equations.

2.2.4. Action on the process. When a failure has been detected or during the thorough cyclical verification, extra signals are injected onto the inputs of the process. So it is possible to identify correctly the parameters of the proceeding's models. The study of the mean and the standard deviation of the news residus : identified parameters - parameters of reference gives a new characterization of the fault :
- variation of the mean and the standard deviation : bias, gain error

or sensor breakdown, actuator locking, deperdition.
- variation of the mean : evolution of the characteristics of the process.
- variation of the standard deviation : modification of the noise.

For the same equation, the comparision of the characterizations obtained by the estimation of the state and by identification, the two realized simultaneously, should permit a characterization of higher precision in certain cases.

Eventually the characterizations can be classified in 4 types :
- bias, constant deperdition,
- actuator locking, non constant deperdition, gain error or sensor breakdown,
- evolution of the process' characteristics,
- modification of the noise.

2.3. Central level

At the central level we are interested in the diagnosis of the origin of the failures having been detected at the local level or at the global level, but whose failures causes could not be determined in absolute certainty. This level centralizes the results of all the realized tests, executes a synthesis of all this information and proposes a diagnosis. To improve a partial diagnosis, the "knowledge based system" realizing this synthesis may be led to propose some actions to the user : quick verification of a part of the installation, use of extra signals at certain points of the circuit, measurement of variables normally not measured.

3. APPLICATION ON A GRINDING-CLASSIFICATION CIRCUIT

3.1. Description of the circuit

Figure 4 gives an outline of a process of grinding-classification currently used in the field of mineral industry for the preparation of the minerals. The role of this circuit is to reduce a mineral to a size small enough to free the useful mineral from its gangue.

The studied circuit is composed of :
- a ball - mill (4) feeded with ore (stream 1) and with water (stream 2)
- a recorvery sump for the grinded products (8), this sump receives an addition of water (stream 6)
- a hydrocyclone (10) feeded with pulp by the stream 9. This appliance assues the separation between big particles, which have to be recycled (stream 11) and the fine particles (stream 12).

The following sensor are installed :

stream 1 : sensor of ore flowrate (W1)
stream 2 : sensor of water flowrate (Q2)
stream 6 : sensor of water flowrate (Q6)
sump 8 : sensor of level (h8) and density (d8)
stream 9 : sensor of pulp flowrate (Q9)
stream 12 : sensor of pulp flowrate (Q12) and density (d12)

The installed actuators are the following ones :
stream 1 : conveyer with adjustable speed
stream 2 and 6 : pneumatic valves
stream 9 : pump with adjustable speed

The use of analogical modules of regulation PID has enabled the setting-up of the following local regulations :
- regulation of ore flowrate (W1)
- regulations of water flowrate (Q2 and Q6)
- regulation of pulp level (h8)

3.2. Realized decomposition

The circuit is split up into 5 subsystems :
- the subsystem 1 includes the whole of mill-sump-hydrocyclone, it uses the totality of the measured variables appart from the measurement h8. Depending on the case, this subsystem is described by a static or a dynamic model.
- the subsystems 2, 3 and 4 correspond to the local regulations of W_1, Q_2 and Q_6. These subsystems are represented by a dynamic model linking the measurement to the set-point.
- the subsystem 5 is represented by a dynamic model linking the level h8 to the inputs and ouputs of the sump.

3.3. Implementation of the detection strategy

3.3.1. Test of steady state. Figure 5 shows the evolution of Q6, the evolution of the corresponding test variable, u, and the result of the test of steady state. This test is realized on a window of 50 measurements, that is to say 2.5 mn. The discrimination threshold of the test variable is taken equal to 6.

3.3.2. Drift test. Figure 6 shows the results obtained by the realization of a drift test on the variable Q9.
On figure 6 you can see the evolution of the non filtered variable Q9, the mean of this variable realized on 50 measurements, the test variable u, calculated from a window of 50 measurements and from a window of 300 measurements. The threshold of static-dynamic discrimination is taken equal to 3 for the first window and equal to 2 for the second. For the narrow window, we notice that the variable u is inferior to its discrimination threshold : the variable Q9 is therefore considered in static state for that window. For the window of the large width, the variable u is very quickly superior to its threshold : so we are in presence of a slow drift.

3.3.3. Standard deviation test. We have create several "visible" failures on the variable Q9 in order to prove the interest of a standard deviation test.
Figure 7 shows :
- the evolution of the variable Q9. The functioning of the installation is normal except at the following points :

. from 5.5, to 10.5 mn : changing of the coefficient P in the regulation of level

. from 15.75 to 21.75 mn : suppression of level regulation, the pump works at a constant speed

. from 26 to 31.5 mn : addition of a action I on the level regulation

. from 35 mn to the end : false intermittent contacts on the current output of the sensor Q9.

- the value of the standard deviation calculated on a window of 50 measurements. The arbitrarily negative values of the standard deviation correspond to the points in dynamic state

- the value of the variable "result of the tests". This variable has 3 states :

It is zero when the variable Q9 is in static state, the functioning being normal

It is negative when the variable Q9 is in dynamic state

It is positive when the variable Q9 is in static state and when the standard deviation test has detected an anomaly.

3.3.4. Evolution of the diagnosis obtained on an example. The realized failure is a bias of the value $\Delta Q6 = 50$ ℓ/h on the measurement of Q6 (see Figure 8). This failure appeared earlier at the starting phase. The failure is detected in the subsystem 1 on a window of 300 measurements. A quick visual inspection of the installation does not allow the determination of the origin of the defect. For each variables of subsystem 1, the computed means are :

$Q_1 = 51$ ℓ/h ; $Q_2 = 115{,}6$ ℓ/h ; $Q_6 = 681$ ℓ/h ; $Q_9 = 1528{,}6$ ℓ/h
$Q_{12} = 786{,}2$ ℓ/h $d_9 = 1{,}242$ kg/ℓ ; $d_{12} = 1{,}10$ kg/ℓ

For detection and localization, the following equations are used :

$$Q_1^* + Q_2^* + Q_6^* - Q_{12}^* = 0$$

$$Q_1^* d_1 + Q_2^* + Q_6^* - Q_{12}^* d_{12}^* = 0$$

$$Q_{12}^* (d_1 - d_{12}^*) + Q_6^* (1 - d_1) + Q_2^* (1 - d_1) = 0$$

$$(d_1 - 1) Q_1^* + Q_{12}^* (1 - d_{12}^*) = 0$$

$$Q_1^* (d_1 - d_{12}^*) + Q_2^* (1 - d_{12}^*) + Q_6^* (1 - d_{12}^*) = 0$$

In this system, only the first two equations are independant. The thresholds of these equations have been chosen equal to :

$$c_1 = 24 \quad c_2 = 27 \quad c_3 = 38 \quad c_4 = 8 \quad c_5 = 6$$

The residus obtained are :

$$b_1 = 61{,}4 \quad b_2 = 68{,}1 \quad b_3 = 97{,}8 \quad b_4 = 6{,}7 \quad b_5 = 0{,}4$$

If we are limited to the sensors' failures (2 simultaneous failures maximum), we get the following classification :

1 : Q_2, Q_6, Q_{12} } distance dEF = 1
2 : $Q_2 + Q_6$ }

3 : $Q_1 + Q_2$, $Q_1 + Q_6$, $Q_1 + Q_{12}$, $Q_2 + Q_{12}$, $Q_2 + d_{12}$ } distance
$Q_6 + Q_{12}$, $Q_6 + d_{12}$, $Q_{12} + d_{12}$ } dEF = 2

4 : Q_1, d_{12} distance dEF = 3

After the calculations, the following provisional diagnosis is obtained :

The failure may be caused by :
- an error on Q2 or Q6 or Q12
- an error on Q2 and Q6 (in this case the separated estimation of the errors in not possible).

At this stage of the diagnosis, nothing allows to affirm that it ist a matter of bias (or several) because this result is obtained from a steady state.

In order to test the 2 regulations acting on Q2 and Q6 without disturbing the whole process, we apply low amplitude extra-signals on the concerned set-points. The behaviour of the identified parameters of the models representing the subsystem 3 and the subsystem 4 allows to verify the normal functioning of the 2 regulations : I there is a failure in these regulations, it can only be a matter of bias.

With large amplitude extra-signals, on compute a estimation of the output variables for subsystem 1. A difference between estimated and measured confirms the presence of a bias.

In conclusion : The failure is caused by a bias on the sensors measuring Q_2 or Q_6 or Q_{12} or Q_2 and Q_6 or by a bias on the regulators.

In order to determine the exact origin of this bias, a human intervention is necessary.

4. CONCLUSION

The recommended strategy enables the detection of failures of all kinds. The efficiency of the localization and the precision of the diagnosis depend on the structure of the installation and on the knowledge of its functioning (models and quantity of information).

Another extension concerns the study of decision-talking, that can be considered from different aspects :
- proceed to a thorough study of the behaviour of suspect elements.
- correct defective measurements (in case of a sensor failure).
- compensate control signals (in case of actuators' failures).
- reconfigure the control structure (in the case of loss of essential

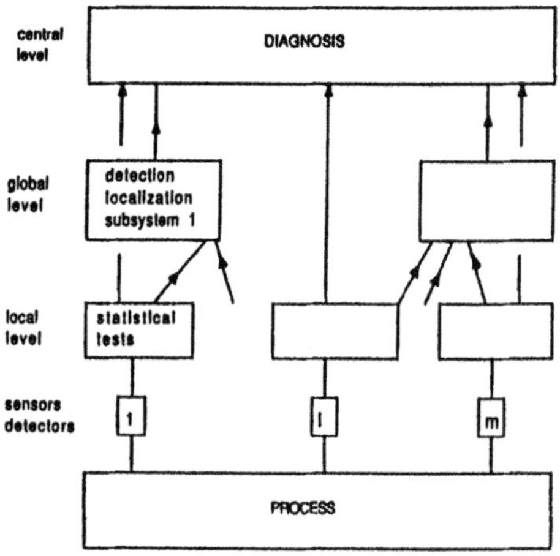

Figure 1. Representation of the strategy.

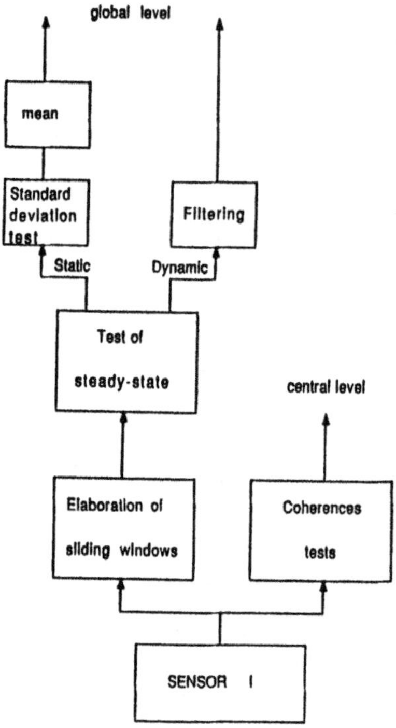

Figure 2. Local level.

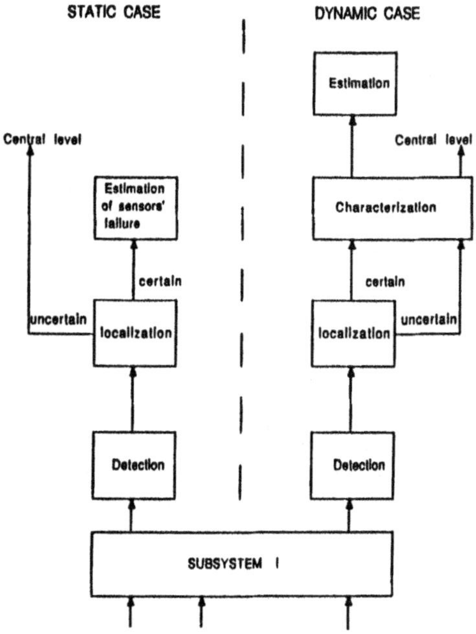

Figure 3. Global level.

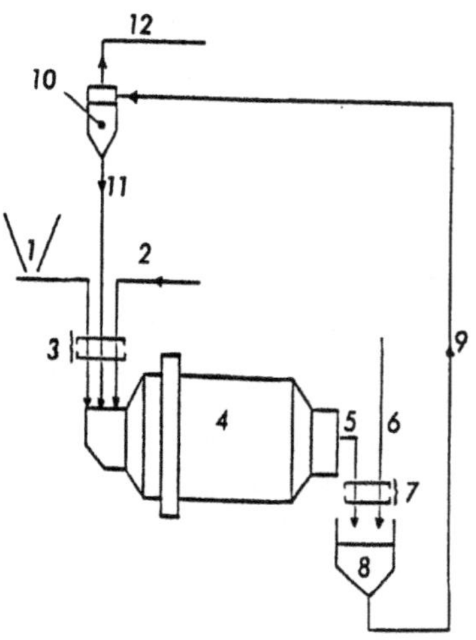

Figure 4. Grinding-classification circuit.

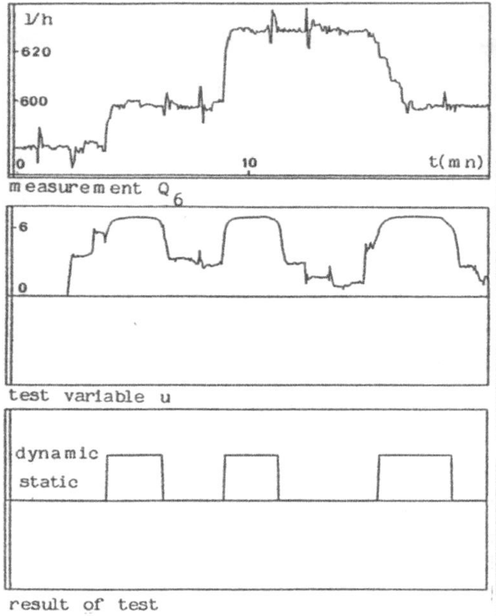

Figure 5. Test of steady state applied on Q_6.

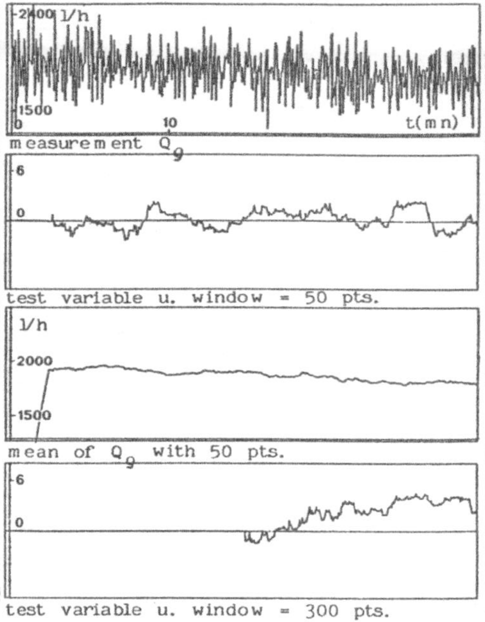

Figure 6. Drift test applied on Q_9.

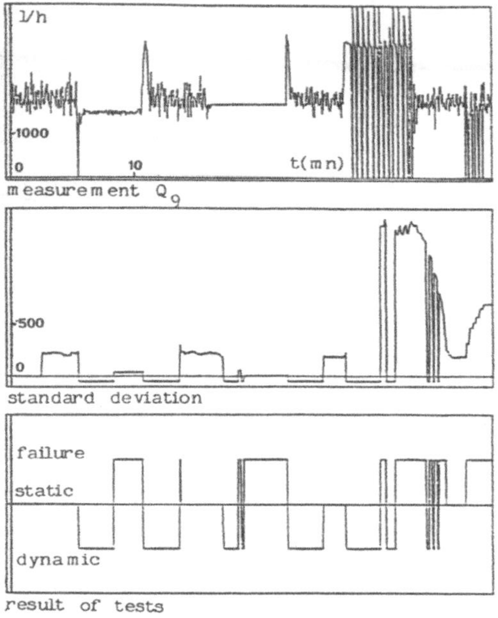

Figure 7. Standard deviation test applied on Q_9.

INFORMATION	DIAGNOSIS	ACTION
Detection on subsystem 1	Failure on subsystem 1	
Localisation and estimation of possible biaise	Diagnosis 1 : Failure on Q_2 or Q_6 or Q_{12} or Q_2 and Q_6	Quick inspection of process
Result of inspection	Diagnosis 2 : No leakage	Low amplitude extra-signals on set points Q_{2c} and Q_{6c}
Evolution of identified parameters on subsystems 3 and 4	Diagnosis 3 : Biase on sensors measuring Q_2 or Q_6 or Q_2 and Q_6, or biase on regulations acting on Q_2 or Q_6 or Q_2 and Q_6, or failure on Q_{12}	Large amplitude extra-signals on Q_{2c} and Q_{6c}
Evolution of estimation residus on subsystem 1	Diagnosis 4 : Biase on sensors measuring Q_2 or Q_6 or Q_{12} or Q_2 and Q_6 , or biase on regulateurs acting on Q_2 or Q_6 or Q_2 and Q_6	Off line measurement of Q_2 and Q_6
Flowrate measurement Q_2 and Q_6	Diagnosis 5 : Biase on Q_6	

Figure 8. Evolution of the diagnosis obtained on an example.

organs) and proceed to substitutions and reparation of organs.
- put installation in "emergency state" (in case of a non-satisfying reconfiguration).
- shut down production.

BIBLIOGRAPHY

(1) A.S. WILLSKY (1976)
'A survey of design methods for failure detection in dynamic systems' Automatica. Vol 12. Pergamon Press.

(2) G. ZWINGELSTEIN, T. ADAM (1984)
'Diagnostics of failures of sensors of nuclear plants with the parity space methodology' IMEKO Symposium on Measurement and Estimation. Bressanone.

(3) B. DUBUISSON, P. MALVACHE, D. GRENIER (1985)
'The use of pattern recognition in order to improve fast nuclear reactor monitoring' Patt. Rec. in Pratice II. Amsterdam.

(4) Session 6 C - diagnostic technique (1985)
Les systèmes experts et leurs applications. 5ème journées internationales d'Avignon.

(5) J.P. PARCY (1982)
'Un système expert en diagnostic sur réacteurs à neutrons rapides'. Thèse 3ème cycle en intelligence artificielle. Université d'Aix-Marseille II.

(6) B. PAPIN, P. BERNARD, J.L. ARTAUD, M. GILLET (1984)
'Un système de diagnostic par analyse de processus appliqué aux réacteurs à eau présurisée'. Journée d'étude AFCET sur la validation des mesures dans les systèmes complexes. Nancy.

(7) T. CECCHIN (1985)
'Conception d'une usine pilote minéralurgique automatisée. Elaboration et application d'une stratégie de diagnostic d'état de fonctionnement de procédés industriels en temps réel'. Thèse de doctorat de l'Institut National Polytechnique de Lorraine.

(8) Commissariat à l'énergie atomique (1978)
Statistique appliquée à l'exploitation des mesures. Ed. Masson.

(9) J. RICHALET, A. RAULT, M. POULIQUEN (1971)
Identification des processus par la méthode du modèle. Théorie des systèmes, Vol 4, Gordon & Breach.

(10) A. RAULT, D. JAUME, M. VERGE (1982)
'On line identification and fault detection on industrial processes'. IFAC.

DESIGN OF FAIL-SAFE TUNABLE DIGITAL SET-POINT TRACKING CONTROLLERS FOR MULTIVARIABLE PLANTS

B Porter and R N Fripp
Centre for Instrumentation and Automation
University of Salford
Salford M5 4WT
England

ABSTRACT It is shown that tunable digital set-point tracking controllers for multivariable plants can be readily designed to 'fail safe' in the event of an actuator failure associated with any one of the manipulated variables. The effectiveness of such a tunable digital controller is illustrated by designing a 'fail-safe' tunable set-point tracking controller for a multivariable plant with two outputs and three manipulated variables.

1. INTRODUCTION

The techniques for the design of tunable set-point tracking controllers proposed by Davison [1], Pentinnen and Koivo [2], Porter [3,4,5], and Porter and Jones [6,7] can be regarded as attempts to extend to multivariable industrial plants the classical techniques of Ziegler and Nichols [8] for the design of controllers for monovariable plants. In order to circumvent the need for mathematical models of multivariable plants expressed in either state space or transfer function matrix form, both the proportional and integral controller matrices embodied in the tunable digital controllers introduced by Porter and Jones [6,7] can be directly determined from open-loop tests performed on asymptotically stable plants. This use of step-response matrices greatly facilitates the choice of sampling period in the design of such digital set-point tracking controllers, and the effectiveness of the resulting design methodology has been illustrated by the excellent set-point tracking behaviour of tunable digital controllers for a distillation column [6] and an advanced gas-cooled nuclear reactor [7].

In this paper, it is shown that such tunable set-point tracking controllers for multivariable plants 'fail safe' in the event of an actuator failure associated with any one of the manipulated variables. The effectiveness of such a tunable digital controller is illustrated by designing a 'fail-safe' tunable digital set-point tracking controller for a multivariable plant with two outputs and

S. Tzafestas et al. (eds.), System Fault Diagnostics, Reliability and Related Knowledge-Based Approaches, Vol. 1, 235–245.

three manipulated variables. In the case of an actuator failure associated with any one of the three manipulated variables of this plant, it is shown that the tunable digital controller exhibits excellent 'fail-safe' set-point tracking behaviour.

2. ANALYSIS

The plants under consideration are assumed to be governed on the continuous-time set $T = [0,+\infty)$ by state and output equations of the respective forms

$$\dot{x}(t) = Ax(t) + Bu(t) \tag{1}$$

and

$$y(t) = Cx(t) \quad , \tag{2}$$

where the state vector $x(t) \in R^n$, the input vector $u(t) \in R^m$, the output vector $y(t) \in R^p$, $m \geqslant p$, and all the eigenvalues of the plant matrix $A \in R^{n\times n}$ lie in the open left half-plane C^-. Furthermore, it is assumed that the introduction of integral action preserves stabilisability and therefore that [9][10]

$$\text{rank } G = p \quad , \tag{3}$$

where the plant transfer function matrix

$$G(s) = C(sI_n-A)^{-1}B \tag{4}$$

and

$$G = G(0) = -CA^{-1}B \in R^{p\times m} \quad . \tag{5}$$

Finally, it is assumed that the input-output decoupling is achievable and therefore that [11]

$$\text{rank } F = p \quad , \tag{6}$$

where the plant decoupling matrix

$$F = \begin{bmatrix} c_1^T A^{d_1}B \\ c_2^T A^{d_2}B \\ \cdots\cdots \\ \cdots\cdots \\ c_p^T A^{d_p}B \end{bmatrix} \in R^{p\times m} \tag{7}$$

and the d_i $(i=1,2,\ldots p)$ and the c_i^T $(i=1,2,\ldots,p)$ are, respectively, the decoupling indices of the plant [11] and the rows of the output matrix. In the case of such plants it is important to note that

$$F = \lim_{t\to 0} \Lambda^{-1}(t)H(t) \quad , \tag{8}$$

where

$$\Lambda(t) = \mathrm{diag}\{t^{d_1+1}/(d_1+1)!, t^{d_2+1}/(d_2+1)!, \ldots$$
$$\ldots, t^{d_p+1}/(d_p+1)!\} \in R^{p\times p} \tag{9}$$

and

$$H(t) = CA^{-1}(e^{At}-I_n)B \in R^{p\times m} \tag{10}$$

is the plant step-response matrix [6].

In order to design digital error-actuated set-point tracking controllers for plants governed by state and output equations of the respective forms (1) and (2), it is convenient to consider the behaviour of such plants on the discrete-time set $T_T = \{0,T,2T,\ldots,kT,\ldots\}$. This behaviour is governed by state and output equations of the respective forms [5]

$$x_{k+1} = \Phi x_k + \Psi u_k \tag{11}$$

and

$$y_k = \Gamma x_k \quad , \tag{12}$$

where $x_k = x(kT) \in R^n$, $u_k = u(kT) \in R^m$, $y_k = y(kT) \in R^p$,

$$\Phi = e^{AT} \quad , \tag{13}$$

$$\Psi = \int_0^T e^{At}B dt \quad , \tag{14}$$

$$\Gamma = C \quad , \tag{15}$$

and $T \in R^+$ is the sampling period.

The state and output equations of such plants under the action of error-actuated digital controllers governed on $T_T = \{0,T,2T,\ldots,kT,\ldots\}$ by control-law equations of the proportional-plus-integral form [5]

$$u_k = TK_1 e_k + TK_2 z_k \tag{16}$$

clearly assume the respective forms

$$\begin{bmatrix} x_{k+1} \\ z_{k+1} \end{bmatrix} = \begin{bmatrix} \Phi - T\Psi K_1 & , & T\Psi K_2 \\ -T\Gamma & , & I_p \end{bmatrix} \begin{bmatrix} x_k \\ z_k \end{bmatrix} + \begin{bmatrix} T\Psi K_1 \\ TI_p \end{bmatrix} v \tag{17}$$

and

$$y_k = [\Gamma \; , \; 0] \begin{bmatrix} x_k \\ z_k \end{bmatrix} \quad , \tag{18}$$

where the error vector $e_k = v - y_k \in R^p$, the set-point input vector $v \in R^p$, the digital integral-of-error vector

$$z_k = z_{k-1} + Te_{k-1} \in R^p \quad , \tag{19}$$

and the controller matrices $K_1 \in R^{m \times p}$ and $K_2 \in R^{m \times p}$. Therefore, provided only that T, K_1, and K_2 are such that all the eigenvalues of the closed-loop plant matrix in equation (17) lie in the open unit disc D^-,

$$\lim_{k \to +\infty} \Delta z_k = \lim_{k \to +\infty} (z_{k+1} - z_k) = 0 \tag{20}$$

and consequently

$$\lim_{k \to +\infty} e_k = 0 \tag{21}$$

so that set-point tracking occurs.

The closed-loop characteristic equation can be readily expressed in the form [5]

$$\phi_c(z) = \phi_1(z)\phi_2(z) \tag{22}$$

by invoking the block-diagonalisation procedure of Kokotović [12], where [5]

$$\phi_1(z) = |zI_n - I_n - TA - ½T^2A^2 + T^2BK_1C + T^2BK_2CA^{-1} + 0(T^3)| \tag{23}$$

and

$$\phi_2(z) = |zI_p - T^2CA^{-1}BK_2 + 0(T^3)| \tag{24}$$

However, in case

$$K_1 = H^T(T)[H(T)H^T(T)]^{-1}\Lambda(T)\Pi \tag{25}$$

where

$$\Pi = \mathrm{diag}\{\pi_1, \pi_2, \ldots, \pi_p\} \quad , \quad \pi_i \in R^+ \; (i=1,2,\ldots,p) \tag{26}$$

and

$$K_2 = G^T(0)[G(0)G^T(0)]^{-1}\Sigma \tag{27}$$

where

$$\Sigma = \mathrm{diag}\{\sigma_1,\sigma_2,\ldots,\sigma_p\} \quad , \quad \sigma_i \in R^+ \ (i=1,2,\ldots,p) \quad , \tag{28}$$

it follows from equations (22), (23), (24), (25), and (27) that $Z_c = Z_1 \cup Z_2$ is the set of closed-loop characteristic roots where

$$Z_1 = \{z \in C: |zI_n - I_n - TA + O(T^2)| = 0\} \tag{29}$$

and

$$Z_2 = \{z \in C: |zI_p - I_p + T^2\Sigma + O(T^3)| = 0\} \tag{30}$$

These expressions indicate that, provided T is sufficiently small, both $Z_1 \subset D^-$ and $Z_2 \subset D^-$ since the open-loop plant is asymptotically stable on the continuous-time set $T = [0,+\infty)$ and since $T^2\Sigma$ is a positive diagonal matrix.

The introduction of error-actuated digital controllers governed by equations (16), (25), and (27) accordingly ensures that, in the absence of actuator failures, set-point tracking occurs when the sampling period $T \in (0,T^*]$, where $T^* = T^*(\Pi,\Sigma)$ can be readily obtained by simple 'on-line' tuning [5]. Moreover, it can be shown [13] that such set-point tracking behaviour also occurs in the event of an actuator failure associated with any one of the manipulated variables in case the controller tuning matrix defined in equation (28) assumes the scalar form

$$\Sigma = \sigma I_p \quad , \quad \sigma \in R^+ \tag{31}$$

These tunable digital set-point tracking controllers therefore 'fail safe' provided only that the tuning matrix associated with the integral component of the proportional-plus-integral control law (16) is chosen in accordance with equation (31).

3. ILLUSTRATIVE EXAMPLE

These general results can be conveniently illustrated by designing a 'fail-safe' tunable digital set-point tracking controller for an asymptotically stable plant with transfer function matrix

$$G(s) = \begin{bmatrix} \dfrac{1}{3s+3} , & \dfrac{2s+4}{(s+1)(s+3)} , & \dfrac{1}{s+3} \\ \\ \dfrac{1}{s+2} , & \dfrac{3s+8}{(s+2)(s+3)} , & \dfrac{2s+5}{(s+2)(s+3)} \end{bmatrix} \tag{32}$$

which remains functionally controllable in the event of the failure

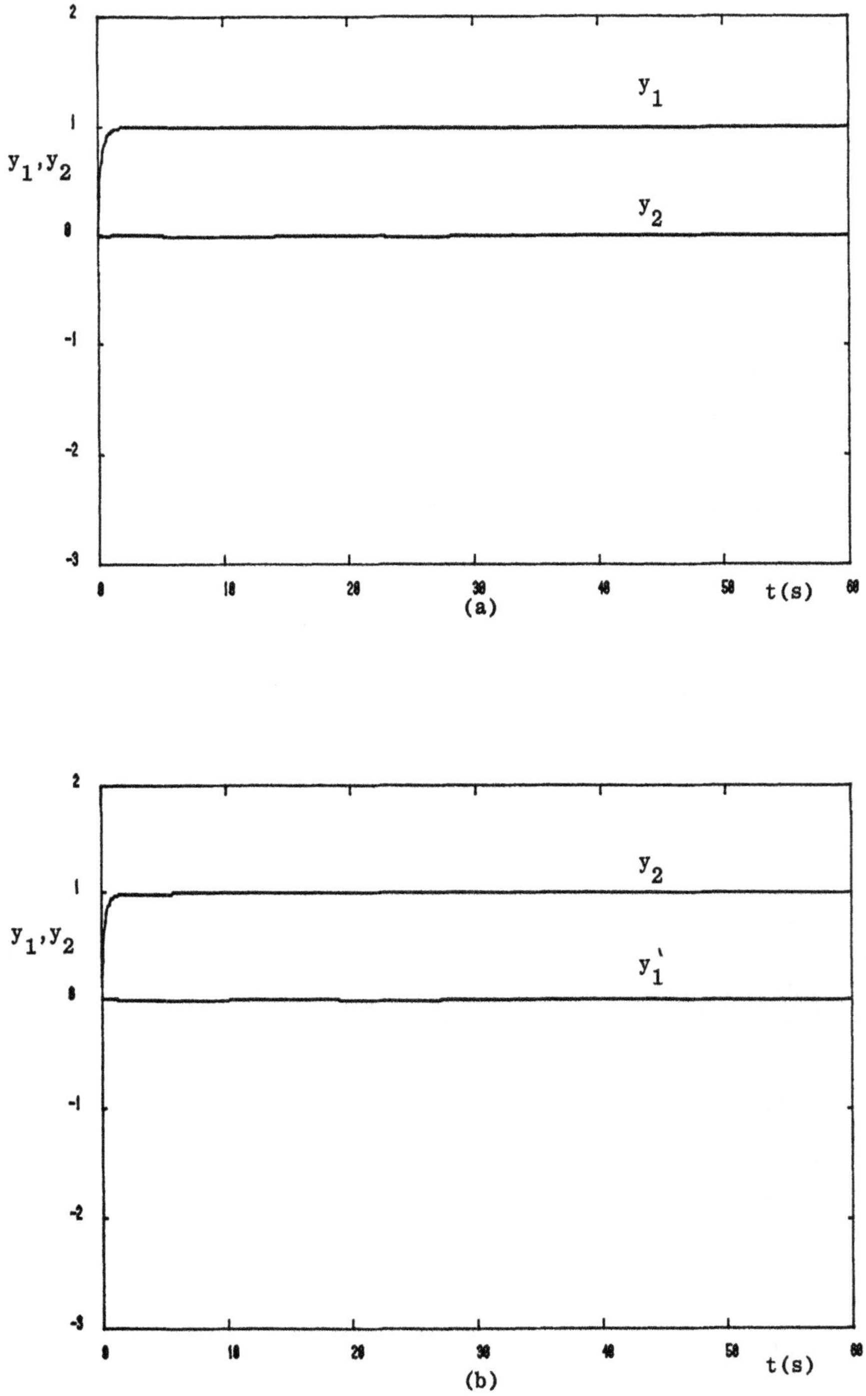

Fig 1: Set-point tracking behaviour:
no actuator failure

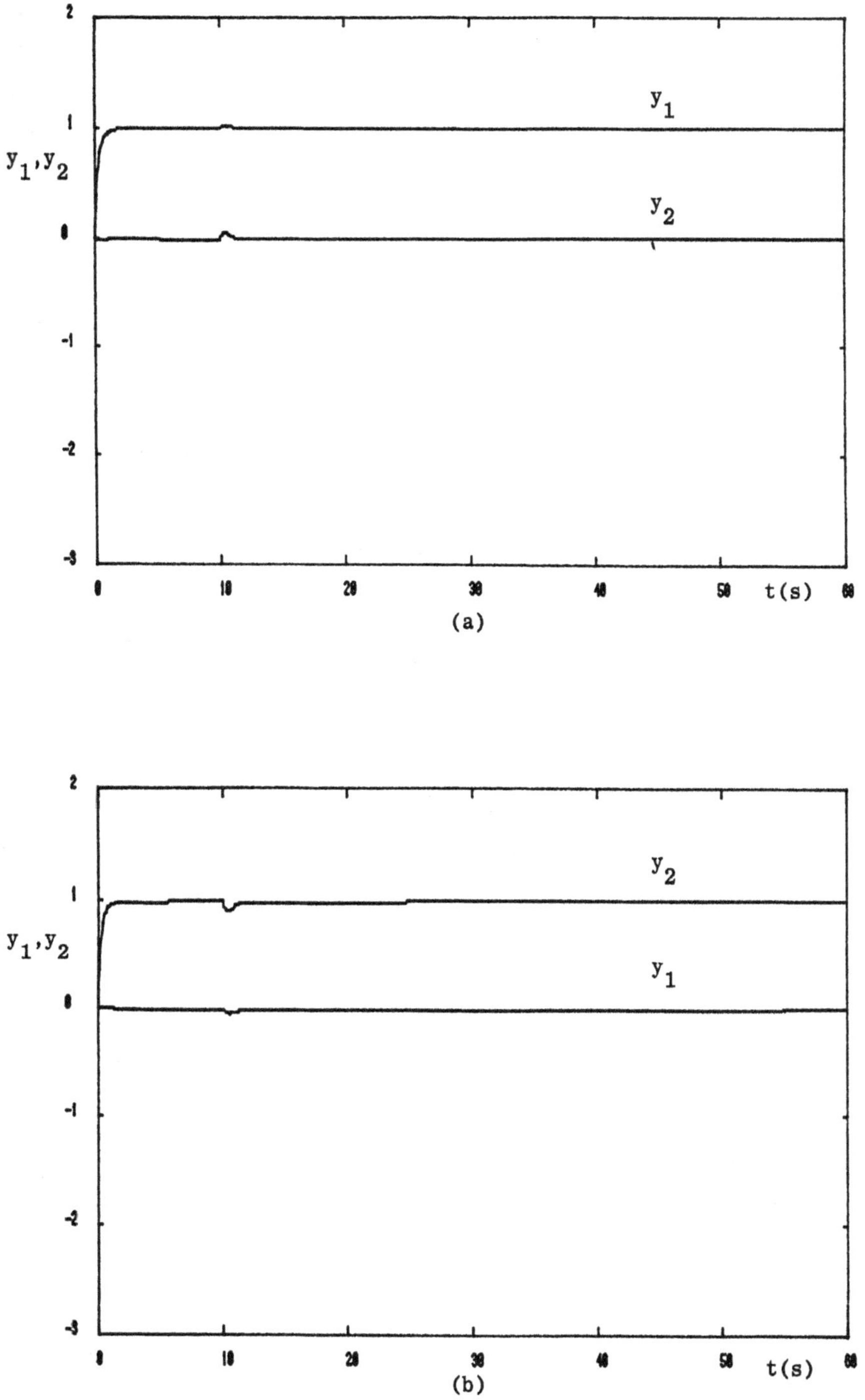

Fig 2: Set-point tracking behaviour: first actuator failure

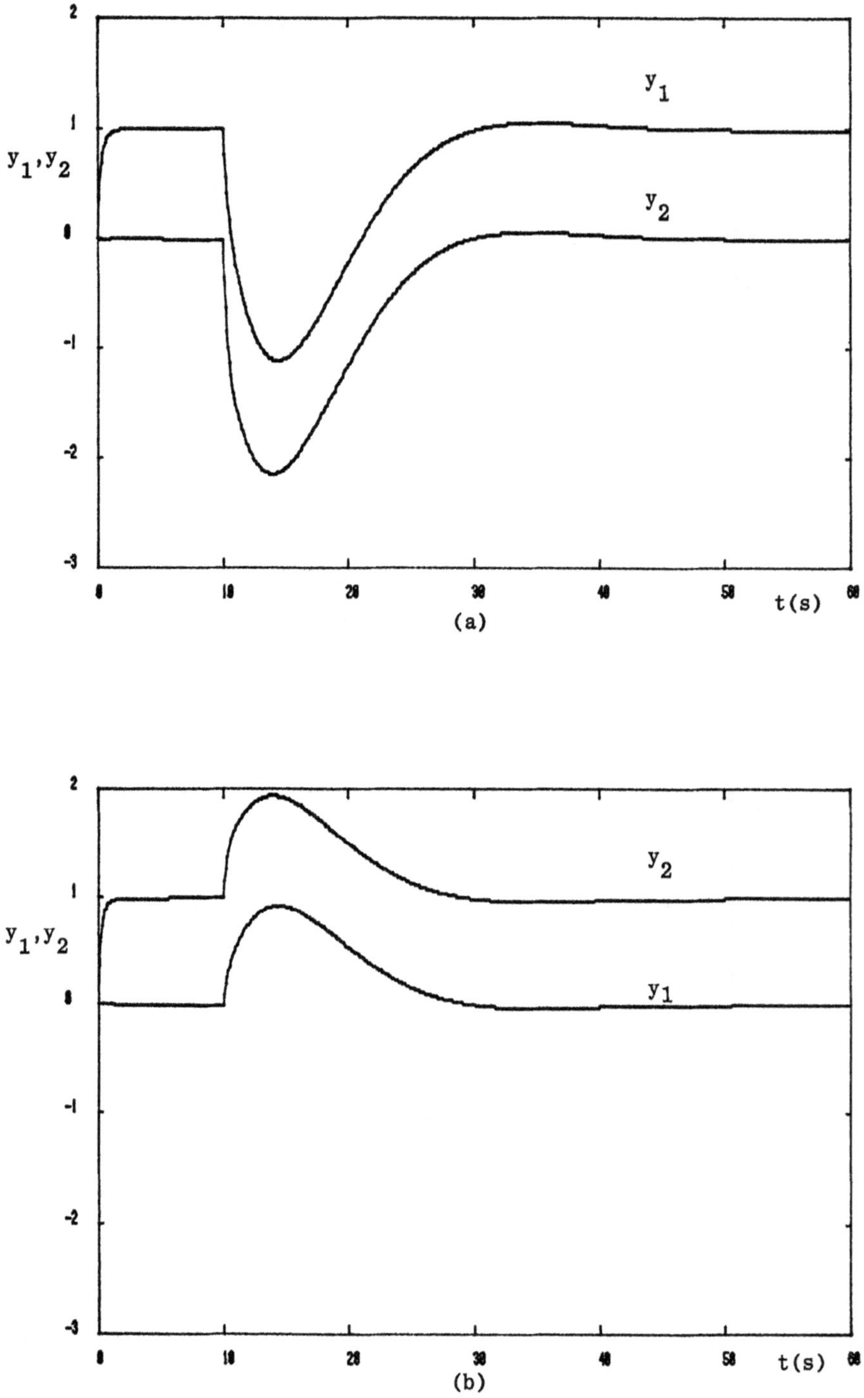

Fig 3: Set-point tracking behaviour:
second actuator failure

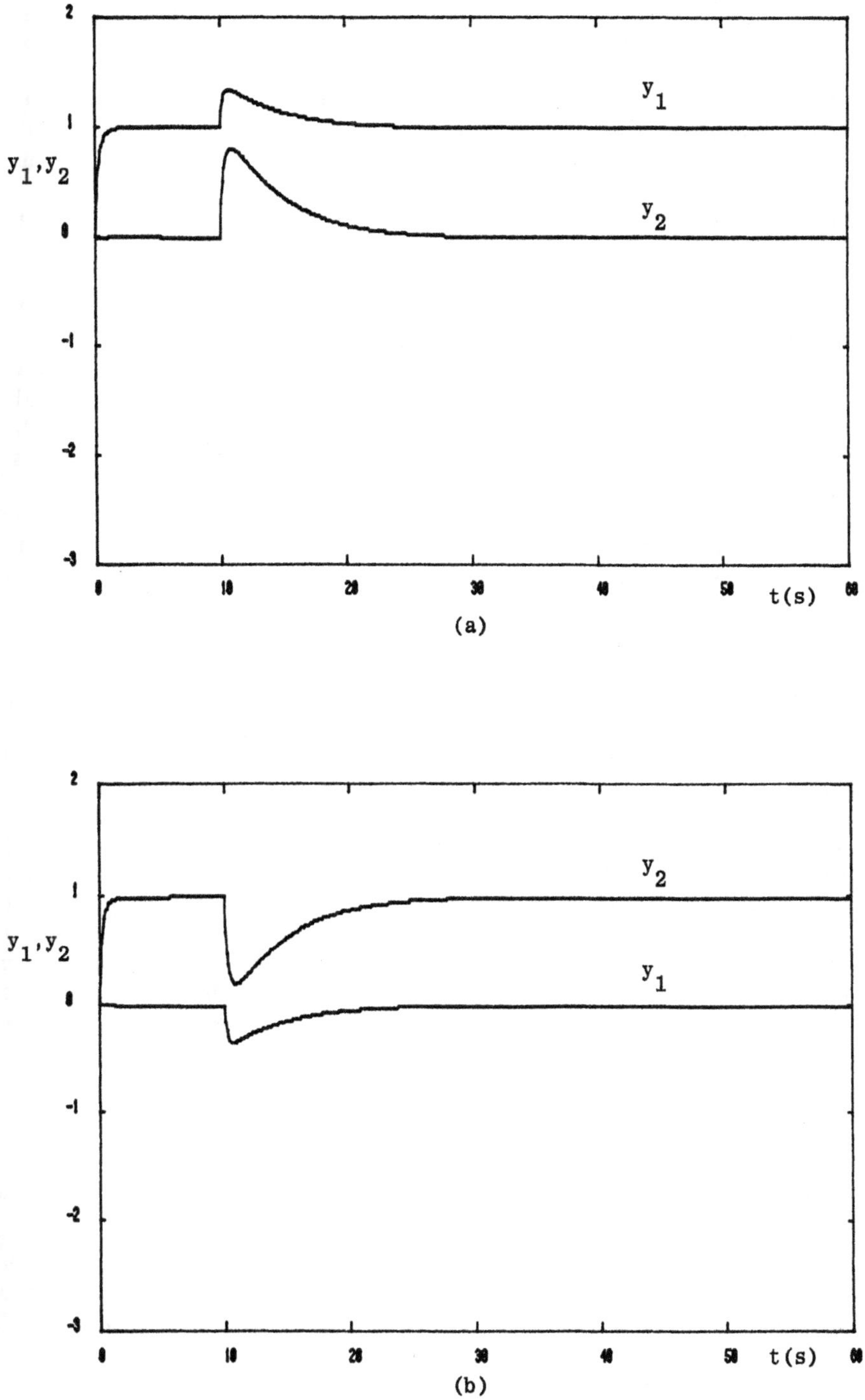

Fig 4: Set-point tracking behaviour:
third actuator failure

of any single actuator. In this case, it can be readily verified that $\sigma_{min}[H(0.05)]$ (ie, the minimum singular value of H(0.05)) is not small and therefore that H(0.05) is well-conditioned, and furthermore that $\sigma_{min}[H(+\infty)]$ (ie, the minimum singular value of $H(+\infty)$) is not small and therefore that G is well-conditioned. The set-point tracking behaviour of the initially quiescent plant in the absence of actuator failure, when the error-actuated digital controller is tuned such that $T = 0.05$, $\Lambda(0.05)\ \Pi = \text{diag}\{4,4\}$, and $\Sigma = \text{diag}\{70,70\}$, is therefore excellent as shown in Fig 1(a) and in Fig 1(b) when $v = [1, 0]^T$ and $v = [0\ ,\ 1]^T$, respectively.

However, since $\Sigma = \text{diag}\{70,70\}$ is clearly a scalar matrix, this controller is also guaranteed to 'fail safe' as shown in Figs 2, 3, and 4 in the event of the failure after 10 secs of the actuator associated with the first, second, and third manipulated variable, respectively. The set-point tracking behaviour of the 'failed' plant is shown in Figs 2(a), 3(a), and 4(a) and in Figs 2(b), 3(b), and 4(b) when $v = [1\ ,\ 0]^T$ and $v = [0\ ,\ 1]^T$, respectively. It is evident that the tunable digital controller exhibits excellent 'fail-safe' set-point tracking behaviour even though, in the case of the actuator failure associated with the second manipulated variable, the resulting 'failed' plant possesses non-minimum phase characteristics.

4. CONCLUSION

It has been shown that tunable digital set-point tracking controllers for multivariable plants can be readily designed to 'fail safe' in the event of an actuator failure associated with any one of the manipulated variables. The effectiveness of such a tunable digital controller has been illustrated by designing a 'fail-safe' tunable set-point tracking controller for a multivariable plant with two outputs and three manipulated variables. In the case of an actuator failure associated with any one of the three manipulated variables of this plant, it has been shown that the tunable digital controller exhibits excellent 'fail-safe' set-point tracking behaviour.

ACKNOWLEDGEMENT

This work was undertaken as part of the activities of the DIGICON (Digital Industrial Control) Unit which is funded by a grant from the CAMPUS Venture and Enterprise Fund of the University of Salford. The financial support of the Science and Engineering Research Council of the United Kingdom is also gratefully acknowledged.

REFERENCES

[1] E J Davison, 'Multivariable tuning regulators: the feedforward

and robust control of a general servomechanism problem', *IEEE Trans*, AC-25, pp 361-375, 1976.

[2] J Pentinnen and H M Koivo, 'Multivariable tuning regulators for unknown systems', *Automatica*, **16**, pp 393-398, 1980.

[3] B Porter, 'Design of error-actuated controllers for unknown multivariable plants', *Electron Lett*, **17**, pp 106-107, 1981.

[4] B Porter, 'Design of set-point tracking and disturbance-rejection controllers for unknown multivariable plants', *Proc IEE Workshop on Theory and Applications of Self-tuning and Adaptive Control*, Oxford, 1981.

[5] B Porter, 'Design of tunable set-point tracking controllers for linear multivariable plants', *Int J Control*, **35**, pp 1107-1115, 1982.

[6] B Porter and A H Jones, 'Design of tunable digital set-point tracking controllers for linear multivariable plants using step-response matrices', *Proc 3rd IASTED International Symposium on Modelling, Identification, and Control*, Innsbruck, 1984.

[7] B Porter and A H Jones, 'Design of tunable digital set-point tracking controllers for linear multivariable plants with multiple time-scale reaction rates', *Proc 2nd IMC Symposium on Application of Multivariable System Techniques*, Plymouth, 1984.

[8] J G Ziegler and B N Nichols, 'Optimum settings for automatic controllers', *Trans ASME*, **64**, pp 759-768, 1942.

[9] B Porter and H M Power, 'Controllability of multivariable systems incorporating integral feedback', *Electron Lett*, **6**, pp 689-690, 1970.

[10] H M Power and B Porter, 'Necessary and sufficient conditions for controllability of multivariable systems incorporating integral feedback', *Electron Lett*, **6**, pp 815-816, 1970.

[11] P L Falb and W A Wolovich, 'Decoupling in the design and synthesis of multivariable systems', *IEEE Trans*, **AC-12**, pp 651-659, 1967.

[12] P V Kokotović, 'A Riccati equation for block-diagonalisation of ill-conditioned systems', *IEEE Trans*, AC-20, pp 812-814, 1975.

[13] R N Fripp, 'Design of digital controllers for linear multivariable plants with actuator failures', *Report USAME/DC/101/86*, Department of Aeronautical and Mechanical Engineering, University of Salford, April 1986.

The Design and Assessment of a Sliding Mode Controller for the Lateral Motion Model of an Aircraft

S.K.Mudge and R.J.Patton,
The Department of Electronics,
University of York,
Heslington,
York, YO1 5DD.
England.

Abstract The design of a multivariable sliding mode controller for an unmanned aircraft is described and an emphasis on robust eigenstructure assignment is given. The linear and non-linear model responses are compared when subjected to the same variable structure control design. It is shown that, after sliding has commenced the response of the objective system is tracked by the corresponding non-linear system. A measure of sensitivity is defined as the proximity of a sub-space of the actual system response from the designed objective.

1. INTRODUCTION

Variable Structure Control Systems (VSCS) have received considerable attention in recent years as a possible solution to the problem of controlling systems for which statistics on parameter variations, noise, etc. are not available. A VSCS is characterised by a state feedback control structure which is switched as the system state crosses certain discontinuity surfaces in the state-space. The central feature of this control structure is the sliding mode, where the system state is constrained to lie within a neighbourhood of all the switching surfaces. The sliding mode technique for control system design gained acceptance as a result of early work by Utkin[1] and Itkis[2]. Amongst other results, this work has shown that by changing the control structure a system of variable structure was produced which could possess new properties not present in any of the composite structures. For example, an asymptotically stable system may consist of two unstable systems. Thus a variable structure system may be constructed by switching between a combination of two sub-systems, each with a fixed structure and each operating in a specified region of the state-space. One of these sub-systems can actually be unstable[1]. Another important property of VSCS is

S. Tzafestas et al. (eds.), System Fault Diagnostics, Reliability and Related Knowledge-Based Approaches, Vol. 1, 247–264.

that after the commencement of the sliding mode, the system is unaffected by parameter variations and disturbances occurring within a sub-space of the full state-space. In fact, the more recent work of Ryan and Corless[3] has shown that variable structure control may be used to establish almost cetain convergence to a neighbourhood of the origin for a class of uncertain systems[4]. In recent years generalised theory and design techniques have been developed for sliding mode controllers, and these have been applied in the areas of model following, model-reference adaptive control, tracking and observers[5,6,7,8,9,10].

The aircraft is a complex, highly non-linear system whose aerodynamic parameters change considerably during flight resulting in variations in the dynamic performance over the flight envelope. In this paper variable structure systems design techniques are used to formulate a non-linear control law for the lateral motion model of an unmanned aircraft or RPV (remotely piloted vehicle). The control law is assessed in its ability to compensate for parameter variations and non-linearities using a non-linear simulation of the vehicle. The robustness of the system is investigated with respect to the definitions of Frank[11] which state that a system is called robust if the system property of interest remains in a bounded region in the face of a class of finite bounded perturbations. Frank also defines a system to be sensitive to parameter variations if $\underline{S}$ is unlike $\underline{0}$. If a measure of $\underline{S}$ is small the system is termed insensitive. The "bounded region" is the manifold intersection in state-space corresponding to the null space of C where C is a weighting matrix. A measure of robustness may be considered to be the "distance" of the response from the desired null-space objective.

2. THE AIRCRAFT DYNAMIC MODEL

The equations of motion of the fixed wing aircraft are normally decoupled into two sub-systems each describing the longitudinal and the lateral motions. The longitudinal motions are concerned with forward velocity and pitching excursions whilst the lateral motions relate to roll, yaw and sideslip velocity. In this work a controller is designed for the lateral motion of a particular remotely piloted vehicle for which the equations of motion and the aerodynamic parameters are well defined and published elsewhere[12]. The lateral and longitudinal motions are assumed decoupled as far as the control system design is concerned. A "stick-fixed" linearisation of the non-linear system gives the following linear model:

$$\dot{\underline{x}} = A\underline{x} + B\underline{u} \tag{2.1}$$

where:

$$\underline{x} = \begin{bmatrix} v \\ p \\ r \\ \phi \\ \psi \\ \zeta \\ \xi \end{bmatrix} \quad \begin{array}{l} \text{-sideslip velocity} \\ \text{-roll rate} \\ \text{-yaw rate} \\ \text{-roll angle} \\ \text{-yaw angle} \\ \text{-rudder angle} \\ \text{-aileron angle} \end{array} \tag{2.2}$$

$$\underline{u} = \begin{bmatrix} \zeta_c \\ \xi_c \end{bmatrix} \quad \begin{array}{l} \text{-rudder angle demand} \\ \text{-aileron angle demand} \end{array} \tag{2.3}$$

and the state A and B matrices are given by:

$$A = \begin{bmatrix} y_v & 0 & y_r - U_0 & g & 0 & y_\zeta & 0 \\ l_v & l_p & l_r & 0 & 0 & l_\zeta & l_\xi \\ n_v & n_p & n_r & 0 & 0 & n_\zeta & n_\xi \\ 0 & 1 & 0 & 0 & 0 & 0 & 0 \\ 0 & 0 & 1 & 0 & 0 & 0 & 0 \\ 0 & 0 & 0 & 0 & 0 & \mu_1 & 0 \\ 0 & 0 & 0 & 0 & 0 & 0 & \mu_2 \end{bmatrix} \quad B = \begin{bmatrix} 0 & 0 \\ 0 & 0 \\ 0 & 0 \\ 0 & 0 \\ 0 & 0 \\ K_1 & 0 \\ 0 & K_2 \end{bmatrix} \tag{2.4}$$

The aerodynamic derivatives for the aircraft are denoted by $y_v, n_v, \ldots$etc, U_0 is the nominal forward air speed and μ_1, K_1, μ_2, K_2 are the rudder and aileron actuator parameters, the two being modelled as simple first-order lags.

The lateral aircraft model is characterised by three modes; roll, Dutch roll and the spiral mode. These modes are such that the spiral mode has a very slow divergence, the roll mode is fast and stable and the Dutch roll mode has 0.7 damping ratio. The unmanned aircraft under consideration here has the following trim flight linear lateral motion model matrices corresponding to a 33ms^{-1} airspeed:

$$A = \begin{bmatrix} -0.277 & 0.000 & -32.900 & 9.810 & 0.000 & -5.432 & 0.000 \\ -0.103 & -8.325 & 3.750 & 0.000 & 0.000 & 0.000 & -28.640 \\ 0.365 & 0.000 & -0.639 & 0.000 & 0.000 & -9.490 & 0.000 \\ 0.000 & 1.000 & 0.000 & 0.000 & 0.000 & 0.000 & 0.000 \\ 0.000 & 0.000 & 1.000 & 0.000 & 0.000 & 0.000 & 0.000 \\ 0.000 & 0.000 & 0.000 & 0.000 & 0.000 & -10.000 & 0.000 \\ 0.000 & 0.000 & 0.000 & 0.000 & 0.000 & 0.000 & -5.000 \end{bmatrix} \tag{2.5}$$

$$B = \begin{bmatrix} 0.000 & 0.000 \\ 0.000 & 0.000 \\ 0.000 & 0.000 \\ 0.000 & 0.000 \\ 0.000 & 0.000 \\ 20.000 & 0.000 \\ 0.000 & 10.000 \end{bmatrix}$$

The eigenvalues of A are as follows:

$-0.502 \pm j3.508$	Dutch roll	
-8.360	roll	
0.122	spiral	(2.6)
$0.$	yaw attitude mode	
$-10.$	rudder mode	
$-5.$	aileron mode	

It should be noted that the open-loop model has a spiral divergence mode at $s = +0.122$. The model is also strictly non-minimum phase with positive zeros (not given here) in addition to the one positive real open-loop pole. This has obvious implications on stability margin as the possibility always exists of achieving a non-minimum phase closed-loop design even when using all the degrees of freedom available in full state feedback.

As well as tailoring the stability margins and time response of the closed-loop system it is necessary to consider the weighting of the various aircraft modes. For example, it is necessary to decouple roll from yaw and Dutch roll from both roll and aileron control actuation. These motions are normally coupled via the airframe with roll giving rise to sideslip and vice-versa. This essentially amounts to assigning a prescribed eigenstructure for the closed-loop system. In this way the control system design can be tailored to satisfy MIL-F8785C specifications on damping, frequency response and modal decoupling[13]. The work in this area is thus complementary to the work of Sobel et al[14,15] with the additional robustness properties exhibited by the sliding mode design approach.

3. VARIABLE STRUCTURE CONTROL SYSTEMS AND THE SLIDING MODE

The essential feature of a variable structure control system (VSCS) is that the state feedback control structure utilised is switched as the system state crosses certain discontinuity surfaces in the state-space. The so-called **sliding mode** behaviour is an idealised concept which might occur when n-m of the directions of motion in the state-space become invariant (i.e. become linearly related to n-m eigenvectors). If the interval between switching instants becomes infinitessimally small then the n-m sub-system becomes a free system with direction in state-space given by the intersection of n-m eigenvectors. In the case of parameter variations the system can be described as having uncertain dynamics and for this case ideal sliding motion can only be approached asymptotically. However, the VSCS design provides a very powerful deterministic design approach for uncertain systems and has strong robustness properties. The VSCS can be thought of as a closed-loop system with an adaptively varying state-feedback gain. Thus a feedback control law of the form:

$$\underline{u} = -K^T\underline{x}$$
$$\text{with:} \quad K^T = k^T + \Delta k^T \tag{3.1}$$

is used where the control matrix K^T is the sum of the fixed and switched matrices k^T and Δk^T. The switching surfaces are usually intersecting hyperplanes S_j passing through the state-space origin. An m input multivariable system has m such manifolds and the switching boundary is defined by the sub-space spanned by the intersection of these manifolds:

$$S_j = \{ \underline{x} : \underline{h}^j\underline{x} = 0 \} \quad j=1,..,m \tag{3.2}$$

Assembling the rows $\underline{h}^j$ into a full rank (mxn) matrix H:

$$H\underline{x} = 0 \tag{3.3}$$

This motion results in the system state repeatedly crossing and re-crossing a switching manifold defined by eigenvector directions in the state-space. All motions in the neighbourhood of this switching manifold must be directed towards the manifold itself. This requires that:

$$s_i\dot{s}_i <= 0 \tag{3.4}$$

In the ideal case (ie. with no system uncertainties) the discontinuous motion actually results in a response known as **sliding motion** which is characterised by an invariant direction within the manifold intersection. The motion of the system is then effectively constrained to lie within a certain subspace of the full state-space and the system becomes equivalent to a system of lower order termed the **equivalent system**[1,5]. During sliding the system is rendered insensitive to some parameter variations because the system trajectory is maintained within the sliding regime. It is also possible to detect parameter variations by considering some measure of the "distance" of the state vector from the sliding manifold by using a standard norm of the orthogonal projection of the state vector into the null space N(C).

4. DESIGN OF A SLIDING MODE CONTROLLER

The design of a VSCS entails the choice of discontinuity surfaces, specification of the discontinuous control and determination of the associated switching logic. The switching surfaces are usually fixed hyperplanes in state-space passing through the state-space origin, the intersection of which forms the sliding subspace. The objective of the design is to drive the state of the system from an arbitrary initial condition to the intersection of the switching surfaces. Once the state is sliding it must be maintained in a neighbourhood of the intersection manifold.

The aim of the design is to regulate the state vector to the equilibrium with an invariant response. The equivalent system must therefore be asymptotically stable. Stable sliding motion is assured by determining a set of hyperplanes such that the sliding mode defined by their intersection gives a specified performance to the equivalent lower order system. This is termed the **existence problem.** The choice of control must ensure that the desired sliding mode is attained. This is the second stage of the design process termed the **reachability problem**[1].

The design may be formulated by considering the linear time-invariant system defined by:

$$\dot{\underline{x}} = A\underline{x} + B\underline{u} \qquad \underline{x} \in R^n, \ \underline{u} \in R^m$$
$$\underline{y} = C\underline{x} \qquad \underline{y} \in R^m \tag{4.1}$$

It is assumed that r(B)=m and that (A,B) are a controllable pair. Without loss of generality the switching vector $\underline{S}$ is chosen as:

$$\underline{S} = C\underline{x} \tag{4.2}$$

A similarity transformation is used to map the system pair (A,B) into a controllable form[4] with:

$$TB = \begin{bmatrix} 0 \\ B_2 \end{bmatrix} \tag{4.3}$$

where B_2 is (mxm) and non-singular. The transformation matrix can be orthogonal to simplify the numerical calculation (as $T^T = T^{-1}$ for this case) and hence the change of basis takes the form:

$$\underline{z} = T\underline{x}$$

The transformed state equations then become:

$$\dot{\underline{z}} = TAT^T\underline{z} + TB\underline{u}$$
$$\underline{S} = CT^T\underline{z} \qquad (4.4)$$

with $\underline{z}$ now partitioned as:

$$\underline{z} = \left[z_1, z_2\right] \quad z_1 \in R^{n-m}, z_2 \in R^m \qquad (4.5)$$

and by partitioning the matrices in (4.4) accordingly:

$$TAT^T = \begin{bmatrix} A_{11} & A_{12} \\ A_{21} & A_{22} \end{bmatrix} \quad , \quad TB = \begin{bmatrix} 0 \\ B_2 \end{bmatrix} \qquad (4.6)$$

and:

$$CT^T = [C_1 \quad C_2]$$

The partitioned state equations then give:

$$\dot{\underline{z}}_1 = A_{11}\underline{z}_1 + A_{12}\underline{z}_2 \qquad (4.7)$$

$$\dot{\underline{z}}_2 = A_{21}\underline{z}_1 + A_{22}\underline{z}_2 + B_2\,\underline{u} \qquad (4.8)$$

$$\underline{S} = C_1\underline{z}_1 + C_2\underline{z}_2 \qquad (4.9)$$

During sliding the switching function $\underline{S}$ is identically equal to zero ie. the state lies simultaneously in each of the hyperplanes $S_j, j=1,..,m$. Equation (4.9) thus gives:

$$\underline{z}_2 = -C_2^{-1}C_1\underline{z}_1 \qquad (4.10)$$

The ideal sliding mode is governed by equations (4.7) and (4.10), an (n-m)th order system where $\underline{z}_2$ has the role of a state feedback control. Closing the loop in (4.7) with the state feedback from (4.10):

$$\dot{\underline{z}}_1 = (A_{11}-A_{12}C_2^{-1}C_1)\underline{z}_1 \qquad (4.11)$$

An asymptotically stable system must be ensured. This existence problem now becomes that of fixing $(C_2^{-1}C_1)$ to give (n-m) left-hand-half-plane eigenvalues of $(A_{11} - A_{12}C_2^{-1}C_1)$. For the aircraft problem eigenstructure assignment is used in the following way. During sliding:

$$C\underline{x} = \underline{0} \qquad (4.12)$$

Differentiating (4.12) with respect to time:

$$C\dot{\underline{x}} = \underline{0} \tag{4.13}$$

Substituting from equation (4.1):

$$C(A\underline{x} + B\underline{u}) = 0$$

$$\text{ie.} \quad \underline{u} = \underline{u}_{eq} = -(CB)^{-1}CA\underline{x} \tag{4.14}$$

where $\underline{u}_{eq}$ is defined as the equivalent control which governs the response during sliding[1,5]. Substituting back into equation (4.1):

$$\dot{\underline{x}} = A_{eq}\underline{x} = (A-BK)\underline{x} \tag{4.15}$$

where $\quad K = (CB)^{-1}CA$

A_{eq} is thus the equivalent system matrix during sliding. The n eigenvalues and associated eigenvectors of this matrix will determine the responses during sliding. Equation (4.12) expresses m of the states in terms of the remaining (n-m) and hence a reduction in system order from n to (n-m) is possible. It follows that m of the eigenvalues of A_{eq} will necessarily be zero valued. Now:

$$C\underline{x} = \underline{0} \quad \Longleftrightarrow \quad C(A-BK)\underline{x} = \underline{0}$$

$$\Longleftrightarrow \quad R(A-BK) \subseteq N(C) \tag{4.16}$$

Let the eigenvalues of (A-BK) be $\lambda_i, i=1,\ldots,n$ with corresponding eigenvectors $\underline{w}_i$. Then, from (4.16):

$$C(A-BK)\underline{w}_i = \lambda_i C\underline{w}_i = \underline{0}$$

$$\text{ie.} \quad \lambda_i = 0 \text{ or } \underline{w}_i \in N(C) \tag{4.17}$$

Assume $\lambda_i, i=1,\ldots,n-m$ are distinct and non-zero. Then by assigning the (n-m) corresponding eigenvectors the null space of C is fixed, although C is not unique:

$$CW = \underline{0} \qquad W = [\underline{w}_1,\ldots,\underline{w}_{n-m}]$$

$$\text{and so} \quad [C_1 \; C_2]\begin{bmatrix} V_1 \\ V_2 \end{bmatrix} = \underline{0} \quad \text{where} \quad \begin{bmatrix} V_1 \\ V_2 \end{bmatrix} = TW \tag{4.18}$$

$$\Longrightarrow C_2[C_2^{-1}C_1 \quad I_m] = 0 \tag{4.19}$$

$$\text{or} \quad (C_2^{-1}C_1)V_1 = -V_2$$

Therefore, provided V_1 is nonsingular, (C_2^{-1}) is uniquely determined and by choosing $C_2 = I_m$ it follows that:

$$C = [C_1 \quad I_m]T \tag{4.20}$$

and the existence problem is solved.

The variable structure control law generally consists of two parts; a linear control law $\underline{u}_L$ and a non-linear control signal $\underline{u}_{NL}$ which are added to form $\underline{u}_{TOTAL}$. The linear control is merely a state feedback controller while the non-linear feedback controller incorporates the discontinuous elements of the control law. The control structure used in the examples which follow has the **unit vector** form with:

$$\underline{u}_{TOTAL} = \underline{u}_L + \underline{u}_{NL}, \tag{4.21}$$

$$\underline{u}_L = L\underline{x}$$

and with:

$$\underline{u}_{NL} = \frac{N\underline{x}}{\|M\underline{x}\|}$$

The null-spaces of N, M and C are coincident as described by Ryan and Corless[3]. The role of the linear control is to force the range space states to zero asymptotically to attain the sliding mode. To attain the null-space of C in finite time the non-linear control component is required. This non-linear component must be continuous whenever $\underline{S}$ is non-zero, but is discontinuous (or zero) during sliding. The control system design is accomplished in the following way. Starting from the transformed state $\underline{z}$ given by equation (4.5) a second non-singular transformation matrix is defined by:

$$\underline{y} = T_2\underline{z} \tag{4.22}$$

This is such that:

$$\underline{y}_1 = \underline{z}_1 \tag{4.23}$$

$$\underline{y}_2 = C_2^{-1}C_1\underline{z}_1 + \underline{z}_2 \tag{4.24}$$

Comparing with equations (4.9) and (4.10) it is seen that the points of the state-space at which $\underline{S}$ equals zero are precisely those points at which $\underline{y}_2$ is equal to zero. Thus to obtain the sliding mode it is necessary to force $\underline{y}_2$ and $\dot{\underline{y}}_2$ to become identically zero. In this way the linear part of the control law (4.21) is formulated. Similarly the non-linear control component is found by considering that whenever the system is sliding, ie. $\underline{y}_2$ is equal to zero, the control must be discontinuous, but for all other $\underline{y}_2$ the control must be continuous. This calculation involves finding the positive definite solution of a Lyapunov equation. For more details of this particular control system development see, for example, Dorling and Zinober[4].

5. AIRCRAFT SLIDING MODE DESIGN

The development of a sliding mode controller for the lateral motion aircraft example requires the selection of an eigenstructure for the sliding or "null space" part of the control. A set of (n-m) null space eigenvalues for the closed-loop modes of the lateral motion aircraft example (2.6) are chosen according to MIL-F8785C[13,14,15] as:

$$\begin{aligned} \lambda_1, \lambda_2 &= -2.0 \pm j1.5 && \text{(roll mode)} \\ \lambda_3, \lambda_4 &= -1.5 \pm j1.5 && \text{(Dutch roll mode)} \\ \lambda_5 &= -0.05 && \text{(spiral mode)} \end{aligned} \tag{5.1}$$

During sliding these correspond to the (n-m) roots of the characteristic polynomial of A_{eq} where A_{eq} is given by equation (4.15). In the idealised sliding case these roots are also the poles which cancel (n-m) transmission zeros of the closed-loop system. To ensure that the various aircraft modes appear in the correct weighting in particular motions the desired eigenvectors are:

$$\underline{v}_1, \underline{v}_2 = \begin{bmatrix} 0 & 0 \\ 1 & v \\ 0 & 0 \\ v & 1 \\ v & v \\ v & v \\ v & v \end{bmatrix}, \quad \underline{v}_3, \underline{v}_4 = \begin{bmatrix} 1 & v \\ 0 & 0 \\ v & 1 \\ 0 & 0 \\ v & v \\ v & v \\ v & v \end{bmatrix}, \quad \underline{v}_5 = \begin{bmatrix} 0 \\ v \\ v \\ 1 \\ v \\ v \\ v \end{bmatrix} \tag{5.2}$$

where v denotes that the magnitude of the element is unimportant[16]. The roll mode should thus show up dominantly on roll rate but not on sideslip velocity or yaw rate ; banking manoeuvres should not incur sideslip buildup. Since $\phi = \int p \, dt$ some mode content must be expected on roll angle and also on the aileron and rudder states if the surfaces are controlling the mode. Similar arguments apply to the Dutch roll mode. There must be no coupling of Dutch roll into either roll rate or roll angle to satisfy the handling quality criteria. The spiral mode is associated with roll angle only and should not appear on sideslip velocity in order to avoid sideslip in steady turns. These desirable eigenvectors are projected in a least-squares sense into the allowable sub-space[17] and a C matrix providing the desired null-space dynamics is found. The non-linear controller design of equation (4.21) then follows with the discontinuous control selected according to actuator limitations.

Figure 1 shows the state responses for a linear model of the aircraft at $33ms^{-1}$ airspeed with initial conditions of $0.1ms^{-1}$ in v and $0.5rads^{-1}$ in p. One non-linear control component and plot of $\underline{S}$ are illustrated. From the switching on the non-linear control component which begins very quickly and the small value of $\underline{S}$ it can be seen that the desired sliding regime is reached almost immediately. The roll mode which is associated dominantly with the roll rate p and thus also with the bank angle ϕ has the desired fast, first order response. The Dutch roll mode appears only in the sideslip v and yaw rate r responses and has the required damping. The spiral mode appears on the bank angle state ϕ . The actuator demands are acceptable, the "chatter" being due to the discontinuous nature of the non-linear control component given by equation (4.21). The discontinuities can occur at each simulation sample point due to the fact that a sampled-data approach is applied to an otherwise continuous time system design.

6. ROBUSTNESS ASSESSMENT OF THE SLIDING MODE CONTROLLER

When the state begins to slide down the intersection of the switching surfaces the system will exhibit the same dynamics as a system given by:

$$\dot{\underline{x}} = (A + BL)\underline{x} \tag{6.1}$$

with L as defined by equation (4.21). Thus sliding may be demonstrated by running a full simulation, storing the state vector when sliding has commenced and using this as the initial condition for system (6.1). This is necessary as the system response is "piecewise" linear only between switching instants. The responses of the full simulation and system (6.1) should match for $t>t_s$ where t_s is the time at which sliding begins.

To establish when sliding begins one may look either at the non-linear control component to see when this becomes discontinuous or at the magnitude of S. A more sophisticated method to determine whether or not a system is essentially following the specified dynamics is to look at the orthogonal projection of the state vector into the null space of C at each simulation sample point. The orthogonal projection into the null-space of C is achieved using the Singular Value Decomposition method[18]. Let:

$$C = [\ U_r\ \bar{U}_r\ ]\ \Sigma\ [\ V_r\ \bar{V}_r\ ]^T \tag{6.2}$$

where Σ is the diagonal matrix of singular values of C and if r = rank(C), then U_r and V_r are (n x r) and $\bar{U}_r$, $\bar{V}_r$ are (n x (n - r)) respectively. The projection matrix is formed as:

$$P = V_r \overline{V}_r^T \tag{6.3}$$

The state vector $\underline{x}$ is then projected into the null-space of C using:

$$\underline{x}_p = P\underline{x} \tag{6.4}$$

The norm of $\underline{x}_p$ then gives the "distance" of the point in state-space from the switching boundary. By testing the ratio of the actual value of each component of the state vector to the equivalent component after projection, information about the position of each state relative to the null space may be determined. This ratio should be seen to converge to unity for all n states; the system is then sliding. Also, if the norm of the projection vector is small the dynamics of the system must be close to those of the null space of C. The norm used here is the 2-norm (or Euclidean norm) defined by:

$$\|\underline{x}\| = (x_1^2 + \ldots\ldots + x_n^2)^{1/2} \tag{6.5}$$

A non-linear simulation of the unmanned aircraft has been used to provide a realistic assessment of the performance of the VSS controller in the presence of parameter variations. The simulation incorporates both the full force and moment lateral and longitudinal dynamics together with cross-coupling effects. Figure 2 shows the responses of the lateral motion states. An initial comparison of the sliding mode controller's performance using the linear and non-linear system models shows that the responses are broadly similar. Figure 3(a) shows the variation of the 2-norm of the orthogonal projection of the state vector into the null space of C at every sample instant for both the linear and non-linear models. The null space is achieved at approximately 2.5 seconds by both systems. The response of the system (6.1) is shown tracking the non-linear model response for one of the system states after this time in figure 3(b).

In order to demonstrate the lack of robustness resulting from the use of simple linear state feedback control a second non-linear simulation was run. The control utilised here was purely the linear state feedbak defined by $\underline{u}_L = L\underline{x}$ ie. the VSS controller minus the switched component. Figure 4(a) shows the responses of the roll rate p and yaw rate r for this simulation. These should be compared with the equivalent responses in figure 2(a). The results show clearly that the modal objectives expressed in equations (5.1) and (5.2) are not maintained when using this linear feedback alone. This illustrates that the linear feedback design does not maintain the correct eigenstructure during parameter variations and is therefore not robust in this sense. Figure 4(b) shows a plot of S_1 (the first component of $\underline{S}$) and it can be seen that this is not close to zero and this further demonstrates the lack of eigenstructure robustness in the sense defined in section (1). The definitions of robustness given here correspond to the eigenstructure of the null space of C as defined by equation (4.12).

7. CONCLUSION

The design and assessment of a sliding mode controller for the lateral motion sub-system of an unmanned aircraft has been considered and the work of Aslin et al extended[18]. The primary objective has been to illustrate the degree of insensitivity achieved by this form of variable structure (non-linear) design. Figures 1 and 2 show that the VSS controller provides acceptable performance with both linear and non-linear aircraft simulations. Figure 3 demonstrates that the sliding regime may be attained by a non-linear system.

A common criticism of switched designs is the "chatter" inherent in a discontinuous control structure. However it is demonstrated in[10] that switching need only be shown at the design stage to ensure sliding. A smoothing control parameter δ can be used when the controller is applied. In this case equation (4.21) becomes:

$$\underline{u}_{TOTAL} = L\underline{x} + \frac{N\underline{x}}{\|M\underline{x}\| + \delta} \tag{7.1}$$

The effect of δ is to provide a compromise between the true sliding condition and motion in a close proximity to the hyperplane intersections. At present there is no systematic way of calculating δ for a given problem. A suitable value must be found from trial and error using simulation studies.

8. ACKNOWLEDGEMENTS

This work has been supported by an SERC CASE Studentship to Miss. S.K.Mudge in collaboration with British Aerospace, Aircraft Group, Brough, North Humberside, UK. The authors wish to acknowledge the guidance of Mr.D.Walker of BAe (Brough). Thanks are also due to Dr.A.S.I.Zinober (University of Sheffield) and Dr.C.M.Dorling (Topexpress LTD, Cambridge) for the use of the V.S.C.S. design package.

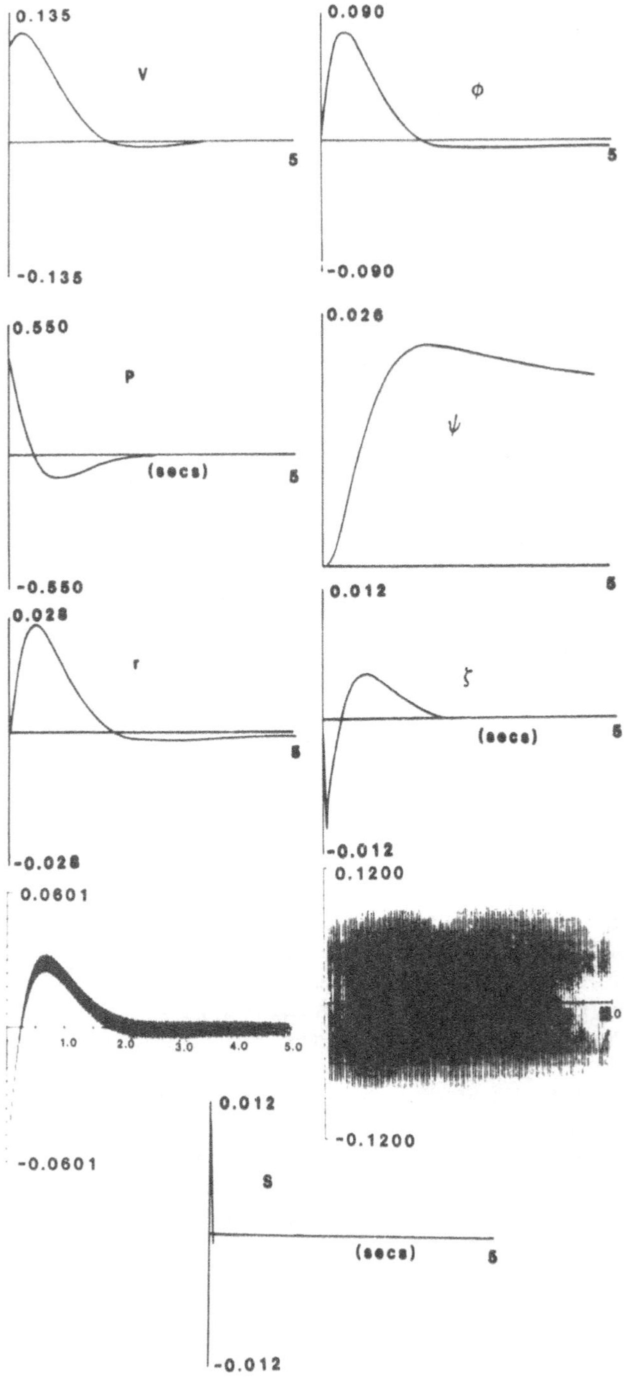

Fig 1 Lateral motion responses for the nominal linear system

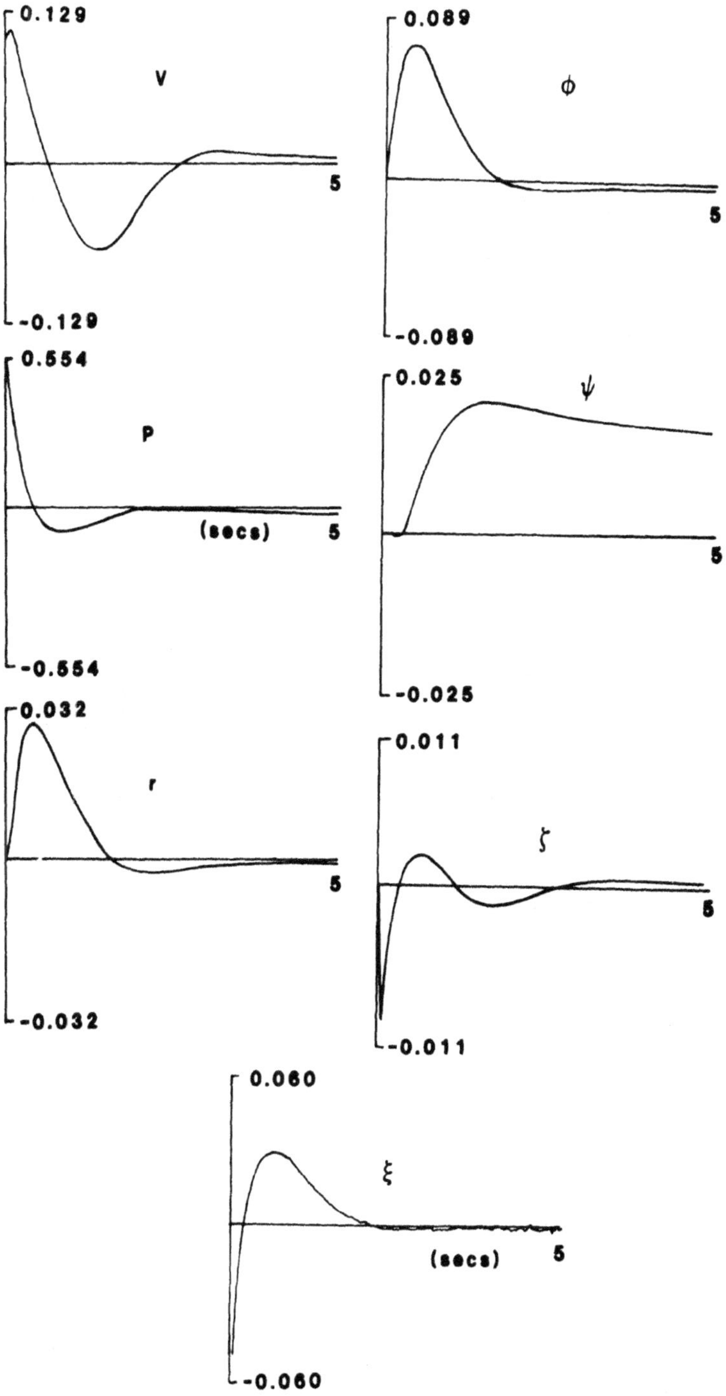

Fig 2 Lateral motion state responses for the full non-linear simulation

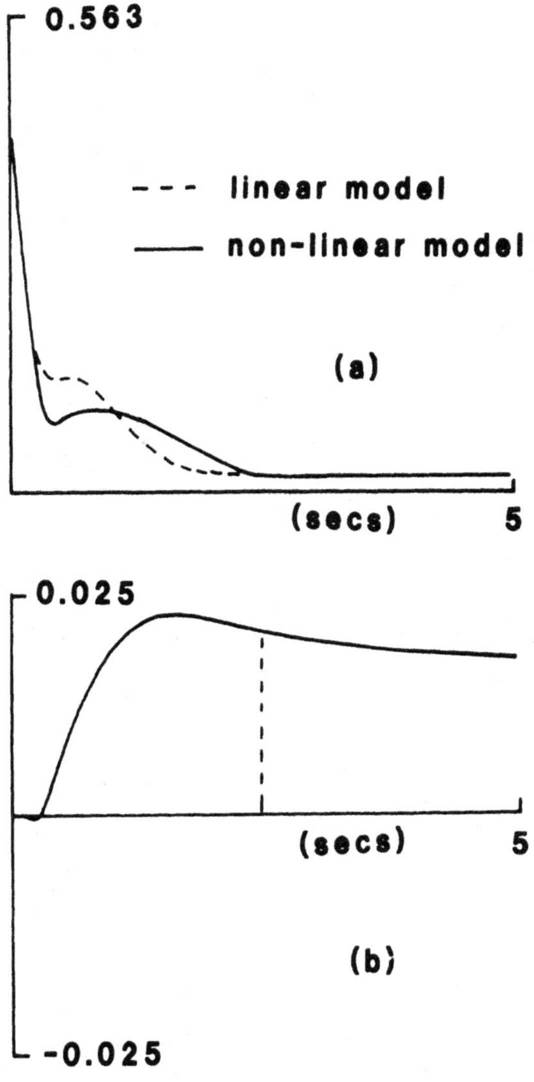

Fig 3 Norm variation

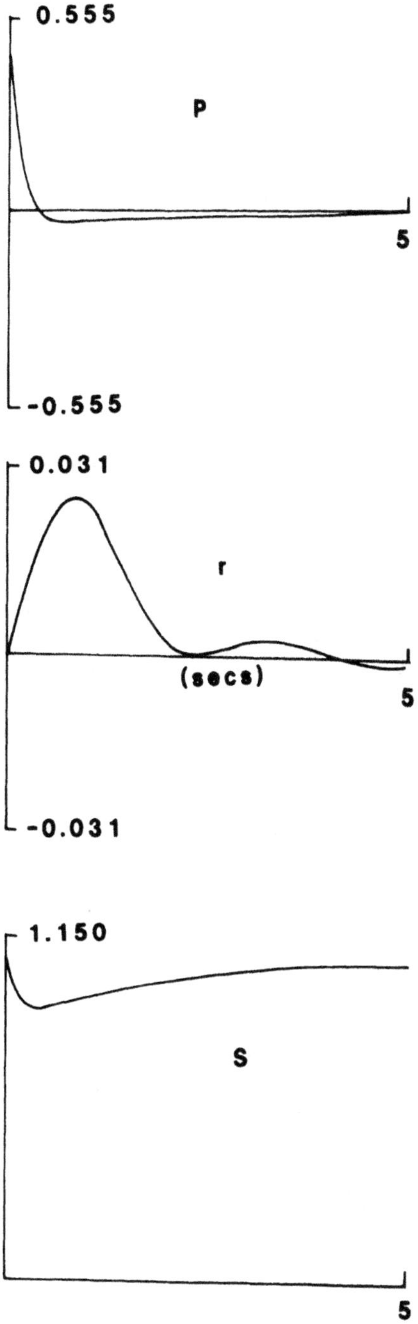

Fig 4 Full non-linear simulation controlled by the linear component of the VSS controller only

9. REFERENCES

1.Utkin,V.'Variable Structure Systems with Sliding Modes',IEEE Trans.AC-22,2,1977.
2.Itkis,U.'Control Systems of Variable Structure',J.Wiley,New York,1976.
3.Ryan,E.P.,and Corless,M.'Ultimate Boundedness and Asymptotic Stability of a class of Uncertain Dynamical Systems via Continuous and Discontinuous Feedback Control',IMA.J.Math.Control Information,1984.
4.Dorling,C.M., and Zinober,A.S.I.'Two Approaches to Hyperplane Design in Multivariable Variable Structure Systems',Int.J.Control,44,65-82,1986.
5.Dorling,C.M.'The Design of Variable Structure Control Systems:A Manual for VASSYD Package',1985.
6.Zinober,A.S.I.,El-Ghezawi,O.M.E.,and Billings,S.A.'Multivariable Variable Structure Adaptive Model-following Control Systems',IEE.Proc,Vol.129,Pt.D,No.1,1982.
7.Young,K-K.D.'Design of Variable Structure Model-following Control Systems',IEEE.Trans.AC-23,1079-1085,1978.
8.Ambrosino,G.,Celentano,G.,and Garofalo,F.'Variable Structure Model-reference Adaptive Control Systems',Int.J.Control.39,6,1339-1349,1984.
9.Slotine,J.J.,and Sastry,S.S.'Tracking Control of Nonlinear Systems using Sliding Surfaces, with Application to Robot Manipulators',Int.J.Control,38,2,465-492,1983.
10.Dorling,C.M.,and Zinober,A.S.I.'A Comparative Study of the Sensitivity of Observers',Proc.IASTED Symp.on Applied Control and Identification, Copenhagen,1983
11.Frank,P.M.'Robustness and Sensitivity:A Comparison of the two Methods',ACI Conference,Copenhagen,1983.
12.Mcruer,D.,Ashkenas,I.,and Graham,D.'Aircraft Dynamics and Automatic Control',Princeton University Press,1973.
13.MIL-F8785C.'Flying Qualities of Piloted Vehicles',1980.
14.Andry,A.N.,Shapiro,E.Y.,and Chung,J.C.'Eigenstructure Assignment for Linear Systems',IEEE Trans.AES-19,711-729,1983
15.Sobel,K.M.,and Shapiro,E.Y.'Flight Control Synthesis via Eigenstructure Assignment;the discrete version',28th Israel Conference on Aviation and Astronautics,1986.
16.Harvey,C.A.,and Pope,R.E.'Design Techniques for Multivariable Flight Control Systems',Control and Dynamic Systems,18,Academic Press,1982.
17.Harvey,C.A.,and Stein,G.'Quadratic Weights for Asymptotic Regulator Properties',IEEE Trans.AC-23,3,378-387,1978.
18.Golub,G.,and Van Loan,C.'Matrix Computations',North Oxford Academic,1983.
19.Aslin,P.P.,Patton,R.J.,and Dorling,C.M.'The Design of a Sliding Mode Controller for Dutch Roll Damping in a nonlinear Aircraft System',Proceedings of the IEE Control 85 Conference, University of Cambridge, England published by Peter Peregrinus Press,1985.

HIGH INTEGRITY SOFTWARE FOR MULTIVARIABLE PID CONTROLLERS

S E Burge and A H Jones
Department of Aeronautical and Mechanical Engineering
University of Salford
Salford M5 4WT
England

ABSTRACT The prominent software engineering techniques used in the development of high integrity software are discussed. The application of these techniques is illustrated by presenting a review of a high integrity software project in the case of a fixed structure multivariable proportional-plus-integral-plus-derivative controller.

1. INTRODUCTION

The digital implementation of real-time control strategies requires the development of not only the hardware but also the software. In those environments where a failure is undesirable or unacceptable great efforts must be made to ensure that failures do not occur or are tolerated (see for example [1]).

In recent years the software industry has developed several techniques to assist in the design of high integrity software (HIS). This paper briefly discusses some of the more prominent of these methods and illustrates their application by presenting a review of a high integrity software project.

The high integrity software project is illustrated in the case of a fixed structure multivariable proportional-plus-integral-plus-derivative controller. This type of controller structure is in widespread use and therefore the points highlighted by this paper are of practical relevance.

2. GENERAL ASPECTS OF HIGH INTEGRITY SOFTWARE (HIS)

When developing HIS, whether the integrity is required for safety or security considerations, there are several phases and stages of the development cycle which can be generally identified. Furthermore, there are several recognised techniques which can

S. Tzafestas et al. (eds.), System Fault Diagnostics, Reliability and Related Knowledge-Based Approaches, Vol. 1, 265–279.

assist in achieving the desired goal of high integrity. This section briefly outlines these and introduces the importance of software quality assurance.

2.1 HIS Development Cycle

In general, HIS systems takes a considerable amount of time and expense to develop Management and quality assurance are the two essential cornerstones of any project.

Figure 1 shows diagrammatically the basic, and absolute minimum, project phases and stages together with the communication paths necessary to achieve HIS.

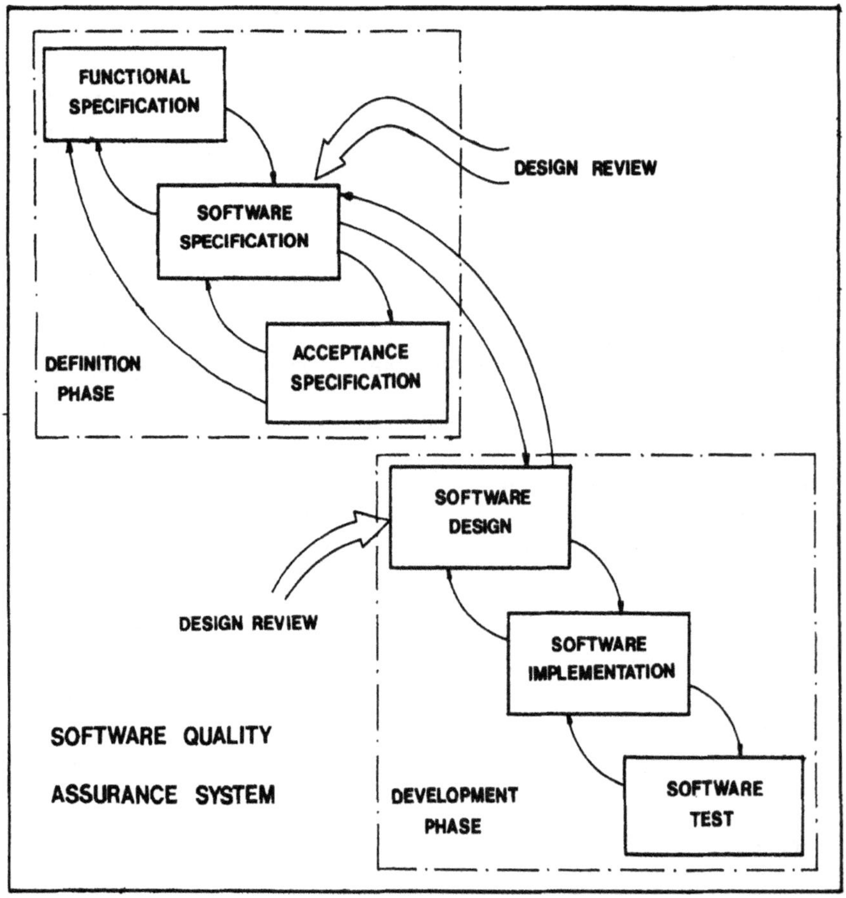

Figure 1: HIS project phases and stages

The first point to be made about Figure 1 is that the quality assurance system encompasses the entire project. It is not something that is applied at the end. The success of the project is dependent upon the operation of, and the adherance to, a good QA system. Such a system must at minimum state responsibilities, standards, codes of practice, and project management methods. Quality must be made the number one objective of all involved staff.

It is also of note that the project has been split into two phases; definition and development. This division is the only natural division. The development phase cannot be started until the definition phase has been performed. This is often used as a management aid because the results of the definition phase allow a more detailed project plant to be generated and thus give a more accurate estimate of time scales and costs. The latter of which may affect the design.

Within the definition phase three stages have been identified. The first results in a document compiled by the customer; the functional specification. It states precisely what is required of the final software. Using this document the software team analyse the requirements to produce their own version of the specification; the software specification. This apparent repetition serves two vital functions. Firstly, it ensures that the software team understands the customer's requirements. Indeed, simple mistakes, such as a misinterpreted sentence, in the functional specification can be identified. Secondly, the customer may not be a software expert and therefore the functional specification is likely to not be in a suitable form for the software development team. The software specification performs a 'jargon' transformation from that of the customer to a form the software team understands.

The next stage, again something that has to be agreed with the customer, is the development of a test schedule to which the final software is accepted.

Provided that the definition phase is successfully completed the development phase will progress in a logical and straightforward manner. The design stage should be a logical extension of the software specification and similarly the testing stage should be a logical extension of the acceptance specification.

Communication between stages and phases of the project is important. Errors, omissions, and other bad practices should be corrected and fed back to modify the earlier stages. One point is that the sooner such anomalies are detected the less effect they will have on the overall project timescale and cost.

The final point is documentation. All the stages should be fully documented, preferably to some standard such as DO-178 [2] or JSP-188 [3]. Documentation has two benefits, firstly it greatly assists in the identification of anomalies and secondly, it facilitates the operation and maintenance of the software once it is implemented.

2.2 Techniques for HIS Development

There are numerous software engineering techniques which can be employed to achieve high software integrity. In general, these methodologies fall broadly into three categories: fault avoidance, fault resistance, and fault correction.

2.2.1 Fault Avoidance

By adopting the attitude of 'right first time' when developing software, many non-deliberate errors can be avoided. To assist in this task several methods have been devised, the major of which include:

Formal Specification It has already been discussed that transforming a customer's specification into a software specification is a method by which non-deliberate errors can be avoided. There are several formal specification methods available, such as MASCOT [4], DeMarco [5], VDM [6], etc. These all allow a formal specification to be generated and in some cases (VDM,Z,EUCLID) the specification may be mathematically tested for correctness.

Structured Techniques The consistent use of structured methods, from analysis through design, coding, and testing, has brought about a revolution in the software industry. The methods permit large and complex software systems to be partitioned into small and easily handled sub-systems. The benefits are simplified project management, improved communication, improved code production and readability, improved testability, and simplified operational maintenance.

Validation and Verification By demonstrating that a piece of software matches its functional requirement implies that it has been validated. Ensuring that the software development stages are complete, consistent, and testable, indicates that the software is verified. HIS requires both validation and verification by extensive testing and the adherence to the quality system.

Assured Code Libraries Some software has been extensively used or has been through an extensive validation-verification programme (for example, NAG and GINO library routines and the final Ada compiler). This software is tried and trusted and can be assumed to have a high integrity. However, although it is expedient to use such routines where possible they can introduce inflexibility into the overall software design because of the tight input/output specifications.

2.2.2 Fault Resistance

Accepting that fault avoidance will never by completely successful, there are several methods which can be used to minimise the effects of errors if they occur.

Defensive Programming The concept of defensive programming is based on the infamous Murphy's law, "*if anything can go wrong it will*", and the corollary, "*the impossible always happens*". If a piece of software receives an incorrect input it will produce an incorrect output. To overcome this, defensive checks are placed at the beginning of the software to test the input data. This is the mechanism of defensive programming.

Software Diversity or N-Version Programming By running several versions of software, each satisfying a common specification, in parallel, then the results of each can be used to detect errors in one or more of the other versions. This technique has been used successfully in many applications, however, because of increased hardware complexity and increased software development it is exceedingly costly.

Recovery Blocks In its simplest form a recovery block comprises two alternative algorithms and an acceptance test. If one version of the algorithms does not pass the acceptance test then the system is reset and the alternate algorithm executed.

2.2.3 Fault Removal

This technique usually complements fault avoidance. Simply, fault removal is the identification and removal of software errors before the software enters operation. There are two categories of fault removal; static and dynamic.

Static fault removal does not require the execution of the software but is based around man-intensive tasks such as code inspection, walk-throughs, syntatic checkers, etc.

Conversely, dynamic fault removal requires the execution of the software. In general, dynamic fault removal follows any static analysis. There are several dynamic analysis methods including:

Structural Testing This requires the testing of all software statements, branches and paths. Even for very small programs the number of possible branch and path permutations makes this method prohibitive.

Black Box Testing This testing method assumes that a piece of software will perform some known transformation on the input data to produce a predictable output.

Reliability Modelling By assuming that the software agrees with some model, it is possible to predict its performance in terms of

reliability or mean time to failure. The difficulty here is in selecting the appropriate model. Moreover, the analysis can only tell you how many errors are still undetected.

Testing Philosophies Testing can be performed in two opposed ways, either 'top down' or 'bottom up'. In top down testing the higher code modules are tested in isolation with the lower modules that are called represented by a 'harness'. Once these modules have been validated the lower modules are tested in a similar fashion. Bottom up testing reverses this process.

The disadvantage of top down testing is that the harness of code required to simulate the lower modules can be complex and subject to errors. This is avoided in bottom up testing because any harness required to test the lowest modules is likely to be very simple. Furthermore, the higher modules can be tested using the already validated lower modules.

The disadvantage of bottom up testing is that no demonstrably working software is available until the highest module has been tested.

In general, neither philosophy has a clear advantage and most software development utilizes a mixture of both.

3. HIGH INTEGRITY SOFTWARE FOR MULTIVARIABLE PID CONTROLLERS

The second section is concerned with describing some steps taken to develop HIS for a multivariable PID control scheme. It is evident from the previous section that in a correctly managed project the necessary documentation is quite substantial. Consequently, the following can only be restricted to presenting the major points and points of interest.

To achieve high integrity, a mixture of the various methods described in the previous section have been used. Structured techniques have been used throughout, since it is considered that these offer a significant contribution to improved integrity. Defensive programming is also used extensively.

The following is presented in three sections; specification, design, and testing.

3.1 Specification

3.1.1 Functional Specification

The general customer requirements are specified in the functional specification as follows:

The design and development of resident software to provide a real-time control function for multivariable systems. The software will provide the facilities:

(a) Receive raw unscaled, unlimited, data which will be scaled and limited as specified. This data will then be processed by a linear fixed structure proportional-plus-integral-plus-derivative control-law to produce scaled and limited control demands.

(b) No target machine is specified and software versions will be written in Pascal, Ada, and Occam.

(c) The software will be tolerant of raw data error and control saturation.

(d) The software will provide the specified self-testing and active testing facilities.

(e) The documentation will be to the specified standards.

Each of these points is further expanded in the functional specification, giving precise details of what is required.

3.1.2 Software Specification

The software specification for the controller HIS has been generated using the DeMarco technique [5]. Whilst other methods may have been employed with equal success, DeMarco presents the specification in a structured format and its graphical representation simplifies communication with the customer. It is a top down method and provides three basic tools:

(a) The Data Flow Diagram (DFD) which is network representation of the system; it has the convention.

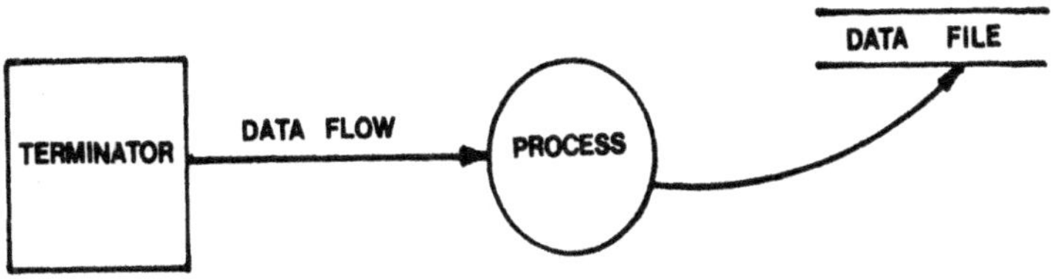

(b) The Data Dictionary (DD) compliments the DFD by defining all the data flows.

(c) The Process Specification (PS) which specifies the processes using a semi-formal structured English.

The importance of this technique is that it describes the system from the viewpoint of data flow rather than control flow. Thus, the usually troublesome interfaces are highlighted. Furthermore, the top down structuring philosophy of the method allows very complex systems to be specified. This can be illustrated in the case of the multivariable PID controller HIS, where the top level DFD is simply

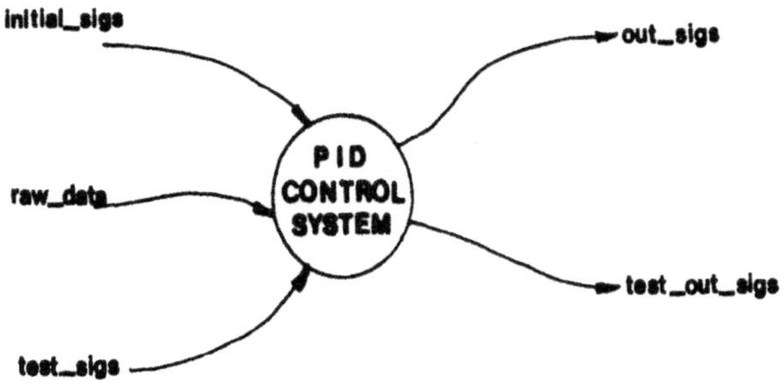

CONTEXT DIAGRAM

This is subsequently decomposed to generate the next level of DFD

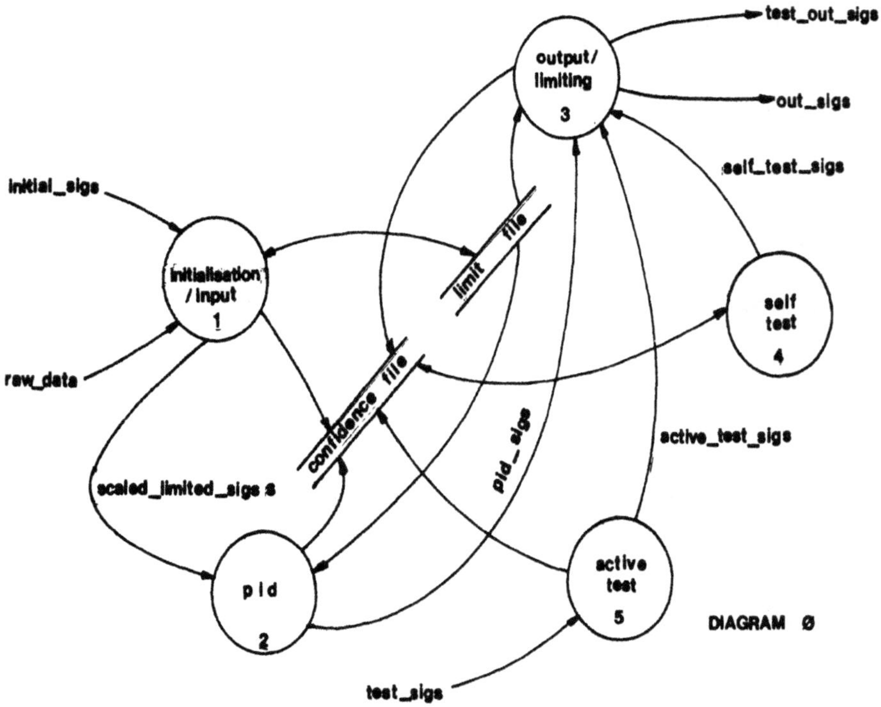

DIAGRAM Ø

Each process on diagram ⌀ is then decomposed until the system has been specified. The specification is completed by constructing the DD and providing a PS for the lowest level processes.

The software specification should not contain any information not present in the functional specification. However, it is frequently found that the software specification highlights deficiences in the functional specification which can then be modified accordingly. The published software specification is only arrived at after several iterations, at this point all parties should have come to an understanding about the requirements of the final software.

3.2 Software Design

The design of the HIS for the multivariable PID controller was documented to JSP-188 [3]. The adoption of this standard dictated the design process. JSP-188 calls for the design to be structured into four levels, each becoming progressively more detailed with the last, level 4, presenting the actual code.

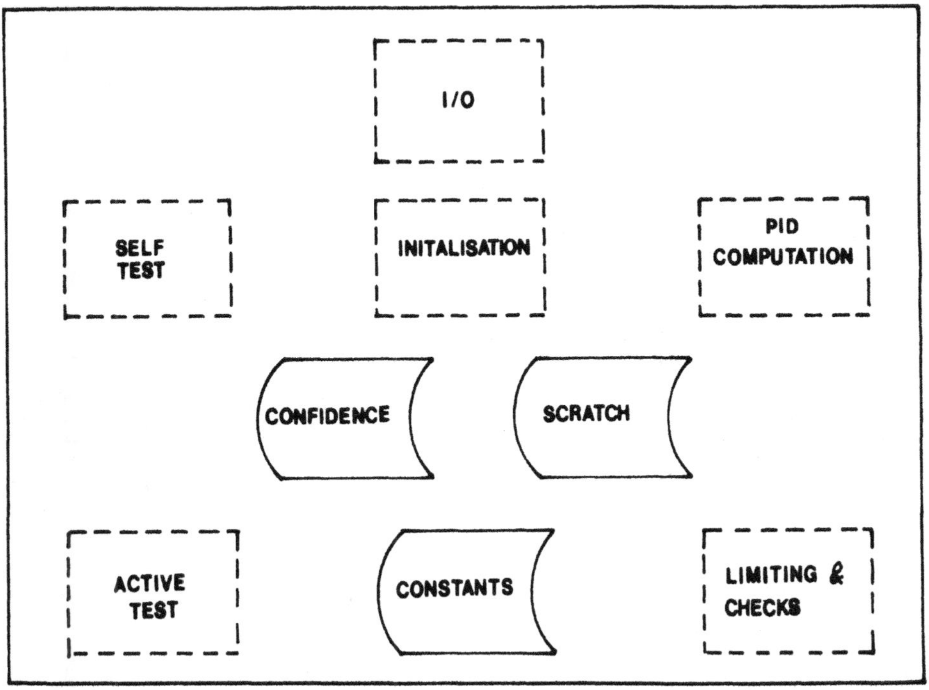

Software Task and Global Data Area Diagram

3.2.1 Level 1 Design

The level 1 design requires the software to be described as a set of TASKS and GLOBAL DATA AREAS dynamically linked to provide the software FACILITIES.

In the case of the HIS controller design the only FACILITY is to provide the control demands. The following diagram shows how this is to be provided in terms of TASKS and GLOBAL DATA AREAS.

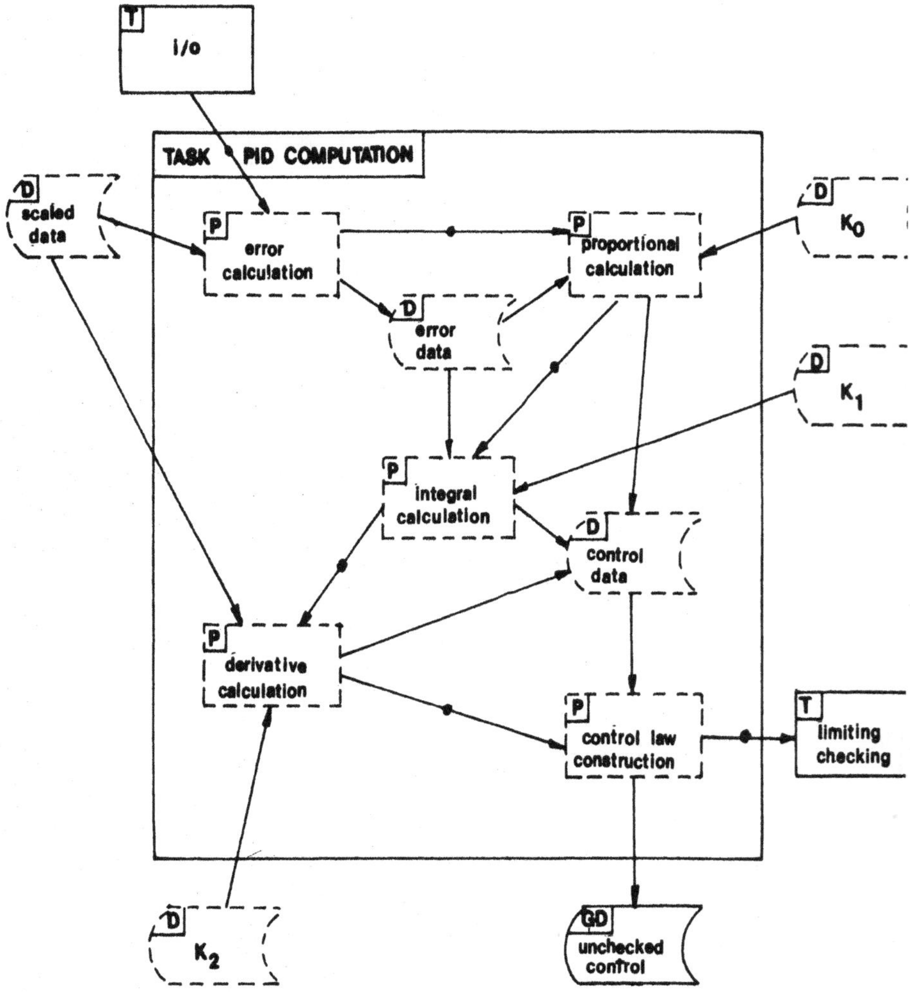

Level 2 Control and Data Interconnection Diagram

Comparing this diagram with diagram 6 of the software specification, it is clear that the two are similar. Indeed, the natural structure which arose from the specification should follow through into the design.

3.2.2 Level 2 Design

The level 2 design decomposes the TASKS and GLOBAL DATA AREAS of the level 1 design into PROCESSES and DATA AREAS respectively. These are documented diagrammatically as Data and Control interconnection diagrams with supporting text as necessary. For the TASK main computation of the level 1 design the Data and Control interconnection diagram was generated

It is clear from this diagram that the detail is increasing. However, by adhering to the guidelines of JSP-188 the increase in detail is proceeding in a controlled manner: the temptation to overdetail too early is avoided.

3.2.3 Level 3 Design

The next stage in the design process is to decompose the PROCESSES and DATA AREAS into BASIC CODED UNITS, DATA, respectively. Once again this is presented graphically if necessary. In the case of the controller HIS this is not necessary since in many cases the PROCESSES are already sufficiently decomposed.

The level 3 design also requires detailed descriptions of the BCU's and DATA. These descriptions can be presented in several formats, for example, PDL [7], structure charts [7], etc.

PDL has been used for describing the BCU's of the multivariable PID controller HIS, an example is presented below.

BCU PROPART - Proportional Action Calculation PROPART limit checks the controller error signals and calculates the proportional action of the multivariable PID controller. If a limit error occurs the relevant flag in the status word is set and the status flag is set to indicate that an error has occurred. On completion the node count is incremented.

```
BEGIN(error_sig,prop_act,lmax,status,status_flag,node_count)

      initial_lim:=-lmax
      DO WHILE initial_lim<0
           prop_act:=0
           FADD(initial_lim,1,initial_lim)
      ENDDO
```

```
        col:=1
        col_lim:=-lmax
        DO WHILE col_lim<0
             FSUB(error_sig{col},low_err_lim{col},lim_flag)
             IF lim_flag<0
             THEN
                  error_sig{col}:=low_err_lim{col}
                  status_flag:=true
             ENDIF
             FSUB(high_err_lim{col},error_sig{col},lim_flag)
             IF lim_flag<0
             THEN
                  error_sig{col}:=high_err_lim{col}

                  status_flag:=true
             ENDIF
             row:=1
             row_lim:=-lmax
             DO WHILE row_lim<0
                  FMUL(kpgain{row,col},error_sig{col},part_sum{row})
                  FADD(prop_act{row},part_sum{row},prop_act{row})


                  FADD(row,1,row)
                  FADD(row_lim,1,row_lim)
             ENDDO
             FADD(col,1,col)
             FADD(col_lim,1,col_lim)
        ENDDO
        IF status_flag true
        THEN
             status:=status.AND.lim_prop_mask
        ENDIF
        FADD(node_count,1,node_count)
   END
```

Immediately clear from this BCU description is the significant proportion of the PDL which is dedicated to defensive action. However, even more defensive programming is included in the addition, subtraction, and multiplication functions FADD, FSUB, and FMUL, respectively. These functions include code for under- and over-flow checking. It is also clear from this BCU description that it is almost high-level code.

3.2.4 Level 4 Design

The level 4 design transforms the level 3 BCU descriptions into the target language code. Since the original specification required the target code to be a high-level language, there is no point in presenting an example because the PDL description is so similar.

3.3 Testing

In the case of the controller HIS the testing was performed in the following stages:

(i) code inspection,
(ii) compilation,
(iii) unit testing,
(iv) integration testing,

and

(v) acceptance testing.

The first two are obvious and will identify most logic and syntax errors. In the case of unit testing, the lowest level BCU's were tested in isolation against their own specification. The tests were performed with the assistance of a test plan developed from the software specification and design documents. For example, in the case of the proportional action BCU the following presents an extract from the test plan/results document.

BCU PROPART - Test Plan/Results

Case	Input	Expected Results	Actual Results	Mod
1	lmax 1 error_sig -1 low_err_lim -0.5 kpgain 1	status_flag true status_word set prop_act -0.5	✓ ✓ -0.5	

It is evident from just this single test case that the testing phase is time-consuming and requires carefully designed test plans.

The integration and acceptance testing follows a similar pattern, with the test plan designed before testing was initiated.

4. CONCLUSIONS

This paper has presented some general comments on the development of high integrity software. In particular, it has identified those techniques which are presently viewed as the most appropriate for developing high integrity software.

Some of these techniques have been illustrated, albeit briefly, by presenting the development of high integrity software for a real-time multivariable PID control scheme. It is shown that the development of high integrity software is an expensive, time-consuming and complex task and one which must not be taken lightly if the desired integrity levels are to be achieved

REFERENCES

[1] C T Sheridan, 'Space Shuttle Software', *Datamation*, July 1978.

[2] 'DO-178A Software Considerations in Airborne Systems and Equipment Certification', *Radio Technical Commission for Aeronautics (RTCA)*, 1981.

[3] 'JSP-188 Requirements for the Documentation of Software in Military Operational Real-Time Systems', *UK MOD*, March 1980.

[4] 'The Official Handbook of MASCOT', *Mascot Suppliers Association*, Royal Signals and Radar Establishment, 1980.

[5] T, DeMarco, 'Structured Analysis and System Specification', *Yourdon Press*, 1978.

[6] C B Jones, 'Software Development of Rigorous Approach', *Prentice-Hall*, 1980.

[7] R S Burgess, 'An Introduction to Program Design Using JSP', *Hutchinson*, 1984.

ADAPTIVE AND COLOURED STRUCTURED PETRI-NETS FOR DESCRIPTION, ANALYSIS AND SYNTHESIS OF HIERARCHICAL CONTROL AND RELIABILITY OF FLEXIBLE CELLS IN MANUFACTURING SYSTEMS

J.P. Bourey, D. Corbeel, E. Craye, J.C. Gentina
Institut Industriel du Nord (I. D. N.) - L. A. I. I.
B.P. 48
59651 Villeneuve d'ascq Cédex - FRANCE
UA CNRS 370 - D.U.S.V.A.

ABSTRACT. We present a methodological approach to the analysis and synthesis of the control of Flexible Cells in Manufacturing (F.C.M.). An F.C.M. can be described by a three levels modelization : the production tools, the direct automatic-control of the operative system and the hierarchical level to parameter the working conditions. Adaptive coloured and structured Petri-nets are used to model the "Control System" with high level of parallelism and its different communications with the "Operative System" and with the hierarchical level by means of a top-down approach. Modelling the changes of working modes, taking the failures into account, adding degraded modes or priorities, parametrizing a flexible production, all these notions could be added to the "Control System" but the command graph would be more cumbersome and unreadable. The notion of an upper hierarchical level enables this problem to be solved. It is more adequate to model this last level in a declarative way which guarantees any "a posterieuri" modification of this level by the user giving the correct requirements of flexibility to the "Operative System".

INTRODUCTION

This paper presents a methodological approach to the analysis and synthesis of the hierarchical control of **F**lexible **C**ells in **M**anufacturing (F.M.C.). The main feature of the methodology consists in the use of both structured and **A**daptive **C**oloured **P**etri-**N**ets (A.C.P.N.).

In order to introduce the different levels of structuration, let us first introduce the three levels of description of an F.M.C. :

(i) the production tools generally called "Operative System" (OS),
(ii) the direct automatic-control of the operative system called "Control
System" (CS),
(iii) The hierarchical level which is used to parameter the working conditions of both the "operative system" and the "control system".

S. Tzafestas et al. (eds.), System Fault Diagnostics, Reliability and Related Knowledge-Based Approaches, Vol. 1, 281–295.

The operative system consists of a small number of production machines. A carriage between the different machines, input and output parts and stocking areas will be assured by a flexible transport system.

By optimisation of the strategy of communication between its different elements, such a flexible cell enables the user to :

(i) increase the operating time of each machine,

(ii) manufacture a large variety of different products according to the requirement specifications.

The potential flexibility of an F.M.C. results from both the architecture of the OS and from the structuring of the CS. Description of the interfaces between OS, CS and the hierarchical level may also be precisely defined according to the different models used for the description of these three levels.

In this paper, we propose the description of an F.M.C. by means of a top-down structured approach from both the point of view of "control" and "error processing".

The presentation is organized in 3 parts.

* A first part deals with the definition of the primitive functions used in structured A.C.P.N.. The model obtained from such description is really convenient to represent the CS and its communications with both the OS and the hierarchical level. From this first structured analysis, we propose in addition a class of primitives devoted to the description of reliability control.

* A second part will concern an illustrative example of parallel control of F.M.C.. In this example, the F.M.C. performs 4 classes of different products. The advantages of structuring in control and reliability will appear in the last section of this part.

* In the last part, we try to characterize the hierarchical level. Thus the example of part 2 will give the basis of our approach. We suppose that production is organized in 2 modes : the automatic normal mode using 2 autonomous resources/units for transport and the degraded automatic mode which uses a single resource when a fault occurs in one of the 2 resources.

The hierarchical level will then decide on error-recovering conditions after identification of the fault and control of transient conditions from mode 2 to mode 1. Hierarchical level description uses the declarative form for description of the different priority rules of each working mode and the different classes of rules used for diagnosis of fault and control of restart in degraded mode.

1. HOMOGENEOUS MODELIZATION OF INDUSTRIAL PROCESS CONTROL SYSTEM GRAPH

1.1. Structured Petri-nets (S P-N)

1.1.1. Introduction. Petri-nets represent an important tool for the description of industrial processes. A particular class of these nets, more restrictive but more systematic has been studied /CORBEEL, 1979/. This tool, (Structured Petri-Nets or S P-N), which adds the power of description and verification of usual Petri-nets to the strict nature of structured programming, will enable a method of analysis and synthesis of industrial processes to be defined (characterized by an important rate of parallelism). Needs of industrial users led us to complete this method by the definition of exception handling's mechanisms /CORBEEL, 1980/.

1.1.2. Methodology of description of processes'systems. a) Introduction.
A system of n processes is defined by the data of :
(i) each process,
(ii) the connections between these processes.

The decomposition of an industrial system into n processes takes into account such criteria as :
(i) parallelism,
(ii) functional coherence,
(iii) minimization of the number of processes.

Each process is then described independently by a process net. The interactions between the processes will enable us to define next the link net (graph whose extremities belong to the process nets) /CORBEEL, 1983-a/.

b) Process net. A process is a sequential chain of tasks and may be described by a structured Petri-net (one initial token), by means of three basic structures (Fig. 1).

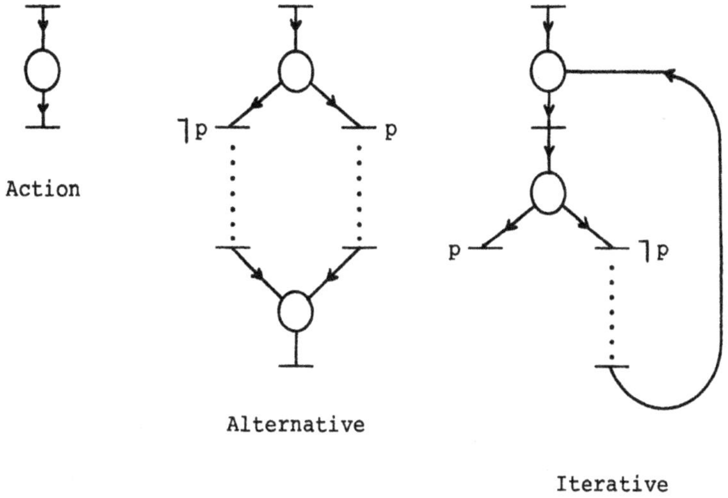

Figure 1. Basic structures of S P-N

The notion of block was then introduced to initially enable a sequential chain of tasks to be grouped under a same name. This procedure appears naturally when we study a problem with sequential steps. This notion is also used to define the link nets and the recovery points /CORBEEL, 1983-b/.

c) Link net. The connections represent the interactions between the processes. Essentially, they will be of three types :

(i) synchronisation,
(ii) mutual exclusion,
(iii) producer/consumer.

d) Conclusion. The use of the three basic structures enables us to ensure that the process's Petri-net will be safe, live and reinitialisable.

We could also easily verify the properties of processes because we used only three types of connections. This method which presents the advantages of being easily implemented by means of a structured language /CORBEEL, 1983-a/, enables us to minimize conception's faults and errors during the problem analysis. It represents in that way, the first step for the analysis and the synthesis of a high reliable system.

1.2. Stuctured adaptive Petri-nets (S A P-N)

Structured Petri-nets are indeterminate and not flexible ; moreover important examples of concurrent processes cannot be modelled by ordinary structured Petri-nets. The structured Petri-net model should be extended, for instance, by devices allowing real tests, but on the other hand should be restricted such that the complexity is decreased. A first step in this direction would be a natural and convincing concept which increases the computational power of structured Petri-nets. From the literature, the following extensions are known :

- inhibitor nets /HACK, 1975-a/,
- priority nets /KACK, 1975-a, 1975-b/,
- reset nets /HACK, 1975-b/,
- self-modifying nets /VALK, 1978-a, 1978-b, 1981/.

We have introduced the extension, called Structured Adaptive Petri-Net (S A P-N) which includes the previous ones /CORBEEL, 1985-a, 1985-b/. A "S A P-N" is defined in the same way as an ordinary structured Petri-net, as a bipartite multigraph, having edges of the form :

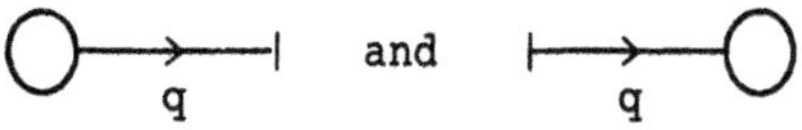

If q = 1, the firing rule of the transition is defined as in the normal case. But q may also be the name of an arbitrary place of the net. In this case, the number of tokens to be moved from or to the place equals the actual number of tokens in the place with name q if and only if this place q is not empty.

Therefore "S A P-N" are able to modify their own firing conditions. Inhibitor nets, priority nets and reset nets are special cases of Structured Adaptive Petri-Nets.

A "S A P-N" can be seen as a structured Petri-net for which dynamic change of the net structure is possible. Hence, structured adaptive

Petri-nets enable the parametrization of the representation of a model of process control command to be added. The S A P-N enable the interpretation of the net to be reduced /CORBEEL, 1985-a, 1985-b/ and the study of the degradate modes of a process control to be carried out.

1.3. The structured coloured adaptive Petri-nets

1.3.1. Introduction. Faced with the growing complexity of industrial control systems, Structured Adaptive Petri-Nets represent an interesting compromise of power and analysis capacity on the models already built. Nevertheless in many cases, the complexity of a S A P-N model may be important even for systems which are functionnally not very complex.

Under these conditions, structured coloured adaptive Petri-nets /CORBEEL, 1985-b/ may make an important contribution. The associations with tokens of colours sets and the definition of labels associated with the nets arcs, allow for a very concise and readable representation of described mechanisms. There is a simplification of the net structure through the transfer of information on the markings (colour) and on the transformations (net label) made on the markings.

1.3.2. Definitions. Let $C = \{C_1, C_2, \ldots, C_n\}$, a finite set of colours. A marking M, is a function defined on P such that :

$$M(p) : C(p) \longrightarrow N \qquad p \in P$$

M(p) give the number of tokens of each colour in place p for the current marking.

The arcs are labelled by expressions which specify the coloured tokens usage. The expressions are :

(i) <K> where K represents a sub-set of C

(ii) <x> where x is a free variable on a domain of colours

(iii) <K,x> where K represents a sub-set of colours and x a free variable on a domain of colour (dom(x)) such as dom(x) and K are two disjoint sets.

All adaptive arcs between a couple of place p and transition t, have the same label.

The first expression associates with the firing of the transition t, the presence of the every colours of K in the input place p :

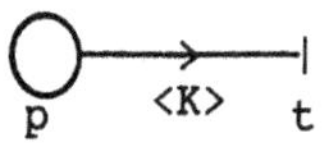

The second expression associates with the firing of the transition t, the existence of one colour of dom(x) in the input place p :

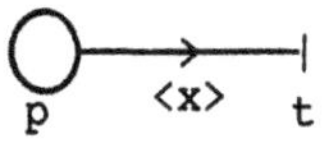

The firing rules are defined more precisely, in /CORBEEL, 1985-b/.

2. EXAMPLE : A FLEXIBLE MANUFACTURING SYSTEM

As an example, let us consider a production workshop which is shown in Figure 2. It is a workshop for mechanical part manufacture.

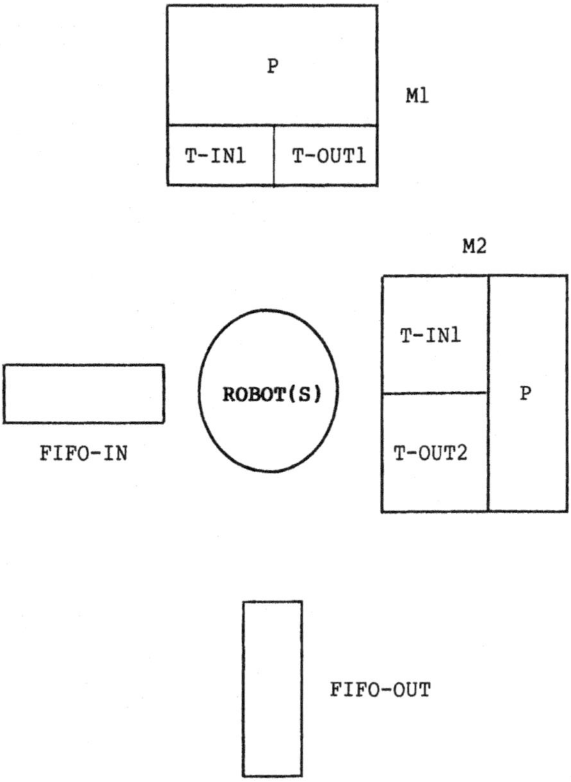

Figure 2. Lay-out of the workshop

This workshop is composed of :

(i) two transfer benches (FIFO-IN and FIFO-OUT tables). FIFO-IN is designed for loading, FIFO-OUT for unloading.

(ii) two stations M1 and M2. Each station is autonomous and is composed of a work post P and two transfer benches T-IN and T-OUT.

(iii) a robot which allows a distribution of parts between the two transfer benches FIFO-IN and FIFO-OUT and the two stations M1 and M2.

We connect a manufacturing program with each part. This manufacturing program is :

either - only station M1,
or - only station M2,
or - station M1 then station M2,
or - station M2 then station M1.

Specification of the workshop : The model of the workshop is shown in Figure 3. We found the next processes :

(FIFO-IN ⟶ M1) : to load a part on the station M1 from the transfer bench FIFO-IN

(FIFO-IN ⟶ M2) : to load a part on the station M2 from the transfer bench FIFO-IN

(M1 ⟶ FIFO-OUT) : to unload a part from the station M1 to the transfer bench FIFO-OUT

(M2 ⟶ FIFO-OUT) : to unload a part from the station M2 to the transfer bench FIFO-OUT

(M1 ⟶ M2) : to transfer a part from M1 to M2

(M2 ⟶ M1) : to transfer a part from M2 to M1

(M1) : to manufacture a part on the station M1

(M2) : to manufacture a part on the station M2

We introduce the 13 colours. The colours LOM1, LOM2, UNM1, UNM2, TRM1M2, TRM2M1, MAM1, MAM2 are associated with the eight processes. The colour TR modelize the communication between the processes. The colours m_1, m_2, m_1m_2, m_2m_1 are associated with each part and define the manufacturing program. Only these colours are represented on the graph of the Figure 3.

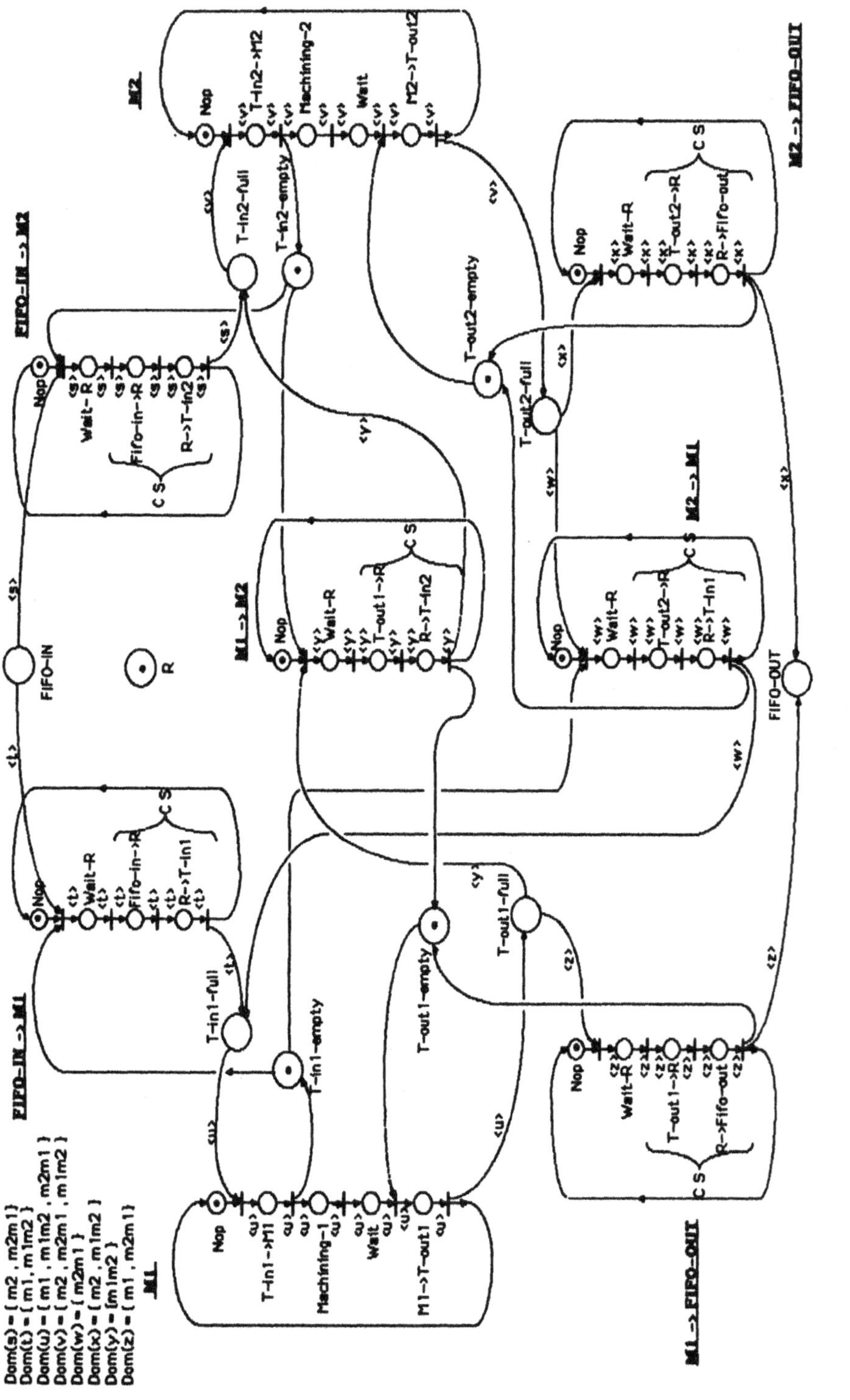

Figure 3. Control graph of the workshop

3. STUDY OF THE HIERARCHICAL LEVEL

3.1. Introduction

The graph of the command part modelizes repetitive working conditions (automatic mode) for a set of industrial production processes. Modelizing the changes of working modes, taking the failures into account, adding degraded modes or priorities, parametrizing a flexible production, all these notions could be added to this model but the command graph would be more cumbersome and unreadable. The notion of an upper hierarchical part enables this problem to be solved by relieving the command part of these functions. The adaptive Petri-nets can be used for modelization of the hierarchical level /CORBEEL, 1985-b/.

A non-procedural declarative language also allows it to be modelized. This representation will be used in the last part of this paper with the following syntax :

R <integer> : IF <predicate(s)> THEN <action(s)> EOR

where the predicates characterize the state of the CS and the actions modify its comportment. The precicates and actions are given in the annex.

Within the framework of our example and with production rules, we aim to show how to :

(i) work out the indeterminism of parallelism,
(ii) modelize the transition from an automatic mode to another,
(iii) analyse misfunctions.

Note : We shall use the following terminology /LAPRIE, 1985/ :
When a fault appears or is made, a latent-error is created which will become effective when it is activated. When the error affects the delivered service, a failure appears. A misfunction stands for any of these four notions.

3.2. Priority rules for determinism of behaviour

Two kinds of access conflict become apparent after a study of the processes :

- Physical impossibility for two processes (for example (FIFO-IN⟶ M1) and (M2 ⟶ M1)) to accede simultaneously to the same transfer bench (T-IN1). We can see it through the simultaneous request of the token of the places T-IN1-EMPTY, T-IN2-EMPTY.
- The mutual exclusion for access to the robot.

To solve the first kind of access conflict, priority was given to the transfer of pieces from a machine to another compared with the introduction of new pieces into the system, thus avoiding the obstruction of the machines. This finds simple expression in two rules.

R1 : IF STATE (T-IN1-EMPTY) ∧ STATE (T-OUT2-FULL,w)
∧ STATE (FIFO-IN,t)
THEN FREEZE (FIFO-IN ⟶ M1) EOR

R2 : IF STATE (T-IN2-EMPTY) ∧ STATE (T-OUT1-FULL,y)
∧ STATE (FIFO-IN,s)
THEN FREEZE (FIFO-IN ⟶ M2) EOR

The definition of such rules is not restrictive : if we want to modify the priorities (for example to introduce dynamic priorities), we merely have to change the second part of the rules.

For the mutual exclusion problem in our example, we have to consider the simultaneous requests of two processes among six parallel processes able to make requests to the robot : C_6^2 conflicts can appear (in fact, only thirteen because two of them are indirectly solves by the rules defined above). The solution that we chose gives priority to the exit of pieces compared with the entering and to certain transfer processes compared with others on account of the lay out of the workshop (R3-R5-R7).

So, we have :

R3 : IF STATE (M1 ⟶ FIFO-OUT.WAIT-R)
∧ STATE (M2 ⟶ FIFO-OUT.WAIT-R)
THEN FREEZE (M1 ⟶ FIFO-OUT) EOR

R4 : IF STATE (X.WAIT-R) ∧ STATE (Y.WAIT-R)
THEN FREEZE (Y) EOR
with X = M1 ⟶ FIFO-OUT OR M2 ⟶ FIFO-OUT
Y = FIFO-IN ⟶ M1 OR FIFO-IN ⟶ M2
OR M1 ⟶ M2 OR M2 ⟶ M1

R5 : IF STATE (M1 ⟶ M2.WAIT-R)
∧ STATE (M2 ⟶ M1.WAIT-R)
THEN FREEZE (M2 ⟶ M1) EOR

R6 : IF STATE (Z.WAIT-R) ∧ STATE (T.WAIT-R)
THEN FREEZE (Z) EOR
with Z = FIFO-IN ⟶ M1 OR FIFO-IN ⟶ M2
T = M2 ⟶ M1 OR M1 ⟶ M2

R7 : IF STATE (FIFO-IN ⟶ M1.WAIT-R)
∧ STATE (FIFO-IN ⟶ M2.WAIT-R)
THEN FREEZE (FIFO-IN ⟶ M2) EOR

This set of rules was given to solve the access conflict between two processes for a single resource. But this set of rules has proved sufficient to solve all the conflicts, whatever the number of simultaneous requests and available resources.

3.3. The different working modes

The automatic mode is completely modelized by the command graph, except for the priority rules which have be given in the upper hierarchical level to get a better flexibility.

Within the framework of our example we shall study the modelization

of two different modifications of the working modes :
- the increase or decrease in the number of operational resources after an operator decision,
- the transition from a production with two machines to a production with only one machine (for example M1).

For the first case, the operator activates two commands (ADD-R) and (RELEASE-R). These commands are the interface between the operator and the hierarchical level. One resource can be added to the system without control. However, the hierarchical level has to hold all the processes during the release and has to verify that at least one resource is free.

Hence, we have two rules for the resources handling.

R8 : IF (ADD-R) THEN ADD(R,1) EOR

R9 : IF (RELEASE-R) ∧ STATE (R)
THEN FREEZE (FIFO-IN ⟶ M1.WAIT-R,
FIFO-IN ⟶ M2.WAIT-R, M1 ⟶ M2.WAIT-R,
M2 ⟶ M1.WAIT-R, M1 ⟶ FIFO-OUT.WAIT-R,
M2 ⟶ FIFO-OUT.WAIT-R)
SUBST (R,1)
UNFREEZE (FIFO-IN ⟶ M1.WAIT-R,
FIFO-IN ⟶ M2.WAIT-R, M1 ⟶ M2.WAIT-R,
M2 ⟶ M1.WAIT-R, M1 ⟶ FIFO-OUT.WAIT-R,
M2 ⟶ FIFO-OUT.WAIT-R)
EOR

For the second case, the operator decides with the command (REQ-STOP-M2), when to stop the machining station M2. We then have to make sure that the machine is waiting for a piece (M((M2).(NOP)) = 1), that the transfer benches are empty (M(T-IN2-EMPTY) = 1, M(T-OUT2-EMPTY) = 1) and that, above the machine M2, the system has no colour that could be unified with variable v (dom(v) = $\{m_2, m_2m_1, m_1m_2\}$). In fact, such colours characterize the need to use the machine M2 and could block the system, if M2 was stopped without precautions.

These restrictions give the following rule.

R10 : IF (REQ-STOP-M2) ∧ STATE (M2.NOP) ∧ STATE (T-IN2-EMPTY)
∧ STATE (T-OUT2-EMPTY) ∧ NOT-STATE (M1,m1m2)
∧ NOT-STATE (FIFO-IN ⟶ M1,m1m2)
∧ NOT-STATE (T-IN1-FULL,m1m2)
∧ NOT-STATE (T-OUT1-FULL,m1m2)
THEN FREEZE (M2) EOR

So, it can clearly be seen that only the colouration enables the moment to stop the machine and to block its graph to be efficiently found. The piece colour in fact specifies its program (for all places we know the colour of the pieces that will pass through these places). So, we only consider the places affected by the studied colours and the scrutiny of the graph becomes partial and therefore faster. In our example, the studied colours are m_2, m_1m_2 and m_2m_1 ; the places T-IN1-

FULL and T-OUT1-FULL and the processes (FIFO-IN ⟶ M1) and (M1) are scrutinized (N.B. : processes (FIFO-IN ⟶ M2) and (M1 ⟶ M2) are not studied, because if the place T-IN2-EMPTY is marked then the places (FIFO-IN ⟶ M2).(NOP) and (M1 ⟶ M2).(NOP) are marked and these processes therefore do not contain the studied colours).

3.4. Error processing

We shall here discuss misfunction analysis and not its detection. Let us suppose therefore that the control-system will give all information about misfunctions detected at this level (for example, a time-out on an action).

We shall hold, within the framework of our example, two kinds of misfunctions, one concerning a resource, the other involving the machine M2.

- Let us first suppose that an impairment of a resource can be detected when not in service (either by usual control or automatic control). A fault is generated which becomes a latent error. To recover it, we just need to release this resource from the system : activation of the rule R9 by assigning the TRUE value to the predicate (RELEASE-R).
- As a second assumption, an impairment of a resource can be detected when used (either by visual control or a time-out controlling the transfert time).

Ex : A failure is generated during the transfert process (FIFO-IN ⟶ M2). The backward recovery rule is the following.

R11 : IF (FAILURE (FIFO-IN ⟶ M2))
THEN RESET (FIFO-IN ⟶ M2)
FORCING (T-IN2-EMPTY,1) EOR

The resource is no longer available : it is out of order and the piece is lost (it can be blocked on the robot).

Note : Since we do not know where the token is (time-out on the critical section), we have to reset the two places of this critical section.

The breakage of a machine M2 tool (this is a fault) happens during the machining :

- the piece is recoverable : this is an effective error,
- the piece is not recoverable : this is a failure,
- the machine M2 is out of order : this is a failure too.

The method retained for the two first assumptions is the following :

- stopping the machine M2,
- operator intervention to decide between a forward recovery (by the command (ACK-EFFECTIVE-ERROR-M2)) and a backward recovery (by the command (ACK-FAILURE-M2)),
- recovery with a new tool.

R12 : IF (ACK-FAILURE-M2)
THEN RESET (M2) EOR

R13 : IF (ACK-EFFECTIVE-ERROR-M2)
THEN RECOVERY (M2) EOR

In the third case, we have to avoid a blocking of all the processes : in fact, a m_1m_2 coloured piece can stay indefinitely in the place T-OUT1-FULL and prevent a degraded mode with the machine M1 only.

The held recovery method is the following :

- the operator has to assume that no more pieces of colour m_2, m_1m_2 and m_2m_1 are introduced into the system,
- we have to extend the <z> called colour domain from $\{m_1,m_2m_1\}$ to $\{m_1,m_2m_1,m_1m_2\}$. Thus the m_1m_2-coloured pieces, which were in the system when the failure appeared can leave the place T-OUT1-FULL without passage on the process M_2.

R14 : IF (ACK-OUT-OF-ORDER-M2)
THEN DOMAIN (<z>,m1,m2m1,m1m2) EOR

It can be seen that the colouration enables the system behaviour to be easily modified.

CONCLUSION

In this paper, we have introduced a 3-level approach for analysis and synthesis of flexible system of production.

Adaptive coloured and structured Petri-nets are used to modelize the "control system" and its different communications with the operative system and the hierarchical level. Structuration has been defined in terms of a restricted list of primitives used to specify the "control system" with a precise and non-interpretative meaning. Colouration is used to mark the different classes of products in the system and also to achieve more rapidly the active links between processes and the active processes. Flexibility is the essential feature of adaptive nets which can be controlled from the hierarchical level.

It is more adequate to model this last level in a declarative way which guarantees any "a posterieuri" modification of this level by the user giving the correct requirements of flexibility to the operative systems.

Different kinds of primitives of detections of faults in the control system well linked to the hierarchical level enable "rules of diagnosis" to be defined as in an expert diagnosis system, in a declarative way and from the experience of the user.

The example of the paper gives an illustration of the interest of such an approach.

This paper introduces a possible modelization of control of transient conditions in flexibility of production and leads the control to wait or to force the system in a given state in order to ensure without fault the change of working conditions.

For flexible systems, the dynamic adaptation of the operative part both must be realized in real time from an efficient adaptive control and decisions upon this control by the hierarchical level.

ANNEX : HIERARCHICAL LEVEL RULES

A.1. HIERARCHICAL LEVEL ⟶ CONTROL SYSTEM

A.1.1. Rules which do not modify the marking

GLOBAL FREEZE ⟺ entire system freeze
FREEZE (<processes identifiers list>)
⟺ processes freeze
FREEZE (<users list>)
⟺ blocks and/or waiting places freeze
GLOBAL UNFREEZE
⟺ entire system unfreeze
UNFREEZE (<processes identifiers list>)
⟺ processes unfreeze
UNFREEZE (<users list>)
⟺ blocks and/or waiting places unfreeze
DOMAIN (<colour domain>,<list of colours>)
⟺ colour domain modification

A.1.2. Rules which modify the marking

FORCING (<connection places identifiers>,<number>)
ADD (<connection places identifiers>,<number>)
REMOVE (<connection places identifiers>,<number>)

A.1.3. Error recovering rules

GLOBAL RESET ⟺ entire system reset
RESET (<processes identifiers list>)
RECOVERY (<process identifier>) ⟺ reset on block header

A.2. CONTROL SYSTEM ⟶ HIERARCHICAL LEVEL

ACTIVITY (<process identifier>)
is true if the process is unfrozen
ACTIVITY (<block identifier>)
is true is the block is unfrozen
ACTIVITY (<place identifier>)
is true is the place is unfrozen
STATE (<process identifier>,<colour>)
is true if the process is marked with the colour
STATE (<block identifier>,<colour>)
is true if the block is marked with the colour
STATE (<place identifier>,<colour>)
is true if the place is marked with the colour

BIBLIOGRAPHIE

CORBEEL D., 1979
"Schéma de cablage et schéma de contrôle. Application à la simulation et à la gestion de processus industriels"
Thèse de Doct. de Spécialité, Lille.

CORBEEL D., VERCAUTER C., GENTINA J.C., 1983-a
"Spécifications et conceptions des systèmes de conduite de processus temps réels"
Proc. of MIMI'83, Lugano, pp. 36-39.

CORBEEL D., VERCAUTER C., GENTINA J.C., 1983-b
"Réductions et propriétés structurelles des réseaux de Pétri structurés"
Proc. of MIMI'83, Lugano, pp. 40-43.

CORBEEL D., VERCAUTER C., GENTINA J.C., 1983-c
"Généralisation des réseaux de Pétri"
AI'83, Lille, Actes III, pp. 201-206.

CORBEEL D., VERCAUTER C., GENTINA J.C., 1984
"Adaptive Petri-nets for real time applications"
Proc. of Digitec'84, IMACS, Patras.

CORBEEL D., VERCAUTER C., GENTINA J.C., 1985-a
"Application of an extension of Petri-nets to modelization of control and production processes"
6th European Workshop on Appl. and Theory of Petri-Nets, Finland, pp. 53-74.

CORBEEL D., GENTINA J.C., VERCAUTER C., 1985-b
"Modelisation homogène du graphe de contrôle d'un système de conduite de processus industriels"
AFCET, Congrès Automatique 1985, Toulouse, pp. 517-533.

HACK M., 1975-a
"Petri-net language"
MIT Computation Structure Group, Memo 124.

HACK M., 1975-b
"Decision problems for Petri-nets and vector addition systems"
MAC Tech. Memo 59, MIT.

LAPRIE J.C., 1985
"Sûreté de fonctionnement des systèmes informatiques et tolérance aux fautes : concepts de base"
Techn. et Science Informatique, **4**, n° 5, pp. 419-429.

VALK R., 1978-a
"Self-modifying nets, a natural extension of Petri-nets"
ICALP, Lect. Notes in Computer Sc., n° 62, Springer, Berlin, pp. 464-476.

VALK R., 1978-b
"On the computational power of Petri-nets"
MFCS, Lect. Notes in Computer Sc., n° 64, Springer, Berlin, pp. 526-535.

VALK R., 1981
"Generalization of Petri-Nets"
MFCS, Lect. Notes in Computer Sc., n° 118, Springer, Berlin, pp. 140-155.

PART 3: SYSTEM RELIABILITY ANALYSIS AND APPLICATIONS

NEW RESULTS ON THE ANALYSIS AND OPTIMIZATION OF A k-OUT-OF-n:F SYSTEM WITH REPAIR

S.Tzafestas*, C.Botsaris†, A.Philippou¶
* National Technical University, Athens, Greece
† Airforce Academy, Tatoi, Greece
¶ University of Patras, Patras, Greece

ABSTRACT. In this paper we review some of the recent results on the reliability of the linearly ordered consecutive-k-out-of-n:F systems. We also examine repairable systems of this type and propose new models for their static and dynamic optimization. Potential applications are also discussed.

1. INTRODUCTION

The consecutive-k-out-of n:F system was first introduced by Chiang and Niu (1981) in connection with telecommunication and oil pipeline systems. Such a system consists of n($\geq$1) linearly or circularly ordered components which functions independently with probability p. The system fails if and only if k ($1 \leq k \leq n$) consecutive components fail.

In this paper we review briefly some results on the reliability of the linearly ordered consecutive-k-out-of-n:F systems. We also present the results of Philippou, Georghiou and Philippou (1983a, 1983b, 1985), Philippou and Makri (1985), and Philippou (1986) on Fibonacci and Fibonacci-type polynomials of order k, in conjunction with longest success runs, used to derive new closed formulas for the reliability of a consecutive-k-out-of-n:F system. Finally, we study the reliability properties, as well as the optimization of consecutive-k-out-of-n:F systems with repair.

2. RELIABILITY OF A CONSECUTIVE-k-OUT-OF-n:F SYSTEM

Let R (p;k,n) be the reliability of a consecutine-k-out-of-n:F system whose components are ordered linearly and function independently with probability p. Chiang and Niu (1981) proved that

$$R(p;2,n)=\sum_{i=0}^{[(n+1)/2]} \binom{n-i+1}{i} p^{n-1}q^{i} \quad ; \quad n \geq 2 \tag{1}$$

S. Tzafestas et al. (eds.), System Fault Diagnostics, Reliability and Related Knowledge-Based Approaches, Vol. 1, 299–307.

and

$$R(p;k,n)=\sum_{i=1}^{n-k+1}\sum_{j=i+1}^{i+k-1} R(p\ k,n-j)p^{i}q^{j-i}+p^{n-k+1},\quad n\geqslant k \tag{2}$$

They also obtained sharp bounds for $R(p;k,n)$ through the inequalities

$$(1-q^{k})^{n-k+1}\leqslant R(p;k,n)\leqslant (1-q^{k})^{[n/k]},\quad n\geqslant k \tag{3}$$

In the above $[x]$ denotes the greatest integer in x.

Derman, Lieberman and Ross (1982) derived the recursive formula

$$R(p;k,n)=\sum_{i=0}^{n} N(i,n-i+1;k-1)p^{n-1}q^{i},\quad n\geqslant k \tag{4}$$

where

$$N(i,r;1)=\begin{cases}\binom{r}{i}, & 0\leqslant i\leqslant r\\ 0, & i>r\end{cases} \tag{5}$$

and

$$N(i,r;m)=\sum_{j=0}^{r}\binom{r}{j}N(i-mj,r-j;m-1),\quad m\geqslant 2 \tag{6}$$

They also considered the case of unequally reliable components.

Bollinger (1982) expressed $R(p;k,n)$ in terms of the number $r_i^{(k)}$ of binary numbers of length n containing i ones, with at least k of these ones consecutive, and the entries of the Pascal triangle of order k. As he proved

$$R(p;k,n)=1-\sum_{i=k}^{n} r_i^{(k)}p^{n-1}q^{i},\quad n\geqslant k \tag{7}$$

where the coefficients $r_i^{(k)}$ are given by

$$r_i^{(k)}=\binom{n}{i}-c_k(n-i+1,i),\quad k\leqslant i\leqslant n \tag{8}$$

and c_k (ℓ,m) is the (ℓ,m) entry in the Pascal triangle of order k.

Finally, Shanthikumar (1982) developed an efficient recursive algorithm for computing the reliability $R(p;k,n)$ of a consecutive-k-out-of-n:F system with unequal component reliabilities. In the case of independent identically distributed components, Shanthikumar's algorithm reduces to

$$R(p;k,n)=R(p;k,n-1)-R(p;k,n-1-k)pq^{k},\quad n\geqslant 2k+1 \tag{9}$$

and

$$R(p;k,n)=1-q^{k}-(n-k)pq^{k},\quad k\leqslant n\leqslant 2k \tag{10}$$

3. DISTRIBUTIONS AND FIBONACCI - TYPE POLYNOMIALS OF ORDER k

Motivated by the work of Philippou and Muwafi (1982), Philippou, Georghiou and Philippou (1983a) introduced the geometric distribution of order k,

$$p(X=n)=p^n \sum_{\substack{n_1,\ldots,n_k \ni \\ n_1+2n_2+\ldots+kn_k=n-k}} \binom{n_1+\ldots+n_k}{n_1,\ldots,n_k} \left(\frac{q}{p}\right)^{n_1+\ldots+n_k}, \quad n\geq k \quad (11)$$

and derived from its study the negative binomial distribution of order k,

$$p(X=n)=p^n \sum_{\substack{n_1,\ldots,n_k \ni \\ n_1+2n_2+\ldots+kn_k=n-kr}} \binom{n_1+\ldots+n_k+r-1}{n_1,\ldots,n_k,r-1} \left(\frac{q}{p}\right)^{n_1+\ldots+n_k}, n\geq kr \quad (12)$$

and the Poisson distribution of order k,

$$p(X=n)=e^{-k\lambda} \sum_{\substack{n_1,\ldots,n_k \ni \\ n_1+2n_2+\ldots+kn_k=n}} \frac{\lambda^{n_1+\ldots+n_k}}{n_1!\ldots n_k!} \quad n\geq 0 \quad (13)$$

For a study of the latter two we refer to Philippou (1983, 1984)Another definition of Philippou, Georghiou and Philippou (1983b) is that of Fibonacci polynomials of order k, $\{f_n^{(k)}(x)\}_{n=0}^{\infty}$, where

$$f_n^{(k)}(x)=\begin{cases} \sum_{i=1}^{n} x^{k-i} f_{n-1}^{(k)}(x), & 2\leq n\leq k+1 \\ \sum_{i=1}^{k} x^{k-i} f_{n-i}^{(k)}(x), & n\geq k+2 \end{cases} \quad (14)$$

with $f_0^{(k)}(x)=0$ and $f_1^{(k)}(x)=1$. In a subsequent paper Philippou, Georghiou and Philippou (1985) defined the sequence, $\{F_n^{(k)}(x)\}_{n=0}^{\infty}$, of Fibonacci-type polynomials of order k, where

$$F_n^{(k)}(x)=\begin{cases} x\{F_{n-1}^{(k)}(x)+\ldots+F_1^{(k)}(x)\}, & 2\leq n\leq k+1 \\ x\{F_{n-1}^{(k)}(x)+\ldots+F_{n-k}^{(k)}(x)\}, & n\geq k+2 \end{cases} \quad (15)$$

and $F_0^{(k)}(x)=0$, $F_1^{(k)}(x)=1$. The authors also derived two closed formulas for computing $F_{n+1}^{k}(x)$, namely:

$$F_{n+1}^{(k)}(x)=\sum_{\substack{n_1+..+n_k \ni \\ n_1+2n_2+..+kn_k=n}} \binom{n_1+..+n_k}{n_1,..,n_k} x^{n_1+..+n_k}, \quad n \geqslant 0 \qquad (16)$$

and

$$F_{n+1}^{(k)}(x)=(1+x)^n \sum_{i=0}^{[n/(k+1)]} (-1)^i \binom{n-ki}{i} x^i (1+x)^{-(k+1)i}$$
$$-(1+x)^{n-1} \sum_{i=0}^{[(n-1)/(k+1)]} (-1)^i \binom{n-1-ki}{i} x^i (1+x)^{-(k+1)i}, \quad n \geqslant 1 \qquad (17)$$

As it is shown, the sequence of Fibonacci-type polynomials is quite appropriate when dealing with longest runs. Indeed, let L_n be the random variable denoting the length of the longest run of successes in $n(\geqslant 1)$ Bernoulli trials. Philippou and Makri (1985) proved that the probability distribution of L_n can be expressed in terms of the Fibonacci-type polynomials of order k, according to the formula

$$p(L_n \leqslant k)=\frac{p^{n+1}}{p} F_{n+2}^{(k+1)}(q/p), \quad 0 \leqslant k \leqslant n \qquad (18)$$

4. NEW RESULTS ON RELIABILITY OF A CONSECUTIVE-k-OUT-OF-n:F SYSTEM

The probability distribution of the length L_n of the longest run of successes in $n(\geqslant 1)$ Bernoulli trials, along with the expansions (16) and (17) for the Fibonacci-type polynomial of order k, was employed by Philippou (1986) to derive new formulas for the reliability of a linearly ordered consecutive-k-out-of-n:F system, two of which are closed. According to these derivations

$$R(p;k,n)=\frac{q^{n+1}}{p} F_{n+2}^{(k)}(p/q), \quad n \geqslant k \qquad (19)$$

$$R(p;k,n)=\frac{q^{n+1}}{p} \sum_{\substack{n_1,..,n_k \ni \\ n_1+2n_2+..+kn_k=n+1}} \binom{n_1+..+n_k}{n_1,..,n_k} \left(\frac{p}{q}\right)^{n_1+..+n_k}, \quad n \geqslant k \qquad (20)$$

and

$$R(p;k,n)=\frac{1}{p}\sum_{i=0}^{[(n+1)/(k+1)]}(-1)^i\binom{n+1-ki}{i}(pq^k)^i$$

$$-\frac{q}{p}\sum_{i=0}^{[n/(k+1)]}(-1)^i\binom{n-ki}{i}(pq^k)^i,\quad n\geq k \qquad (21)$$

Expression (19) combined with Bollinger's formula (7) for $R(p;k,n)$ establishes the relation

$$F_{n+2}^{(k)}(p/q)=(p/q)^{n+1}\sum_{i=0}^{n}c_k(n-i+1,i)(q/p)^i,\quad n\geq k \qquad (22)$$

between the Fibonacci-type polynomials of order k and the entries of the Pascal triangle of order k.

Moreover, using agian (19) Philippou (1986) offered an alternative proof of Shanthikumar's recursions (9) and (10).

5. CONSECUTIVE k-OUT-OF-n:F SYSTEMS WITH REPAIR

The consecutive k-out-of-on-n : F systems considered in this section are repairable in the sense that their components are repairable items. We assume that the times to failure of the operating components are independent identically distributed random variables and have an exponential distribution function with mean $1/\lambda$. Each of the linearly ordered components in the system is associated with a stock of (r-1) similar components in standby, ready to replace failed components (Fig. 1). Following its failure, the defective component is repaired and put back to the stock of standby components. The repair times of the components are independent identically distributed random variables and have an exponential distribution function with mean $1/\mu$. Each component in the system and the associated with stock of standby components will be called "subsystem".

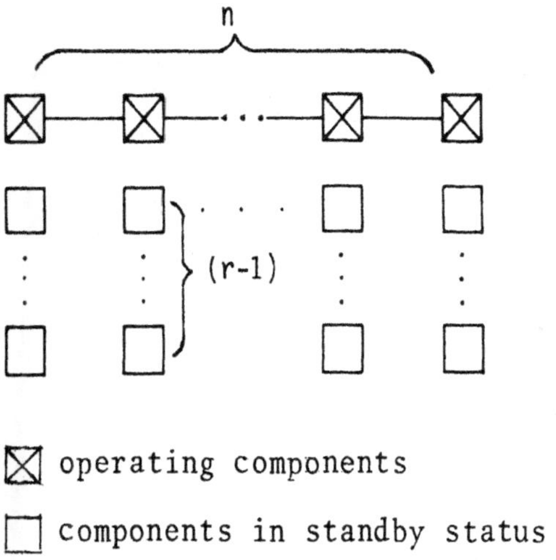

Fig.1.

As may be found in any standard handbook on reliability, e.g. Kozlov and Ushakov (1970), the probability of failure, q,of a subsystem is given in this case by

$$q=\frac{\rho^r}{r!}/\sum_{\ell=0}^{r}\frac{\rho^\ell}{\ell!},\quad \rho=\lambda/\mu \tag{23}$$

Hence, the system reliability is a function of λ,μ,n,k and r. Moreover, the number n of linearly ordered components and the number k of consecutive subsystems required to fail in order to occur a system breakdown, depend both on the quality and the operational characteristics of the components and, consequently, on their price.

It follows now that when the system is to be designed subject to a budget constraint b, we are facing the problem of how to allocate the amount b between the linearly ordered components and the components in standby status.

The above argument suggests that the optimal configuration of the system is determined by solving the non-linear programming problem

$$\left.\begin{array}{rl} \max & R(r,\lambda,\mu;n,k) \\ \text{subject to} & crn\leqslant b \\ & g_j(r,\lambda,\mu;n,k)\leqslant 0,\ j=1,..,J \end{array}\right\} \tag{24}$$

where c is the price (cost) of each component. We recall that not only λ and μ, but n and k are also functions of c. Moreover, k should also depend on the number n of linearly ordered components. Assuming that these functions are given analytically, the decision variables of the problem are r and c. Note that even in this case, a suitable non-linear integer programming algorithm is required in order to derive the optimal solution. The second set of constraints in (24) expresses all other physical constraints imposed to the system, such as operational and lay-out characteristics.

The problem considered above refers to the optimization of the system in equilibrium. Other models of this type, where the decision variables are the reliability parameters λ and μ, and the number r of the components in each subsystem, can also be formulated. Motivated by the work of Tzafestas (1982), Botsaris and Tzafestas (1984, 1985), Tzafestas and Botsaris (1984, 1986), and Botsaris and Fakinos (1986), we are now at the stage of developing static, as well as dynamic optimization models for consecutive repairable systems with components having a modular design and with repair facilities subject to breakdowns.

6. POTENTIAL APPLICATIONS

Let us consider the problem of illuminating a highway segment using electric lamps which are to be selected among various qualities (Fig.2). The lamps are subject to random failures which occur in accordance to a Poisson process of rate λ. Denoting by u the quality of the lamps, it is clear that λ should be some decreasing function $\lambda=\lambda(u)$ of u. A defective lamp is replaced by an operative one. We assume that the replacement time is independent of u and follows an exponential distribution with mean $1/\mu$. In this case, formula (23) reduces to (r=1):

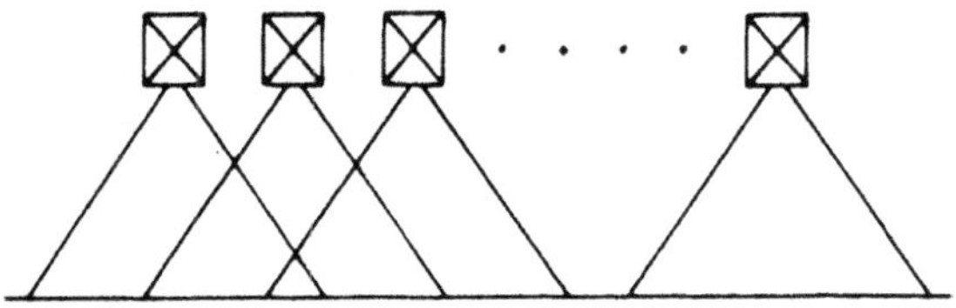

Fig.2.

$$q(u)=\frac{\lambda(u)}{\lambda(u)+\mu} \tag{25}$$

For the system under consideration a breakdown occurs whenever the illumination at any point on the highway drops below a given level. Clearly, the number k of consecutive lamps required to fail so that a system breakdown occurs, is some increasing function $k=k(n,u)$ of the number n of linearly ordered lamps and their quality u.

On the other hand, the minumum number n_{min} of lamps needed to achieve the required illumination is some decreasing function $n_{min}=n_{min}(u)$ of the quality of the lamps.

Recognizing that the cost c of each lamp is an increasing function $c=c(u)$ of this quality, model (24) assumes the form

$$\left.\begin{array}{rl} \max & R(u;n,k) \\ \text{subject to} & c(u)n \leq b \\ & k=k(u,n) \\ & n \geq n_{min}(u) \end{array}\right\} \tag{26}$$

In the following we propose an iterative method for solving the above problem when u is a discrete variable and the functions $c(u)$, $k(n,u)$ and $n_{min}(u)$ are given in a tabular form:

Let $n^*(u)$ be the optimal value of n for a given quality u. Then, we must have

$$R(u;n^*(u),\ k(n^*(u),u)) \geqslant R(u;n^*(u)-1,\ k(n^*(u)-1,u)) \quad (27)$$

and

$$R(u;n^*(u),\ k(n^*(u),u)) \geqslant R(u;n^*(u)+1,\ k(n^*(u)+1,u)) \quad (28)$$

In order to determine $n^*(u)$ for each quality u, $R(u;n,k)$ as given by (21), should be computed for $n=n_{min}(u)$, $n_{min}(u)+1,\ldots$. until the inequalities (27) and (23) are satisfied, or until n attains its upper bound $[b/c(u)]$. Having determined the optimal number of linearly ordered lamps for each quality, the overall optimal values n^* and u^* are chosen to be those for which $R(u;n,k)$ attains its maximum among the $R(u;n^*(u),\ k(n^*(u),u))$'s.

REFERENCES

1. Bollinger, R.C., 1982, Direct Computation for Consecutive-k-out-of-n:F Systems, IEEE Transactions on Reliability, Vol. R-31, No 5, 444-446.
2. Botsaris, C.A. and Fakinos, D., 1986, Analysis and Optimization of a Loss Service System with Servers Subject to Random Breakdowns, Journal of Information and Optimization Sciences, Vol. 7, No 2, 177-185.
3. Botsaris, C.A. and Tzafestas, S.G. 1984, Optimal Design of a Stanbdy System with Repair Facilities Subject to Breakdown, in Methods and Applications of Measurement and Control, Eds. S.G. Tzafestas and M.H. Hanza, Vol. 2, 628-631 (Anaheim: Acta Press).
4. Botsaris, C.A. and Tzafestas, S.G., 1985, Reliability Analysis and Optimization of a Repairable Standby System with Modular Items, Int. J.Systems Sci., Vo. 16, No 3. 313-323.
5. Chiang, D.T. and Niu. S., 1981, Reliability of Consecutive-k-out-of-n:F System, IEEE Transactions on Reliability, Vol. R-30, No 1, 87-89.
6. Kozlov, B.A. and Ushakov, I.A., 1970, Reliability Handbook (New York: Rinehart and Wilson).
7. Philippou, A.N., 1983, Poisson and Compound Poisson Distributions of Order k and Some of Their Properties (in Russian, English summary), Zapiski Nauchnykh Seminarov Leningradskogo Otdeleniya Matematicheskogo Instituta im V.A. Steklova AN SSSR 130, 175-180.
8. Philippou, A.N., 1984, The Negative Binomial Distribution of Order k and Some of Its Properties, Biometrical Journal 26, No. 7, 789-794.

9. Philippou, A.N., 1986, Distributions and Fibonacci Polynomials of Order k, Longest Runs, and Reliability of Consecutive-k-out-of-n:F Systems, in Fibonacci Numbers and Their Applications, Eds. A.N. Philippou, G.E. Bergum and A.F. Horadam, 203-227, (Dordrecht: D.Reidel).

10. Philippou, A.N., Georghiou, C. and Philippou, G.N., 1983a, A Generalized Geometric Distribution and Some of Its Properties, Statistics and Probability Letters, Vol. 1, No. 4, 171-175.

11. Philippou, A.N., Georghiou, C. and Philippou, G.N., 1983b, Fibonacci Polynomials of Order k, Multinomial Expansions, and Probability, International Journal of Mathematics and Mathematical Sciences, Vol. 6, No 3, 545-550.

12. Philippou, A.N., Georghiou, C. and Philippou, G.N.,1985, Fibonacci-Type Polynomials of Order k with Probability Applications, The Fibonacci Quarterly, Vol. 23, No. 2, 100-105.

13. Philippou, A.N. and Makri, F.S., 1985, Longest Success Runs and Fibonacci-Type Polynomials, The Fibonacci Quarterly, Vol. 23, No 4, 338-346.

14. Philippou, A.N. and Muwafi, A.A., 1982, Waiting for the kth Consecutive Success and the Fibonacci Sequence of order k, The Fibonacci Quarterly 20, No 1, 28-32.

15. Shanthikumar, G.J., 1982, Recursive Algorithm to Evaluate the Reliability of a Consecutive-k-out-of-n:F System, IEEE Transactions on Reliability, Vol. R-31, No 5, 442-443.

16. Tzafestas, S.G., 1982, Optimal Control Policies in System Reliability and Maintenance, in Optimization and Control of Dynamic Operational Research Models, Ed.S.G. Tzafestas (Amsterdam: North Holland).

17. Tzafestas, S.G. and Botsaris, C.A., 1984, Optimal Maintenance Policies for Modular Standby Systems, in Lecture Notes in Control and Information Sciences, Ed.P. Thoft. Christenseh, Vol. 59, 618-630 (Berlin: Springer-Verlag).

18. Tzafestas, S.G. and Botsaris, C.A., 1986, Reliability Measures and Optimal Maintenance Policies for a Standby System with Repair Facilities Subject to Breakdown, Journal of the Franklin Institute, Vol. 321, No 6, 309-323.

WEIBULL PARAMETER ESTIMATION ON A PERSONAL COMPUTER

Qiushi Cao and Leif Poulsen
Institute for Engineering Design
Technical University of Denmark
Building 421, DTH, DK-2800 Lyngby, Denmark

ABSTRACT. This paper gives a brief review of 5 basic methods for estimating parameter values in the 3-parameter Weibull distribution. The methods, which include Maximum Likelihood Estimators (MLE), Wyckoff, Bain and Engelhardts simple percentile estimators (WBE), Zanakis estimators (ZE), Modified Moment Estimators (MME) and emperical estimators suggested by Sen and Prabhashanker (SPE), have been evaluated by Monte Carlo Simulation on a mainframe computer. A wide parameter field with sample sizes from 10 to 80 was studied. Based on the simulation results the method with the best overall performance was chosen and a userfriendly parameter estimation program was written in BASIC to run on a Personal Computer.

1. INTRODUCTION

Reliability prediction for a new system is normally based on knowledge of actual life times of the different components or subsystems, which make up the new system. This knowledge can be achieved either by getting life time data from the field of users or by carrying through a special life time test.

By life time testing a certain number of items, a sample of N items, are tested under specific conditions and the individual life time is recorded whenever an item fails. If the test is continued until all items have failed, the sample is called complete. However, some items might be removed from the sample during the test or the test might be terminated before all items have failed; the sample is then called censored.

A normal procedure of censoring is to test N items until some fixed time x_o has elapsed (Type I Censoring) or until some fixed number of failures R has occured (Type II Censoring). In both cases, the test data consists of R life time observations $(x_1, x_2 \ldots\ldots x_R)$, where $(x_1 \leq x_2 \leq \ldots\ldots x_R)$ plus the information that (N-R) items survived beyond the time of termination (x_o or x_R).

S. Tzafestas et al. (eds.), System Fault Diagnostics, Reliability and Related Knowledge-Based Approaches, Vol. 1, 309–326.

In order to use the test results in a reliability prediction model a theoretical probability distribution has to be fitted to the life time data. Practise has shown that the Weibull distribution can fit almost all kinds of life time data. The density function for the 3-parameter Weibull distribution is denoted by:

$$f(x) = \frac{\beta}{\eta}\left(\frac{x-\gamma}{\eta}\right)^{\beta-1} \exp\left\{-\left(\frac{x-\gamma}{\eta}\right)^{\beta}\right\}, \tag{1}$$

$x \geqq 0 \; ; \quad \beta, \eta > 0$

where β is a shape parameter, η is a scale parameter (also called characteristic life time) and γ is a location parameter (also called minimum life time).

Several methods for estimating the parameters in this distribution have been proposed (see list of references). A very popular method used by reliability engineers is the probability paper |7|, which is an easy way of getting a rough estimate of the parameter values. The advantage of using this method is the graphical representation, which gives the reliability engineer opportunity to examine the data and check for out- layers and goodness of fit. However this method is only well suited for the special case where γ equals to zero; i.e. when dealing with a 2- parameter Weibull distribution and major errors may be introduced by the subjectivity in fitting the line to the plotted points. These disadvantages can be avoided by using analytical estimators.

2. ANALYTICAL PARAMETER ESTIMATORS

In this section 5 analytical methods for estimating the parameters in the 3-parameter Weibull distribution are briefly described; i.e. Maximum Likelihood Estimators (MLE) |2,10|, Wyckoff, Bain and Engelhardts simple percentile estimators (WBE) |24|, Zanakis estimators (ZE) |9|, Modified Moment Estimators (MME) |25| and emperical estimators suggested by Sen and Prabhashanker (SPE) |11|.

2.1. Maximum Likelihood Estimators (MLE)

The likelihood function for the 3-parameter Weibull distribution is denoted by

$$L\,(x_1,\ldots.x_N;\ \gamma,\ \beta,\ \eta) = \prod_{i=1}^{N} \frac{\beta}{\eta}\left(\frac{x_i-\gamma}{\eta}\right)^{\beta-1} \exp\left\{-\left(\frac{x_i-\gamma}{\eta}\right)^{\beta}\right\} \tag{2}$$

where x_i is observation number i and N is the sample size. After taking the logarithm of (2), differentiating with respect to β, η and γ in turn and equating to zero we obtain:

$$\frac{1}{\beta} + \frac{1}{N} \quad \Sigma\ln(x_i - \gamma) -\{ \Sigma (x_i - \gamma)^{\beta} \ln(x_i - \gamma) \}\{\Sigma(x_i - \gamma)^{\beta}\}^{-1} = 0 \qquad (3)$$

$$(1 - \beta) \Sigma(x_i - \gamma)^{-1} + \{ N \cdot \beta \quad \Sigma(x_i - \gamma)^{\beta-1} \}\{ \Sigma(x_i - \gamma)^{\beta} \}^{-1} = 0 \qquad (4)$$

$$\eta = \{ \frac{1}{N} \quad \Sigma(x_i - \gamma)^{\beta}\}^{1/\beta} \qquad (5)$$

The equations (3) and (4) can be solved for γ and β by using a modified quasilinearization method |2, 10|. Then the parameter η is estimated from (5).

The MLE-method can be applied not only to complete samples but also to censored samples |2, 10|.

2.2. Simple Unbiased Percentile Estimators (WBE)

In 1980, Wyckoff, Bain and Engelhardt |24| proposed the following simple unbiased percentile estimators:

$$\hat{\beta}_o = 2.989/\ln\{ (x_{[N \cdot P_k]+1} - x_1)/(x_{[N \cdot P_i]+1} - x_1)\} \qquad (6)$$

$$\hat{\gamma} = (x_1 - \bar{x}/N^{1/\hat{\beta}_o})/(1-1/N^{1/\hat{\beta}_o}) \qquad (7)$$

$$\hat{\beta} = N \cdot K_N/\{ - \sum_{i=1}^{S} \ln(x_i - \hat{\gamma}) + \frac{S}{N-S} \sum_{i=S+1}^{N} \ln(x_i - \hat{\gamma})\} \qquad (8)$$

$$\hat{\eta} = \exp\{ K_E/\hat{\beta} + \sum_{i=1}^{N} \ln(x_i - \hat{\gamma})/N\} \qquad (9)$$

where x_i is observation number i, N is the sample size, $P_i = 0.16731$, $P_k = 0.97366$, $[N \cdot P_i]$ and $[N \cdot P_k]$ denote the largest integer less than or equal to $N \cdot P_i$ and $N \cdot P_k$ respectively, $\bar{x}$ is the average of the observations, $S = [0.84 \cdot N]$ denotes the largest integer less then or equal to $0.84 \cdot N$, $K_E = 0.5772$ (Eulers constant) and K_N is a unbiased constant, which is tabulated in |21| as function of N in the range from 1 to 60.

The WBE-method can also be applied to censored samples; then the formulas should have a little change as shown in appendix A.

2.3. Zanakis Estimators (ZE)

Zanakis in 1979 suggested a set of simple estimators |9| as follows:

$$\hat{\gamma} = (x_1 \cdot x_N - x_2^2)/(x_1 + x_N - 2x_2) \tag{10}$$

$$\hat{\eta} = -\hat{\gamma} + x_{[0.63 \cdot N]+1} \tag{11}$$

$$\hat{\beta} = \ln\{ \ln(1-P_k)/\ln(1-P_i) \} / \ln\{ (x_{[N \cdot P_k]+1} - \hat{\gamma}) / (x_{[N \cdot P_i]+1} - \hat{\gamma}) \} \tag{12}$$

where P_i, P_k and the brackets are denoted as in section 2.2 above.

The great advantage of the ZE-method is its simplicity and accuracy when both β and sample size N are small.

2.4. Modified Moment Estimator (MME)

In 1984 Cohen et al |25| proposed the modified moment estimators as follows:

$$\frac{S^2}{(\bar{x} - x_1)^2} = \frac{\Gamma_2 - \Gamma_1^2}{\{(1-N^{-1/\hat{\beta}})\Gamma_1\}^2} \tag{13}$$

$$\hat{\gamma} = (N^{1/\hat{\beta}} x_1 - \bar{x})/(N^{1/\hat{\beta}} - 1) \tag{14}$$

$$\hat{\eta} = (N^{1/\hat{\beta}}(\bar{x} - x_1)/(N^{1/\hat{\beta}} - 1) \cdot \Gamma_1 \tag{15}$$

where $\bar{x} = \sum_{i=1}^{N} x_i/N$ and $S^2 = \sum_{i=1}^{N} (x_i - \bar{x})^2/(N-1)$, $\Gamma_k = \Gamma(1+k/\beta)$ and $\Gamma(\ )$ is the gammafunction denoted by $\Gamma(z) = \int_0^\infty t^{z-1} e^{-t}\, dt$.

Equation (13) has to be solved by some iterative procedure in order to get an estimate for β then subsequently γ and η can be estimated by equation (14) and (15).

2.5. Emperical Method (SPE)

Sen and Prabhashanker |11| have in 1980 described an emperical method as follows:

$$\hat{\eta} = \sigma / \{\Gamma(1+2/\beta) - \Gamma^2(1+1/\beta)\}^{\frac{1}{2}} \qquad (16)$$

$$\hat{\gamma} = U_1 - \hat{\eta}\Gamma(1+1/\beta) \qquad (17)$$

$$\hat{\beta} = Z + C_5 (Z^2 + C_6 Z + C_7) \qquad (18)$$

where $Z = C_1 \exp(C_2 \cdot Y) + C_3 \exp(C_4 \cdot Y)$, $Y = U_3/U_2^{3/2}$, U_2 and U_3 are second and third moment of the mean and σ is the standard deviation. The constants C_i for i= 1,7 are given in |11|.

3. EVALUATION OF ESTIMATORS BY SIMULATION

In order to study the efficiency of the 5 estimation methods described in section 2 a simulation program was set up. Complete samples in the range from 10 to 80 generated from Weibull distributions with typical parameter values were studied. In this section the simulation procedure is briefly outlined.

Let N denote the sample size. A sample of N uniform (0,1) distributed random numbers R_i, i = 1, N is generated. For every random number R_i a random life time from a Weibull distribution with given values of γ, η and β is generated according to inverse transformation by the formula:

$$x(i) = \gamma + \eta(-\ln(R_i))^{1/\beta} \qquad (19)$$

Then the generated life time values are ordered by using the Bobble Sorting method. We now have a complete samlpe which can be used to study the efficiency of different estimation methods.

All samples (size N = 10, 15, 20, 30, 50, 80) were generated from standardized Weibull distributions (as described above) with all the combinations of γ = 100, 200, 600, 1200, 2400; β = 0.5, 1.0, 1.5, 2.0, 3.0, 5.0 and η = 200, 400, 800, 1600, 3600. However the η-values always were bigger than the γ-values according to practical experiences.

The performance of the estimators considered were measured by BIAS and Mean Squared Error (MSE) in 100 replications. For BIAS the following equations were used:

$BIAS_{\gamma} = (\hat{\gamma}-\gamma)/\gamma$, $BIAS_{\beta} = (\hat{\beta}-\beta)/\beta$ and $BIAS_{\eta} = (\hat{\eta}-\eta)/\eta$. As equations for MSE the BIAS-equations were squared.

In order to get a well arranged and clear output from the program, a special index searching subroutine was employed. By using this routine it became easy to see which method were superior to others with respect to different criterias, as f.ex. performance by single parameter estimation (γ or η or β), double parameter estimation (γ and η, γ and β, or η and β) or triple parameter estimation (γ and η and β). The latter corrosponds to overall performance. In Table 1 (Appendix C) an example of output using the first mentioned criteria is given.

Another two types of indicators were used in order to give a better evaluation of the estimars: $\alpha = \eta/\gamma$ and $p = \hat{\eta}/\hat{\gamma}$.

4. RESULTS OF SIMULATION

Since the output files from the simulation content an enormous number of figures every single test result cannot be shown in this paper. However, based on the test results the following conclusions are drawn:

1) The performance of the estimators depend very much on the value of β. Considering overall performence ZE is the best for $\beta < 1$, WBE is the best for $1 < \beta < 3$ and MLE is the best for $\beta \geq 3$. This appereance seems to be independent of sample size (for $N > 10$).

2) Considering estimation of β when $N < 50$, ZE is the best for $\beta \leq 1$, WBE is the best for $1 \leq \beta \leq 3$ and MLE is the best for $\beta \geq 3$. When $N > 50$, ZE is the best for $\beta < 1.5$ otherwize MLE is the best. All methods seem to give an overestimate of β when β is small; for example $\hat{\beta}$ will be approximately 10% overestimated when $\beta = 1$. For larger β values $\hat{\beta}$ will normally be underestimated such as $\hat{\beta} = 2.4$ while $\beta = 3$. It is important to notice that WBE can only be reasonably applied for sample sizes less then 50 since peculiar results showed up for sample sizes above this value.

3) With respect to estimation of γ WBE is the best for $\beta < 3$. For $\beta \geq 3$ WBE will still be the best if $\alpha < 6$. For $\beta \geq 3$ and $\alpha > 6$ all methods perform badly ($\hat{p}$ will be much less than α), but MLE is the less worst. It is notable that MLE gives good estimates of β in this range where $\beta \geq 3$ and $\alpha > 6$.

4) Considering estimation of η WBE is the best for $\beta < 3$ otherwize MLE is the best.

5) Compaired with MLE and WBE, MME seems to be less efficient. However, MME has some advantages under certain conditions, i.e. when $1 \leq \beta \leq 2$ and $N \geq 30$, where it performs quite well.

6) SPE has a bad performance compaired with other methods and is not recommended.

Due to simplicity and the good overall performance of the WBE-method we have chosen this method as the basis for developing a userfriendly parameter estimation program to run on a personal computer.

5. A BASIC PARAMETER ESTIMATION PROGRAM

This section gives a short discription of a program for estimating parameter values in the 3-parameter Weibull distribution based on the estimators proposed by Wyckoff, Bain and Engelhardt |24|. Both complete and censored samples can be treated.

In order to reach a large number of users the program was written in BASICA, i.e. Advanced BASIC including graphical commands. The program, which is listed in Appendix B will run on an IBM PC or an IBM compatible PC.

As shown in the flow chart, fig. 1, the program is structured in two main parts; i.e. one part which can handle the numerical calculations and another which can give a graphical representation of the achieved results. If the user for some reason only wants to use the first part of the program, for example when only BASIC is available, it can be done by ignoring lines 1350 to 3900.

After having loaded the program and typed RUN the user will be prompted for input data, i.e. the number of tested items N, the number of failed items R, the total test time TS, and the individual life times T(I) for the R items, which have failed. These life times can be entered in random order since the program is able to sort the data in ascending order. After having answered all prompts, the input data will be redisplayed in correct ascending order. This gives the user the ability to check the input data. The program will then estimate the 3-parameter values using either equation (6) to (9) (complete samples) or (22) to (24) (censored samples). When using table values linear interpolation is used to get the most precise result. The estimated parameter values are used to calculate the mean and the variance of the distribution according to following equations:

$$\text{MEAN} = \hat{\gamma} + \hat{\eta} \cdot \Gamma(1/\hat{\beta} + 1) \tag{20}$$

$$\text{VAR} = \hat{\eta}^2 \; \{ \Gamma(2/\beta + 1) - \Gamma^2(1/\beta + 1)\} \tag{21}$$

where $\Gamma(\;)$ denotes the gammafunction as defined in section 2.4.

The results will then be displayed and the user will have to answer YES if he wants a density plot of a Weibull distribution with the calculated values. Fig. 2 is an example of such a density plot. The program will provide for a nice layout of the screen (scaling) and characteristic values as the minimum life time, γ and the mean life time, μ will be indicated.

Next and final step is a prompt for a Weibull plot. If the user wants it the program can provide a nice display of a traditional Weibull plot. This will give the user all the advantages of using a graphical method, i.e. possibility for checking for out-layers and for goodness of fit. Again scaling is used to give a nice layout and optimum use of the screen area. Fig. 3 shows an example of a Weibull plot generated by the program.

6. REFERENCES

(1) D.I. Gibbons & L.C. Vance, "A simulation study of estimators for the 2-parameter Weibull distribution". IEEE Transactions on reliability, Vol. R-30, No.1, pp61-66, April 1981.

(2) D.R.Wingo,"Solution of the three-parameter Weilbull equations by constrained modified quasilinearization (progressively censored samples)". IEEE Transactions on reliabillity, Vol.R-22, no.2, pp96-102, June 1973.

(3) M.V. Menon,"Estimation of the shape and scale parameters of the Weibull distribution". Technometrics, Vol. 5, No. 2, pp175-182, May 1963.

(4) K.Fung & A.K.S. Jardine, "Weibull parameter estimation". Microelectron Reliab. Vol. 22, No. 4. pp.681-684. 1982

(5) H.L.Harter & A.H.Moore,"Maxiimum-likelihood estimation of the parameters of Gamma and Weibull populations from complete and from censored samples", Technometrics, Vol. 7, No. 4. pp639-643, November, 1965.

(6) A.C.Cohen, "Maximum likelihood estimation in the Weibull distribution based on complete and on censored samples". Technometrics, VOl. 7, NO. 4. pp579-588. November, 1965.

(7) Nelson, Wayne and Thompson, V.C., "Weibull Probability Papers". Journal of Quality Technology, Vol. 3, NO. 2, pp45-50, April 1971.

(8) L.J. Ringer & E.E. Sprinkle, III, "Estimation of the parameters of the Weibull destribution from multicensored samples". IEEE Transactions on reliability, VOl. R-21, No. 1, pp46-51, February 1972.

(9) S.H. Zanakis, "A simulation study of some simple estimators for the three-parameter Weibull distribution". J. Statist. Comput. Simul.,Vol. 9, pp101-116, 1979.

(10) D.R. Wingo, "Maximum likelihood estimation of the parameters of the Weibull distribution by modified quasilinearization". IEEE Transactions on reliability, Vol. R-21, No. 2, pp89-93,May 1972.

(11) I.Sen & V. Prabhashanker, "A nomogram for estimating the three parameters of the Weibull distribution". Journal of Quality Technology, Vol. 12, NO. 3, pp138-143, July 1980.

(12) S. KuchII, N: Kaio & S. Osaki, "Simulation comparisons of point estimation methods in the 2-parameter Weibull distribution" Microelectron Reliab. Vol. 19. pp333-336. 1979.

(13) N.R. Mann & K.W. Fertig, "Simplified efficient point and interval estimators for Weibull parameters". Technometrics, Vol. 17, No. 3, pp361-368. August 1975.

(14) B.D. Bunday & I.Ai-Mutwali, "Direct optimisation for calculating Maximum likelihood estimates of parameters of the Weibull distribution". IEEE Transactions on reliability, Vol. R-30, No. 4, pp367-369. October 1981.

(15) M. Engelhardt, "On simple estimation of the parameters of the Weibull or extreme-value distribution". Technometrics, Vol. 17, No. 3, pp369-374. August 1975.

(16) G.H. Lemon. "Maximum likelihood estimation for the 3-parameter Weibull distribution based on censored samples. Technometries 17, pp247-254, 1975.

(17) G.R.Herd, "Estimation of the parameters of a population from a multi-censored sample," Iowa state College, Ames, Unpublished doctoral dissertation. 1956.

(18) N.R.Mann,R.E.Schafer & N.D. Singpurvalla, "Methods for statestical analysis of reliability and life data". Wiley, New York, 1974.

(19) K.C.Kapur & L.R. Lamberson, "Reliability engineering design", Wiley, New York, 1977.

(20) N.R.Mann, "Point and interval estimates for reliability parameters when failure times have the two-parameter Weibull distribution". Unpublished doctoral dissertation, Univ. of California at Los Angeles, 1965.

(21) M. Engelhardt & L.J. Bain, "simplified statistical procedures for the Weibull or extreme-value distribution". Technometrics 19, pp323-331, 1977.

(22) K.M.Hassanein, "Simultaneous estimation of the parameters of the extreme value destribution by simple quantiles". Technometrics 14, pp63-70, 1972.

(23) N.R. Mann & K.W. Fertig, "Efficient unbiased quantile estimators for moderate size complete samples from extreme value and Weibull distributions, Tecnometrics 15, pp87-93, 1977.

(24) Wyckoff, J., Bain, L., and Engelhardt, M. "Some complete and censored sampling results for the thre-parameter weibull distribution". Journal of Statistical Computation and simulation 11, pp.139-151, 1980.

(25) A.C. Cohen, B.J. Whitten and Y. Ding, "Modified moment estimation for the three-parameter Weibull distribution". Journal of quality technology, Vol. 16, No. 3, pp159-167, July 1984.

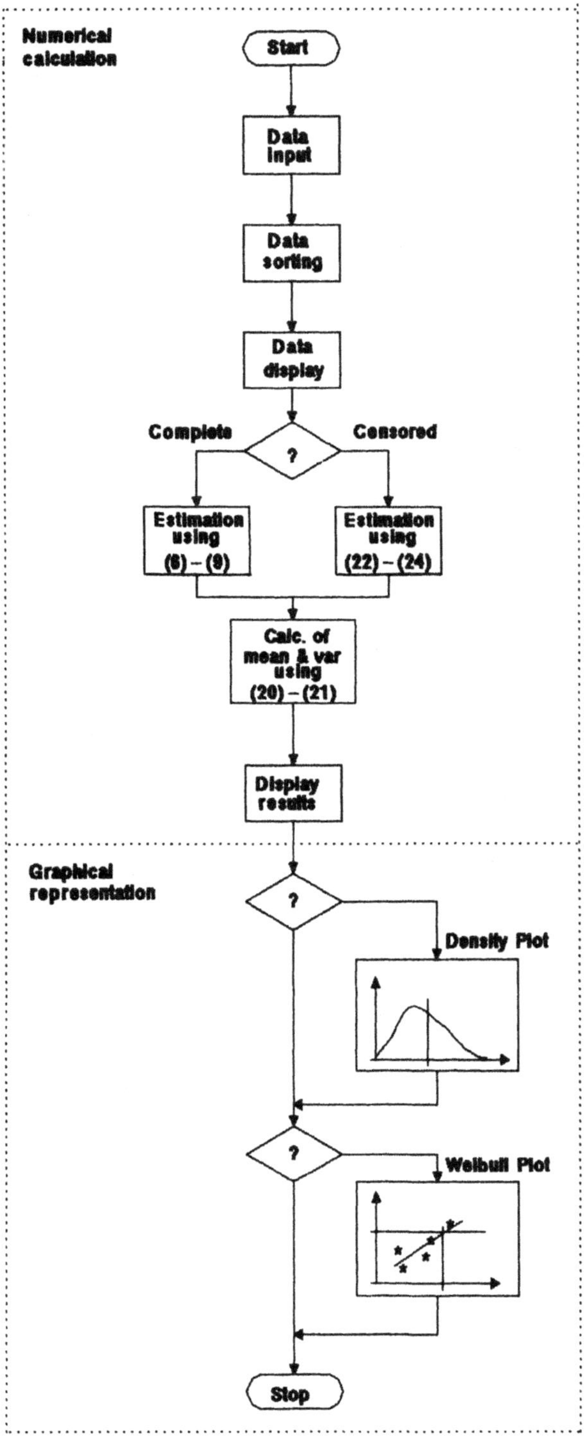

Fig. 1 Program flow chart

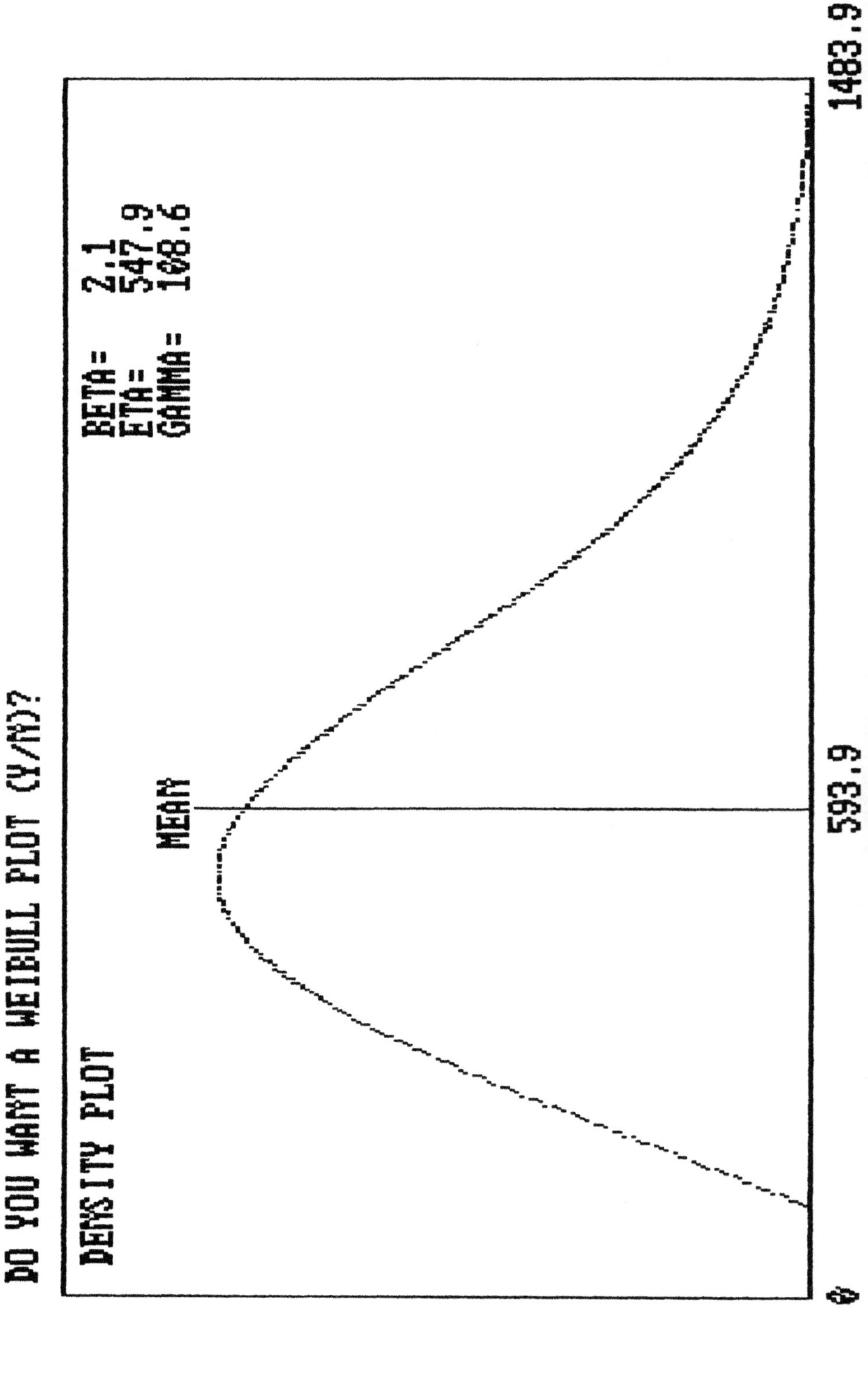

Fig. 2. Example of density plot

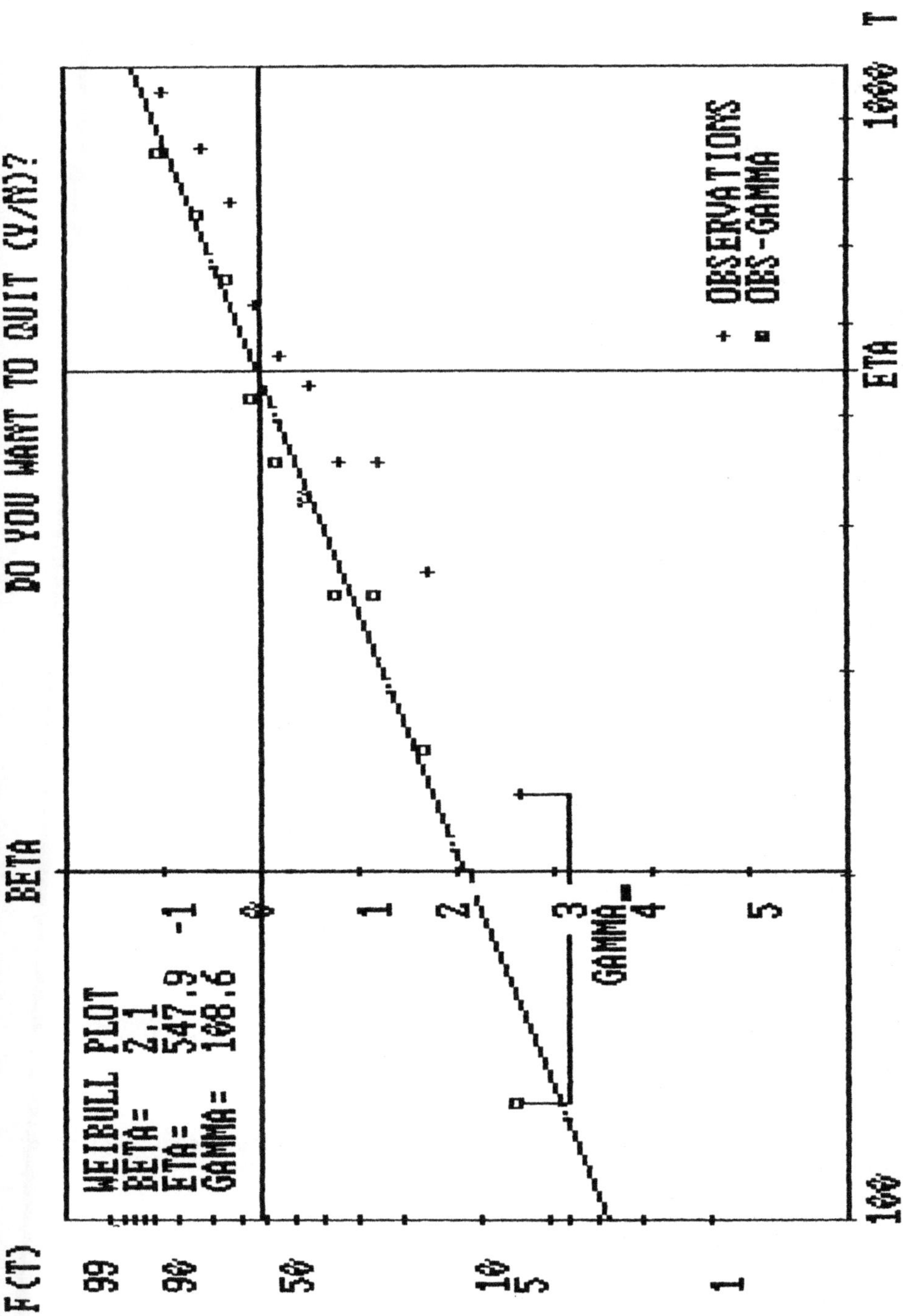

Fig. 3. Example of Weibull plot

Appendix A

WBE-estimators for censored samples.

Wyckoff, Bain and Engelhardt |24| have in 1980 proposed the following simple unbiased percentile estimators for censored samples of Type I or II:

$$\hat{\beta} = (N \cdot K_{R,N} - C_{R,N} + C_{1,N}) / \sum_{i=2}^{R-1} \{\ln(x_R - x_1) - \ln(x_i - x_1)\} \tag{22}$$

$$\hat{\gamma} = \frac{x_1 - x_R \quad (e_1/e_R)^{1/\hat{\beta}}}{1 - (e_1/e_R)^{1/\hat{\beta}}} \tag{23}$$

$$\hat{\eta} = (x_R - \hat{\gamma}) \cdot \exp\,(-C_{R,N}/\hat{\beta}) \tag{24}$$

where N is the sample size, R is the number of failed items and

$$e_i = -\ln(1 - i/(N + 1)),$$

$$C_{R,N} = E\ \{\ \ln\ ((x_R - \gamma)/\ \eta)^{\beta}\},$$

which can be approximately calculated as follows:

$$C_{1,N} = -\ 0.5772 - \ln(N),$$

$$C_{R,N} = c_p + a_1/N + a_2/N^2\ , \text{ and}$$

$$K_{R,N} = k_p + d_1/N + d_2/N^2\ ,$$

where the constants c_p, a_1, a_2, k_p, d_1 and d_2 are tabulated in |24| for ratios of R/N in the range of 0.1 to 0.9.

Appendix B

Program Listing

```
10 REM**********************************************************
20 REM
30 REM        WEIBULL PARAMETER ESTIMATION
40 REM        - 3 Parameter Weibull Distribution
50 REM        - Complete or Single Censored Samples
60 REM          of type I or type II
70 REM        - Sample size max. 60
80 REM
90 REM   Simple unbiased point estimators proposed by
100 REM  Wyckoff, Bain and Engelhardt (WBE) in 1980.
110 REM
120 REM  Programmer: Leif Poulsen    Date: 040286
130 REM
140 REM*********************************************************
200 REM***SCREEN LAYOUT*****************************************
210 CLS:SCREEN 0,0,0:WIDTH 80:COLOR 7,1,1:CLS:KEY OFF
220 PRINT:PRINT "WEIBULL-PARAMETER ESTIMATION":PRINT
230 REM***DATAINPUT*********************************************
240 DIM T(60)
250 INPUT "INPUT N-NO. OF SPECIMENS TESTED";N:PRINT
260 IF N<2 OR N>60 THEN 270 ELSE 280
270 PRINT "INVALID INPUT":PRINT:GOTO 250
280 INPUT "INPUT R-NO. OF SPECIMENS FAILED";R:PRINT
290 IF R<0 OR R>N THEN 300 ELSE 310
300 PRINT "INVALID INPUT":PRINT:GOTO 280
310 INPUT "INPUT TS-TIME TEST STOPPED";TS:PRINT
320 IF TS<0 THEN 330 ELSE 340
330 PRINT "INVALID INPUT":PRINT:GOTO 310
340 FOR I=1 TO R
350 INPUT "INPUT T(I)-FAILURE TIME";T(I)
360 IF T(I)<0 OR T(I)>TS THEN 370 ELSE 380
370 PRINT "INVALID INPUT":PRINT:GOTO 350
380 NEXT I
385 GOSUB 4500
390 REM***DATASORTING*******************************************
400 J=0
410 FOR I=2 TO R
420 IF T(I)<T(I-1) THEN 430 ELSE 440
430 X=T(I):T(I)=T(I-1):T(I-1)=X:J=J+1
440 NEXT I
450 IF J>0 THEN 400
460 PRINT:PRINT "I","T(I)"
470 FOR I=1 TO R
480 PRINT I, T(I)
490 NEXT I
500 PRINT
510 IF R<N THEN 820 ELSE 520
520 REM***COMPLETE SAMPLE***************************************
570 SUM=0
580 FOR I=1 TO N
590 SUM=SUM+T(I)
600 NEXT I
610 ET=SUM/N
620 PI=.16731:PK=.97366:KE=.5772
630 S=INT(.84*N):NPI=INT(N*PI)+1:NPK=INT(N*PK)+1
640 BO=2.989/LOG((T(NPK)-T(1))/(T(NPI)-T(1)))
650 G=(T(1)-ET/N^(1/BO))/(1-1/N^(1/BO))
655 IF G>T(1) THEN 656 ELSE 660
656 G=.9*T(1):PRINT "GAMMA = 0.9 * T(1)":PRINT
660 SUM=0
670 FOR I=1 TO S
680 SUM=SUM+LOG(T(I)-G)
690 NEXT I
700 SUM=-SUM
710 SS=S+1
720 FOR I=SS TO N
730 SUM=SUM+S*LOG(T(I)-G)/(N-S)
740 NEXT I
750 B=N*K(N)/SUM
760 SUM=0
770 FOR I=1 TO N
780 SUM=SUM+LOG(T(I)-G)
790 NEXT I
800 E=EXP(KE/B+SUM/N)
810 GOTO 1040
820 REM***CENSORED SAMPLE***************************************
830  DIM X(6)
854 IX=10*R/N
858 I1=INT(IX)
862 I2=I1+1
863 DX=IX-I1
866 IF I1=0 OR I1=9 THEN 886
870 FOR J=1 TO 6
874 X(J)=Z(I1,J)+DX*(Z(I2,J)-Z(I1,J))
878 NEXT J
882 GOTO 910
886 IF I1=0 THEN I=1
890 IF I1=9 THEN I=9
894 FOR J=1 TO 6
898 X(J)=Z(I,J)
902 NEXT J
910 C1N=-.5772-LOG(N)
920 CRN=X(1)+X(2)/N+X(3)/N^2
930 KRN=X(4)+X(5)/N+X(6)/N^2
940 E1=-LOG(1-1/(N+1))
950 ER=-LOG(1-R/(N+1))
960 J=R-1:SUM=0
970 FOR I=2 TO J
980 SUM=SUM+LOG(T(R)-T(1))-LOG(T(I)-T(1))
990 NEXT I
1000 B=(N*KRN-CRN+C1N)/SUM
1010 Y=(E1/ER)^(1/B)
1020 G=(T(1)-T(R)*Y)/(1-Y)
1030 E=(T(R)-G)*EXP(-CRN/B)
1040 REM***MEAN AND VARIANCE************************************
1050 X=1/B+1
1060 GOSUB 1130
1070 MEAN=G+E*GAMF
1080 Q=GAMF^2
1090 X=2/B+1
1100 GOSUB 1130
1110 VAR=(E^2)*(GAMF-Q)
1120 GOTO 1280
1130 REM***GAMMAFUNCTION****************************************
1140 A1=-.5748646
1150 A2=.9512363
1160 A3=-.6998588
1170 A4=.4245549
1180 A5=-.1010678
1190 Y=INT(X)
1200 Z=X-Y
1210 GAMF=1+A1*Z+A2*Z^2+A3*Z^3+A4*Z^4+A5*Z^5
1220 J=Y-1
1230 IF J=0 THEN 1270
1240 FOR I=1 TO J
1250 GAMF=GAMF*(I+Z)
1260 NEXT I
1270 RETURN
1271 REM***ROUNDING*********************************************
1272 X=INT(X)+CINT(10^DC*(X-INT(X)))/10^DC
1273 RETURN
```

```
1280 REM***DATAOUTPUT*******************************************
1290 PRINT "RESULTS:":PRINT
1295 DC=1:X=B:GOSUB 1271:B=X
1300 PRINT "SHAPE PARAMETER BETA=";B:PRINT
1305 DC=1:X=E:GOSUB 1271:E=X
1310 PRINT "SCALE PARAMETER ETA=";E:PRINT
1315 DC=1:X=G:GOSUB 1271:G=X
1320 PRINT "LOCATION PARAMETER GAMMA=";G:PRINT
1325 DC=1:X=MEAN:GOSUB 1271:MEAN=X
1330 PRINT "MEAN=";MEAN:PRINT
1335 DC=1:X=VAR:GOSUB 1271:VAR=X
1340 PRINT "VARIANCE=";VAR:PRINT
1350 INPUT "DO YOU WANT A PLOT OF DENSITY FUNCTION (Y/N)";A$:PRINT
1360 IF A$="y" OR A$="Y" THEN 1370 ELSE 1380
1370 GOSUB 2000:GOTO 1390
1380 INPUT "DO YOU WANT A WEIBULL PLOT (Y/N)";A$:PRINT
1390 IF A$="y" OR A$="Y" THEN 1400 ELSE 1410
1400 GOSUB 3000
1405 CLS:KEY ON
1410 END
2000 REM***PLOT OF DENSITY FUNCTION**********************************
2010 SCREEN 2,,0,0 :CLS
2020 LOCATE 4,9:PRINT "DENSITY PLOT";
2030 GOSUB 4000
2031 LOCATE 4,55:PRINT "BETA=  ";B;
2032 LOCATE 5,55:PRINT "ETA=   ";E;
2033 LOCATE 6,55:PRINT "GAMMA= ";G;
2040 XMAX=G+E*(-LOG(.001))^(1/B)
2050 DX=XMAX/540
2051 IF B<1 THEN 2052 ELSE 2060
2052 X=DX:GOTO 2070
2060 X = G+E* (1-1/B)^(1/B)
2070 YMAX = B/E* ((X-G)/E)^(B-1)*EXP(-1*((X-G)/E)^B)
2080 KY=.8*160/YMAX
2090 FOR I = 0 TO 540
2100 X = I*DX
2105 IF X<G THEN 2106 ELSE 2110
2106 Y=0:GOTO 2120
2110 Y = KY*B/E*((X-G)/E)^(B-1)*EXP(-1*((X-G)/E)^B)
2120 X=INT(X*540/XMAX)+50
2130 Y = 180-INT(Y)
2140 PSET (X,Y)
2150 NEXT I
2151 X1=INT(MEAN*540/XMAX+50):Y1=180:Y2=45
2152 LINE (X1,Y1)-(X1,Y2)
2153 X=INT(X1/8)-1:Y=6:LOCATE Y,X
2154 PRINT "MEAN";:Y=24:LOCATE Y,X:PRINT MEAN;
2155 LOCATE 24,7:PRINT "0";:XG=INT((G*540/XMAX+50)/8)-2
2156 IF XG>12 THEN 2157 ELSE 2158
2157 LOCATE 24,XG:PRINT G;
2158 LOCATE 24,72:DC=1:X=XMAX:GOSUB 1271:XMAX=X:PRINT XMAX;
2160 LOCATE 2,8:INPUT "DO YOU WANT A WEIBULL PLOT (Y/N)";A$
2180 SCREEN 0,0,0
2190 RETURN
3000 REM***WEIBULL PLOT*********************************************
3020 SCREEN 2,,0,0:CLS
3030 LOCATE 4,9:PRINT "WEIBULL PLOT";
3035 DIM TC(60), LT(60), LTC(60), MR(60)
3040 GOSUB 4000
3041 LOCATE 5,9:PRINT "BETA=  ";B;
3042 LOCATE 6,9:PRINT "ETA=   ";E;
3043 LOCATE 7,9:PRINT "GAMMA= ";G;
3050 FOR I = 1 TO R
3060 TC(I) = T(I)-G
3070 LT(I) = LOG (T(I))
3080 LTC(I) = LOG (TC(I))
3090 MR(I) = (I-.3)/(R+.4)
3100 NEXT I
3110 IF G<0 THEN 3130 ELSE 3120
3120 TMAX = T(R):TMIN = TC(1):GOTO 3140
3130 TMAX = TC(R):TMIN = T(1)
3140 EMAX=INT(LOG(TMAX)/LOG(10))+1
```

```
3150 EMIN = INT(LOG(TMIN)/LOG(10))
3160 ND = EMAX-EMIN
3170 TMAX = 10^EMAX:TMIN = 10^EMIN
3180 REM***X-AXIS AND Y-AXIS*****************************************
3190 FOR I=1 TO ND
3200 FOR J = 1 TO 9
3210 X=(I-1+LOG(J)/LOG(10))*540/ND
3220 X = INT(X)+50:Y = 180
3230 PSET (X,Y):DRAW "BU1 D2"
3240 IF J = 1 THEN 3250 ELSE 3270
3250 X=INT(X/8):Y=24:LOCATE Y,X
3260 X = TMIN*10^(I-1):PRINT X;
3270 NEXT J
3280 NEXT I
3290 LOCATE 24,70:PRINT TMAX;" T";
3300 LOCATE 2,2:PRINT "F(T)";
3420 FOR I = 1 TO 19
3440 Y=LOG(LOG(1/(1-Y(I)/100)))
3450 X = 50:Y = INT(180-20*(Y+6))
3460 PSET (X,Y):DRAW "BL2 R4"
3465 NEXT I
3470 FOR I=1 TO 6
3475 X=2:Y=LOG(LOG(1/(1-Y1(I)/100)))
3476 Y=INT((180-20*(Y+6))/8+1)
3480 LOCATE Y,X:PRINT Y1(I);
3490 NEXT I
3500 REM***PLOT OF POINTS*******************************************
3510 FOR I = R TO 1 STEP -1
3520 X = 540*(LT(I)-LOG(TMIN))/(ND*LOG(10))+50
3530 Y = LOG (LOG(1/(1-MR(I))))
3540 Y1= INT (180-20*(Y+6)):X1= INT(X)
3550 PSET (X1,Y1):DRAW "BL2 R4 BU1 BL2 D2"
3560 X = 540*(LTC(I)-LOG(TMIN))/(ND*LOG(10))+50
3570 X2= INT(X):PSET (X2,Y1):DRAW "BH2 D2 R4 U2 L4"
3580 NEXT I
3581 Y2=Y1+10:LINE(X1,Y2)-(X2,Y2)
3582 LINE (X1,Y1)-(X1,Y2)
3583 LINE (X2,Y1)-(X2,Y2)
3584 D=X1-X2:X=X1-D/2
3585 X=INT(X/8)-1:Y=INT(Y1/8)+3
3586 LOCATE Y,X:PRINT "GAMMA";
3590 PSET (464, 155):DRAW "BL2 R4 BU1 BL2 D2"
3600 LOCATE 20,61:PRINT "OBSERVATIONS";
3610 PSET (464, 163):DRAW "DH2 D2 R4 U2 L4"
3620 LOCATE 21,61:PRINT "OBS-GAMMA";
3630 REM***LINE FIT*************************************************
3640 X1=-6/B+LOG(E):Y1=-6
3645 IF X1<LOG(TMIN) THEN 3650 ELSE 3655
3650 X1=LOG(TMIN):Y1=B*(LOG(TMIN)-LOG(E))
3655 X1=INT(540*(X1-LOG(TMIN))/(ND*LOG(10))+50)
3660 Y1=INT(180-20*(Y1+6))
3665 X2=2/B+LOG(E):Y2=2
3670 IF X2>LOG(TMAX) THEN 3675 ELSE 3680
3675 X2=LOG(TMAX):Y2=B*(LOG(TMAX)-LOG(E))
3680 X2=INT(540*(X2-LOG(TMIN))/(ND*LOG(10))+50)
3685 Y2=INT(180-20*(Y2+6))
3690 LINE (X1,Y1)-(X2,Y2)
3700 REM***BETA SCALE***********************************************
3710 X1 = 540*(LOG(E)-1-LOG(TMIN))/(ND*LOG(10))+50
3720 X2 = INT(X1/8)-2
3730 Y1 = 18:Y2 = 180:X1 = INT(X1)
3740 LINE (X1, Y1)-(X1, Y2)
3750 FOR I = 1 TO 7
3760 Y = 20+I*20
3770 PSET (X1,Y):DRAW "BL2 R4"
3780 Y2 = INT (Y/8)+1
3790 LOCATE Y2,X2: J = I-2:PRINT J;
3795 NEXT I
3800 X2=X2+2:LOCATE 2,X2:PRINT "BETA";
3810 REM***ETA SCALE********************************************
3820 X1 = 50:X2 = 590
3830 Y1 = 60:Y2 = 60
```

```
3840 LINE (X1, Y1)-(X2, Y2)
3850 X1 =INT(540*(LOG(E)-LOG(TMIN))/(ND*LOG(10))+50)
3860 Y1 = 20:Y2 = 182
3870 LINE (X1, Y1)-(X1, Y2)
3880 X =INT(X1/8):Y=24
3890 LOCATE Y,X:PRINT "ETA";
3891 LOCATE 2,44:INPUT "DO YOU WANT TO QUIT (Y/N)";A$
3892 IF A$="y" OR A$="Y" THEN 3900 ELSE 3891
3900 RETURN
4000 REM***STANDARD SCREEN LAYOUT******************************
4010 LINE (50,180)-(50,20)
4020 LINE (50,20)-(590,20)
4030 LINE (590,20)-(590,180)
4040 LINE (590,180)-(50,180)
4080 RETURN
4500 REM***READ DATA*******************************************
4510 DIM K(60),Z(9,6),Y(19),Y1(6)
4520 FOR I=1 TO 60:READ K(I):NEXT I
4530 FOR I=1 TO 9:FOR J=1 TO 6:READ Z(I,J):NEXT J:NEXT I
4540 FOR I=1 TO 19:READ Y(I):NEXT I
4545 FOR I=1 TO 6:READ Y1(I):NEXT I
4550 RETURN
5000 DATA 0.0000, 0.6931, 0.9808, 1.1507, 1.2674
5010 DATA 1.3545, 1.1828, 1.2547, 1.3141, 1.3644
5020 DATA 1.4079, 1.4461, 1.3332, 1.3686, 1.4004
5030 DATA 1.4293, 1.4556, 1.4799, 1.3960, 1.4192
5040 DATA 1.4408, 1.4609, 1.4797, 1.4975, 1.5142
5050 DATA 1.4479, 1.4642, 1.4796, 1.4943, 1.5083
5060 DATA 1.5216, 1.4665, 1.4795, 1.4920, 1.5040
5070 DATA 1.5156, 1.5266, 1.4795, 1.4904, 1.5009
5080 DATA 1.5110, 1.5208, 1.5303, 1.4891, 1.4984
5090 DATA 1.5075, 1.5163, 1.5248, 1.5331, 1.5411
5100 DATA 1.5046, 1.5126, 1.5204, 1.5279, 1.5352
5110 DATA 1.5424, 1.5096, 1.5167, 1.5236, 1.5304
5120 DATA -2.2504, -0.55743, -0.07201, 0.10265, -1.0271, 0.000
5130 DATA -1.4999, -0.30740, -0.01886, 0.21129, -1.0622, 0.030
5140 DATA -1.0309, -0.22859, -0.00767, 0.32723, -1.1060, 0.054
5150 DATA -0.67173, -0.19301, -0.00335, 0.45234, -1.1634, 0.089
5160 DATA -0.36651, -0.17619, -0.00091, 0.58937, -1.2415, 0.145
5170 DATA -0.08742, -0.17114, -0.00111, 0.74274, -1.3540, 0.242
5180 DATA 0.18563, -0.17727, 0.00369, 0.92026, -1.5313, 0.433
5190 DATA 0.47589, -0.20110, 0.00891, 1.1382, -1.8567, 0.906
5200 DATA 0.83403, -0.27773, 0.02825, 1.4436, -2.6929, 2.796
5210 DATA 1,2,3,4,5,10,20,30
5220 DATA 40,50,60,70,80,90,95,96,97,98,99
5230 DATA 1,5,10,50,90,99
```

Appendix C

SMALL MEAN SQUARED ERRORS SEARCHING

IN THE FOLLOWING TABLE, THE NUMBER INDICATE THE NTH METHOD
J=1 INDICATE THE BEST METHOD J=5 INDICATE THE WORST METHOD

INPUT PARAMETERS				F(GAMMA(J))					F(ETA(J))					F(BETA(J))			
GAMMA	ETA	BETA	N	J= 1	2	3	4	5	J= 1	2	3	4	5	J= 1	2	3	4
				NUMBER OF THE METHOD					NUMBER OF THE METHOD					NUMBER OF THE METHOD			
100	200	0.5	30	2	1	4	5	3	1	2	4	5	3	1	2	4	3
100	200	1.0	30	1	4	2	3	5	1	4	2	3	5	2	1	4	3
100	200	1.5	30	1	4	2	5	3	1	4	2	5	3	2	1	4	5
100	200	2.0	30	1	2	5	4	3	1	2	5	4	3	1	2	5	4
100	200	3.0	30	1	5	2	3	4	1	2	5	3	4	1	2	5	3
100	200	5.0	30	1	5	4	2	3	1	5	4	2	3	1	5	4	2
100	400	0.5	30	2	1	4	5	3	2	1	4	5	3	1	2	4	5
100	400	1.0	30	1	4	2	5	3	1	4	2	5	3	1	2	4	5
100	400	1.5	30	1	4	2	5	3	1	5	4	2	3	1	2	4	5
100	400	2.0	30	1	5	2	4	3	1	5	4	2	3	1	2	5	4
100	400	3.0	30	1	5	2	3	4	1	5	2	3	4	1	2	5	3
100	400	5.0	30	5	1	2	3	4	5	1	2	3	4	1	5	2	3
100	800	0.5	30	1	2	4	5	3	1	2	4	5	3	1	2	4	5
100	800	1.0	30	1	4	2	5	3	1	4	2	5	3	1	2	4	5
100	800	1.5	30	1	5	2	4	3	5	1	2	4	3	2	1	5	4
100	800	2.0	30	1	5	2	4	3	1	5	2	4	3	1	2	5	4
100	800	3.0	30	5	1	2	3	4	5	1	2	3	4	1	5	2	3
100	800	5.0	30	5	1	2	4	3	5	1	2	4	3	5	1	2	4
100	1600	0.5	30	2	1	4	5	3	1	2	4	5	3	2	1	4	5
100	1600	1.0	30	1	2	4	5	3	1	4	2	5	3	2	1	4	3
100	1600	1.5	30	1	5	4	2	3	1	5	2	4	3	2	1	5	4
100	1600	2.0	30	1	5	2	4	3	1	5	2	4	3	1	2	5	4
100	1600	3.0	30	1	5	2	3	4	1	5	2	3	4	1	2	5	3
100	1600	5.0	30	5	1	2	3	4	5	1	2	3	4	5	1	2	3
100	3600	0.5	30	2	1	4	5	3	1	5	2	4	3	2	1	5	4
100	3600	1.0	30	5	1	2	4	3	5	4	1	2	3	5	2	1	4
100	3600	1.5	30	1	5	2	4	3	1	5	4	2	3	2	1	5	4
100	3600	2.0	30	1	5	2	4	3	1	5	2	4	3	2	1	5	4
100	3600	3.0	30	5	1	2	3	4	1	5	2	3	4	1	5	2	3
100	3600	5.0	30	5	1	2	3	4	5	1	2	3	4	5	1	2	3
200	400	0.5	30	2	1	4	5	3	1	2	4	5	3	1	2	4	3
200	400	1.0	30	1	4	2	3	5	1	4	2	3	5	2	1	4	3
200	400	1.5	30	1	4	2	5	3	1	4	2	5	3	2	1	4	5
200	400	2.0	30	1	2	5	4	3	1	2	5	4	3	1	2	5	4
200	400	3.0	30	1	5	2	3	4	1	2	5	3	4	1	2	5	3
200	400	5.0	30	1	5	4	2	3	1	5	4	2	3	1	5	4	2
200	800	0.5	30	2	1	4	5	3	2	1	4	5	3	1	2	4	5
200	800	1.0	30	1	4	2	5	3	1	4	2	5	3	1	2	4	5
200	800	1.5	30	1	4	2	5	3	1	5	4	2	3	1	2	4	5
200	800	2.0	30	1	5	2	4	3	1	5	4	2	3	1	2	5	4
200	800	3.0	30	1	5	2	3	4	1	5	2	3	4	1	2	5	3
200	800	5.0	30	5	1	2	3	4	5	1	2	3	4	1	5	2	3
200	1600	0.5	30	1	2	4	5	3	1	2	4	5	3	1	2	4	5
200	1600	1.0	30	1	4	2	5	3	1	4	2	5	3	1	2	4	5
200	1600	1.5	30	1	5	2	4	3	5	1	2	4	3	2	1	5	4
200	1600	2.0	30	1	5	2	4	3	1	5	2	4	3	1	2	5	4
200	1600	3.0	30	5	1	2	3	4	5	1	2	3	4	1	5	2	3
200	1600	5.0	30	5	1	2	4	3	5	1	2	4	3	5	1	2	4
200	3600	0.5	30	2	1	4	5	3	1	2	5	4	3	2	1	4	5
200	3600	1.0	30	1	2	4	5	3	1	4	5	2	3	5	2	1	4
200	3600	1.5	30	1	5	4	2	3	1	5	2	4	3	2	1	5	4
200	3600	2.0	30	1	5	2	4	3	1	5	2	4	3	1	2	5	4
200	3600	3.0	30	1	5	2	3	4	1	5	2	3	4	1	2	5	3
200	3600	5.0	30	5	1	2	3	4	5	1	2	3	4	5	1	2	3
600	800	0.5	30	2	1	4	5	3	1	2	4	5	3	1	2	4	3
600	800	1.0	30	1	4	2	3	5	1	4	2	3	5	2	1	4	3
600	800	1.5	30	1	4	2	3	5	1	4	2	3	5	2	1	4	5
600	800	2.0	30	1	2	4	5	3	1	2	4	5	3	1	2	5	4
600	800	3.0	30	1	2	5	3	4	1	2	5	3	4	1	2	5	3
600	800	5.0	30	1	5	4	2	3	1	5	4	2	3	1	5	4	2
600	1600	0.5	30	2	1	4	5	3	2	1	4	5	3	1	2	4	3
600	1600	1.0	30	1	4	2	3	5	1	4	2	3	5	1	2	4	3
600	1600	1.5	30	1	4	2	5	3	1	4	2	5	3	1	2	4	5
600	1600	2.0	30	1	5	2	4	3	1	4	2	5	3	1	2	4	5
600	1600	3.0	30	1	5	2	3	4	1	5	2	3	4	1	2	5	3
600	1600	5.0	30	5	1	2	3	4	5	1	2	3	4	1	5	2	3
600	3600	0.5	30	1	2	4	5	3	1	2	4	5	3	1	2	4	5
600	3600	1.0	30	1	4	2	5	3	1	4	2	5	3	1	2	4	5
600	3600	1.5	30	1	5	2	4	3	5	1	2	4	3	2	1	5	4
600	3600	2.0	30	1	5	2	4	3	1	5	2	4	3	1	2	5	4
600	3600	3.0	30	1	5	2	3	4	1	5	2	3	4	1	2	5	3
600	3600	5.0	30	5	1	2	4	3	5	1	2	4	3	5	1	2	4
1200	1600	0.5	30	2	1	4	5	3	1	2	4	5	3	2	1	4	3
1200	1600	1.0	30	1	2	4	3	5	1	4	2	3	5	2	1	4	3
1200	1600	1.5	30	1	4	2	5	3	1	2	4	5	3	2	1	4	5
1200	1600	2.0	30	1	2	5	4	3	1	2	5	4	3	1	2	4	5
1200	1600	3.0	30	1	2	5	3	4	1	2	5	3	4	1	2	5	3
1200	1600	5.0	30	1	2	5	3	4	1	2	5	3	4	1	2	5	3
1200	3600	0.5	30	2	1	4	5	3	1	2	4	5	3	2	1	4	5
1200	3600	1.0	30	1	2	4	3	5	4	1	2	3	5	2	1	4	3
1200	3600	1.5	30	1	2	4	5	3	1	4	5	2	3	2	1	4	5
1200	3600	2.0	30	1	2	5	4	3	1	2	5	4	3	2	1	5	4
1200	3600	3.0	30	1	5	2	3	4	1	5	2	3	4	1	2	5	3
1200	3600	5.0	30	5	1	2	3	4	5	1	2	3	4	1	5	2	3
2400	3600	0.5	30	2	1	4	5	3	1	2	4	5	3	1	2	4	3
2400	3600	1.0	30	1	4	2	3	5	1	4	2	3	5	2	1	4	3
2400	3600	1.5	30	1	4	2	5	3	1	4	2	5	3	2	1	4	3
2400	3600	2.0	30	1	2	4	5	3	1	2	5	4	3	1	2	5	4
2400	3600	3.0	30	1	2	5	3	4	1	2	5	3	4	1	2	5	3
2400	3600	5.0	30	1	5	4	2	3	1	5	4	2	3	1	5	4	2

RELIABILITY AND FAULT DIAGNOSIS METHODS OF POWER SYSTEM COMPONENTS

E.N. Dialynas A.V. Machias J.L. Souflis
National Technical University
Department of Electrical Engineering
Energy Systems Laboratory
42, 28th October Street, Athens
GR 106 82 - Greece

ABSTRACT. The service reliability provided by power systems can be measured by using recently published computational methods all of which require the expected values of component outage rates and outage durations to be known. Objective of the paper is to describe computational methods for calculating the required component parameters from field data. The evaluated component parameters of the Hellenic Transmission system through a study of its historic records are presented. An expert system is also described which has been developed for the purpose of fault diagnosis of power system components. The inference engine task is based on knowledge acquired during the study of component field outage data and the Bayes' theorem. An example of applying the developed expert system to faults occurring on power transformers is also included.

1. INTRODUCTION

During the last decade, the development and application of probabilistic modelling and evaluating techniques for power system reliability evaluation have improved considerably and have been the subject of many papers [1,2]. The emphasis has generally been directed towards the system generation and transmission sections due to the possible total collapse of the system that can be caused by certain failure states in these areas. The ability to assess emergency conditions together with the evaluation of load-point reliability indices have added considerable impetus to reliability evaluation. More recently, a few publications [3-5] have devoted to the development of reliability models and evaluation techniques which can be also applied to distribution systems. The main purpose of evaluating the reliability of such systems is to deduce the consumers' load-point indices. The approach of nearly all the evaluating methods is the calculation of system reliability from component parameters such as the expected values of component outage rates and outage durations. The aim of the paper is to describe computational methods for calculating the required component parameters from field data. The faults occurred on each type of

S. Tzafestas et al. (eds.), System Fault Diagnostics, Reliability and Related Knowledge-Based Approaches, Vol. 1, 327–341.

power system components are classified into categories according to the cause of outages, their duration and the weather conditions and in groups according to the component nominal rating or voltage class, the fault nature, the failure initiating cause, the failure contributing cause and the subassembly on which the failure occurred. The paper presents the evaluated parameters and their confidence limits for all these categories and groups of the Hellenic Transmission system transformers through a study of the system historic operating records.

A computer-based expert system seeks to capture enough of the human specialist's knowledge so that it too will solve problems expertly. Over the past ten years various research groups in artificial intelligence have built highly specialised systems containing the expertise needed to solve many scientific and engineering problems [6-8]. Practical applications for expert systems abound. Whenever human experts are in great demand and short supply, a computer-based consultant can help amplify and dissenimate the needed expertise. An expert system can capture the practical experiental knowledge that is hard to be found in bibliography. In equipment repair an expert system may be advantageous because there is a shortage of trained experts, and as technology becomes more complex, this problem will get progressively worse. The objective of this paper is also to describe the development of an expert system which can be used for the purpose of fault diagnosis of power system components. The inference engine task is based on knowledge acquired during the study of component field outage data and the Bayes' theorem. An example of applying the developed expert system to faults occurring on power system transformers is also included.

2. EVALUATING THE COMPONENT RELIABILITY INDICES

The reliability evaluation study of power transmission and distribution systems requires basically the knowledge of two types of component indices. These indices are the average outage rate λ and the average outage duration r. The first step in estimating these indices is the separation of outages into categories according to factors believed to influence them. The component outages can generally be divided into the following four major categories [5,9] :

- permanent outages
- temporary outages
- transient outages
- scheduled (maintenance) outages.

Furthermore, depending on the single or multiple failure effects, each outage category can be divided into independent or common mode subcategories respectively. Computational techniques have been published [9] for evaluating the component reliability indices and their confidence limits from the system historic operating records. These techniques separate the outages of each type of components into groups according to the weather conditions, the voltage rating and the different (if any) constructional properties. The obtained results provide sufficient information for conducting reliability assessment and cost - benefit studies of reinforcement schemes in power transmission and distribution networks [9]. However, the knowledge base of fault diagnosis systems may also need information concerning the component reliability indices (rates, durations, probabilities) for all types of fault nature, failure initiating causes, failure contributing causes, etc. The existing computational algorithm has therefore been modified to provide the required increased set of information and consists of the following main steps :

Step 1 : Assign a code number to each type of power system components to be considered.

Step 2 : For each type of components :
- separate the components into groups according to their nominal power rating, voltage rating and the different constructional properties. Assign a code number to each group.
- divide the component assembly into subassemblies and assign a code number to each of them.
- identify the nature of all categories of failures (mechanical, electrical, etc.) which may lead to a component outage and assign a code number to each of them.
- detect all the different causes which may initiate a component failure and assign a code number to each of them.
- detect all the different causes which may contribute to a component failure and assign a code number to each of them.
- identify the causes that initiate or contribute to faults of each nature category and input the corresponding code numbers.

Step 3 : For each year of the analysis read :
- the average duration of the normal and adverse weather
- the number of similar component units in each group and type of components.

Step 4 : For each year of the analysis read the following characteristics for each outage event to be considered :
- the outage category (permanent, temporary, transient)
- the failure effect (independent, common mode)
- the weather conditions (normal, adverse weather)
- the code number indicating the type of the failed system component
- the code number of the group in which it belongs
- the outage duration in hours or minutes
- the code number indicating the fault nature category
- the code number indicating the failure initiating cause
- the code number indicating the failure contributing cause
- the code number indicating the subassembly in which the fault has occurred.

Step 5 : For each outage category and for each type i and group j of components :

(a) calculate the following reliability indices for normal and adverse weather separately [9] :
   - the average outage rate $\hat{\lambda}_{ij}$
   - the average outage duration $\hat{r}_{ij}$
   - the upper (λ_U, r_U) and lower (λ_L, r_L) limits of indices $\hat{\lambda}$ and $\hat{r}$ for a given confidence probability.

(b) for each category of fault nature, failure initiating cause, failure contributing cause and component subassembly calculate separately :

- the reliability indices $\hat{r}_{ijk}$, $\hat{\lambda}_{ijk}$ of outage code k, where k indicates the code number of the corresponding fault nature category, failure initiating cause, failure contributing cause or component subassembly.
- the probability of occurrence P_{ijk} of outage code from equation (1)

$$P_{ijk} = \hat{\lambda}_{ijk} \cdot \hat{r}_{ijk}/8760 \tag{1}$$

- the conditional probability of occurrence CP_{ijk} of outage code k given that a failure event has occurred from equation (2)

$$CP_{ijk} = P_{ijk} / \sum_{m=1}^{n} P_{ijm} \tag{2}$$

where n is the total number of the corresponding outage codes.

(c) for each failure initiating cause and failure contributing cause calculate separately the conditional probability of occurrence FP_{ijk} of cause code k given that a fault of nature category l has occurred from equation (3)

$$FP_{ijk} = P_{ijk} / \sum_{m=1}^{n} P_{ijm} \cdot \alpha_{ml} \tag{3}$$

where α_{ml} is a coefficient which is either one or zero ; it becomes one when the failure m initiates or contributes to fault nature category l.

Because of scarsity of failure incidents, the sample used for the reliability indices of a particular type and group of components may be small so that the obtained results are not statistically valid. In this case two or more groups of components must be pooled together to provide better basis for results. In order to meet this requirement of the analysis, the developed algorithm offers the facility of pooling together data on a preselected number of groups of similar components and evaluating the parameters of the combined group. These results can be also used by the fault diagnosis methods in the case that the group of the failed component is not known accurately.

3. ANALYSIS OF HELLENIC TRANSMISSION SYSTEM OUTAGE DATA

The analysis reported in this paper has studied the outage data of transmission lines, cables, transformers, breakers, isolators and busbars extracted from the operating records of the Hellenic Transmission system for a period of twelve years. Although the developed algorithm considers two states for weather conditions, the analysis used a model with one state because the available data does not provide sufficient weather information. The Hellenic Transmission system is divided into the 400 KV, 150 KV and 66 KV systems. The 66 KV system exists only in two islands and compared with the other two is smaller in size and older in age.

The power transformers of the Hellenic Transmission system have been divided into four groups according to their nominal power rating as shown in Table I. Applying step (5a) of the developed algorithm, the reliability indices of each transformer group have been evaluated and are presented in Table I together with the indices of a selected number of combined groups. Tables II-V present the reliability indices and the probabilities P and CP which have been evaluated applying step (5b) of the algorithm for all categories of fault nature, failure initiating causes, failure contributing causes and the transformer subassemblies respectively. The outage duration indices ($\hat{r}$, r_L, r_U) are given in hours and the rates ($\hat{\lambda}$, λ_L, λ_U) in failures per year (f/yr) per 100 transformer units. Finally, Table V presents the conditional probabilities FP which have been evaluated applying step (5c) of the algorithm.

Table I Transformer Reliability Indices

Code No.	Transformer Group Description	Number of Units	$\hat{\lambda}$	λ_L	λ_U	$\hat{r}$	r_L	r_U
1	Rating< 25 MVA	154	1.75	1.13	2.60	36.3	25.6	56.1
2	25≤Rating< 50 MVA	81	14.55	11.75	17.83	63.1	52.2	78.1
3	50≤Rating<100 MVA	22	34.70	27.32	43.53	64.2	52.0	81.5
4	100≤Rating	48	11.56	8.45	15.45	119.7	91.9	163.6
1,2	Rating<50 MVA	235	5.72	4.74	6.84	57.4	48.4	69.3
3,4	50≤Rating	70	19.72	16.38	23.57	85.2	72.0	102.7
1-3	Rating<100 MVA	257	8.46	7.32	9.74	60.0	52.5	69.4
2,3	25≤Rating<100 MVA	103	19.64	16.80	22.84	63.6	55.1	74.3
2-4	25≤Rating	151	17.08	14.88	19.52	75.6	66.5	86.8
1-4	All Transformers	305	8.92	7.83	10.12	71.5	63.3	81.4

Table II Reliability Indices - Fault Nature

Code No.	Transformer Group Description	Fault Nature Categories								
		Mechanical-No.1			Electrical-No.2			Magnetic - No.3		
		$\hat{\lambda}$	$\hat{r}$	CP	$\hat{\lambda}$	$\hat{r}$	CP	$\hat{\lambda}$	$\hat{r}$	CP
1	R < 25 MVA	0.78	8.9	0.109	0.88	53.1	0.732	0.10	104.1	0.159
2	25 ≤R < 50 MVA	10.86	60.8	0.719	3.26	65.3	0.232	0.43	103.9	0.049
3	50 ≤R <100 MVA	22.49	46.4	0.469	9.00	129.0	0.521	3.21	7.2	0.010
4	100 ≤R	5.60	10.2	0.041	5.25	31.5	0.120	0.70	1657.4	0.839
1,2	R < 50 MVA	3.90	53.6	0.637	1.61	60.7	0.299	0.20	104.0	0.064
3,4	50 ≤R	11.56	35.0	0.241	6.57	78.5	0.307	1.59	478.7	0.452
1-3	R <100 MVA	5.66	50.9	0.567	2.31	85.9	0.391	0.49	43.5	0.042
2,3	25 ≤R <100 MVA	13.80	54.8	0.606	4.71	96.0	0.362	1.14	34.9	0.032
2-4	25 ≤R	11.20	47.8	0.414	4.88	74.0	0.280	1.00	395.4	0.306
1-4	All Groups	5.65	44.9	0.398	2.75	70.5	0.304	0.52	366.3	0.298

The width of the confidence interval for the mean value of a component parameter is a measure of the reliability of the estimated mean value. If the interval is narrow, confidence that the estimate is close

Table III Reliability Indices - Failure Initiating Cause (All Power Transformers)

Code No.	Cause Description	$\hat{\lambda}$	$\hat{r}$	P $x10^{-4}$	CP
1	Transient overvoltage disturbance (lighting, etc.)	0.52	60.8	0.360	0.049
2	Continuous overvoltage	0.10	44.4	0.052	0.007
3	Overheating	0.31	130.5	0.464	0.064
4	Winding insulation breakdown	0.31	588.8	2.090	0.287
5	Insulating bushing breakdown	0.16	26.6	0.047	0.006
6	Other insulation breakdown	0.26	6.0	0.018	0.002
7	Mechanical breaking,cracking,loosening, abrading, or deforming of static or structural parts	2.70	57.6	1.770	0.244
8	Mechanical burnout, friction, or seizing of moving parts	0.93	24.2	0.258	0.035
9	Mechanically caused damage from foreign source (digging,vehicular accident etc.)	0.41	44.7	0.212	0.029
10	Shorting by tools or other metal objects	0.21	129.4	0.306	0.042
11	Shorting by birds,snakes,rodents, etc.	0.47	26.0	0.138	0.019
12	Malfunction of protective relay control device or auxiliary device	0.57	56.2	0.366	0.050
13	Low voltage	0.0	0.0	0.0	0.0
14	Low frequency	0.0	0.0	0.0	0.0
15	Improper operating procedure	0.57	138.4	0.901	0.124
16	Loose connection or termination	0.47	14.0	0.075	0.010
17	Others	0.93	20.2	0.215	0.030

Table IV Reliability Indices - Failure Contributing Cause (All Power Transformers)

Code No.	Cause Description	$\hat{\lambda}$	$\hat{r}$	P $x10^{-4}$	CP
1	Persistent overloading	0.21	182.9	0.433	0.060
2	Abnormal temperature	0.41	15.2	0.072	0.010
3	Exposure to aggresive chemicals,solvents, dusts, moisture or other contaminants	0.05	3.7	0.002	0.001
4	Exposure to nonelectrical fire or burning	0.05	3.5	0.002	0.001
5	Obstruction of ventilation by foreign object	0.47	16.9	0.090	0.012
6	Normal deterioration from age	1.66	205.9	3.900	0.536
7	Severe wind,rain or other weather condition	0.73	121.5	1.010	0.138
8	Improper setting of protective device	0.26	104.9	0.311	0.043
9	Lack of protective device	0.05	24.0	0.014	0.002
10	Inadequate protective device	0.0	0.0	0.0	0.0
11	Malfuction of protective device	0.36	62.6	0.259	0.036
12	Loss,deficiency,contamination or degradation of oil or other cooling medium	1.35	17.2	0.265	0.036
13	Improper operating procedure or testing error	0.47	8.2	0.044	0.006
14	Inadequate maintenance	1.66	12.2	0.231	0.032
15	Others	1.19	47.5	0.646	0.089

Table V Reliability Indices - Transformer Subassemblies (All Power Transformers)

Code No.	Subassembly	λ	$\hat{r}$	P $\times 10^{-4}$	CP
1	Bushings	0.62	22.0	0.156	0.021
2	Windings	0.26	769.8	2.280	0.313
3	On-load Tap Changer	3.37	89.5	3.440	0.473
4	Leads	0.21	9.0	0.021	0.003
5	Cooling	1.56	9.6	0.170	0.023
6	Auxiliary Equipment	0.47	7.3	0.039	0.006
7	Control and Protection Equipment	0.93	10.3	0.110	0.015
8	Surge Arrester	0.62	15.0	0.106	0.015
9	Bus	0.21	5.0	0.012	0.002
10	Disconnect	0.0	0.0	0.0	0.0
11	Potential Devices	0.0	0.0	0.0	0.0
12	Motor-Operated Ground Switch	0.0	0.0	0.0	0.0
13	Others	0.67	122.4	0.942	0.129

Table VI Conditional Probabilities (All Power Transformers)

Code No.	Failure Initiating Cause - FP			Failure Contributing Cause - FP		
	Fault Nature			Fault Nature		
	Mechanical	Electrical	Magnetic	Mechanical	Electrical	Magnetic
1	-	0.125	-	-	0.063	0.077
2	-	0.018	0.014	-	0.010	0.013
3	0.117	0.161	0.122	0.001	0.001	-
4	-	-	0.552	0.001	0.001	-
5	0.012	-	0.012	0.015	-	-
6	0.005	-	0.005	0.630	0.563	0.698
7	0.448	-	-	0.163	0.145	-
8	0.065	-	-	-	0.045	-
9	0.053	-	-	-	0.002	-
10	-	0.106	-	-	0.0	-
11	-	0.048	-	-	0.037	-
12	-	0.127	-	0.043	-	0.047
13	-	-	0.0	0.007	0.006	0.008
14	-	-	0.0	0.037	0.033	0.041
15	0.227	0.313	0.238	0.104	0.093	0.116
16	0.019	0.026	-	-	-	-
17	0.054	0.075	0.057	-	-	-

to correct is high. If the interval is wide, however, it is possible that future experience may be considerably different than the predicted. The results of the analysis showed that the confidence interval of only a few number of component parameters is significantly wide and therefore they provide accurate information for the knowledge base of fault diagnosis systems.

Table VII Bucnholz Relay Disconnection Procedures

No.	Failure Cause	Fault Nature
1	Overheating due to large flux density because of a particular transformer connection	Magnetic
2	Overheating due to large core flux density because of improper operating procedure	Magnetic
3	Overheating due to large magnetising currents because of flat topped E.M.F.	Magnetic
4	Overheating in cross-over type transformers	Electrical
5	Overheating in coil conductors with several wires in parallel to form each conductor	Electrical
6	Overheating due to carbonization of badly soldered joints between coils and conductor insulation	Electrical
7	Overheating due to core-bolt insulation breakdown	Magnetic
8	Tap changing improper procedure	Mechanical
9	Local overheating due to corona discharge	Electrical
10	Loss of oil due to leakage	Mechanical
11	Ingress of air into oil system	Electrical
12	Auxiliary cooling plant failure	Electrical
13	Others	-

4. FAULT DIAGNOSIS COMPUTATIONAL PROCEDURE FOR POWER SYSTEM COMPONENTS

4.1. General Description

An expert system can be defined [8] as one that handles real world, complex problems requiring an expert's interpretation and solves them using a computer model of expert human reasoning, reaching the same conclusions that the human expert would reach if faced with a comparable problem. Although expert systems typically focus on a very narrow domain, they have achieved dramatic success with real-life problems. This has excited widespread interest outside the research laboratories from which they emerged. Areas successfully tackled so far include medical diagnosis, geological exploration, organic chemistry and fault-finding in electronic equipment. The objective of this paper is to describe the expert system FANC1 (Failure Nature Classification) which has been developed at NTUA and can be used for classifying the nature characteristics of faults occurring on power system components. The developed approach is based on surface knowledge relating the operating conditions of power system components prior the fault incident, the failure contributing causes and the failure initiating causes with the failure nature by using a production rules scheme.

The fully fledged expert system FANC1 consists of the following four essential and critical modules :

- the knowledge base
- the inference engine
- the knowledge-acquisition module

- the explantatory interface

In many expert systems work there has been a concious effort to divide the problem into two parts, the inferencing engine and the knowledge base. The idea is that the inferencing engine is the general purpose thinking machine and the knowledge base is that about which the engine shall think. The developed fault diagnosis computational procedure is a forward heuristic search and in logical terms it is carried out by using inductive reasoning steps. The procedure interrogates its users in order to acquire information enough to reach to a conclusion by accepting or rejecting a hypothesis. Because of the presence of unreliable data or knowledge, the Bayesian theorem is used as the thread for tying together information from disparate sources. More specifically, the Bayes theorem computes the relative likelihoods between competing hypotheses on the strength of the evidence.

4.2. The Knowledge Base

The developed knowledge base for each type of power system components is a simple scheme consisting of the following three basic data formats:

(a) The first format holds the knowledge of a particular category of fault nature (F.N.) :

[(F.N. name), (p), (k), ((j,pyj,pnj), j=1,2,...,k)]

where :

- the first item is the name of fault nature category, i.e. mechanical, electrical, magnetic, etc.
- the second item (p) is the prior probability of that fault nature category occurred in any member of the population of that type of power system components taken at random
- the third item (k) is the number of failure causes which are applicable to that category of fault nature either as an indication or contraindication of the cause
- after these three items, there is a series of triples corresponding to each of the applicable causes in which the first item j is a reference number giving the cause being considered, the second item pyj is the probability that this cause will be observed given that the faulted component suffers the category of fault nature in question, and the third item pnj is the probability that the cause will be observed given that the faulted component does not suffer this category of fault nature.

(b) The second data format is that concerning the particular failure causes :

[(cause number), (cause name), (disconnection procedure number), (question(s) to be asked concerning this cause), (rules explaining the question(s) being asked)]

where :

- the first item is the reference number of the cause given as the corresponding item j in the triples of format (a)
- the second item is the name of the cause
- the third item is the number of the disconnection procedure which follows after an outage of that cause has occurred on the component
- the fourth item determines the question(s) which may be asked of the

user of the system in an attempt to acquire information about the consequences of this particular failure cause
- the fifth item determines the linguistically expressed rules which are used for the explanation of a particular question (answer to the question why is it asked ?)

(c) The third data format is that concerning the particular disconnection procedures :

[(disconnection procedure number), (procedure name), (question(s) to be asked concerning this procedure)]

where :
- the first item is the reference number of the disconnection procedure which is used in format (b)
- the second item is the name of the procedure
- the third item determines the question(s) which may be asked of the user of the system to determine the failure cause and the corresponding category of the fault nature.

4.3. Bayes Posterior Probability Method

The developed expert system implements the Bayes posterior probability method which is probably the most appealing approach from among the statistical theories. If a conclusion is to be selected from a set of mutually exclusive alternatives the Bayesian approach in its most direct form indicates the selection of the conclusion with the highest probability for the given data. If P(H) is the prior probability of some hypothesis H given no evidence, the posterior probability P(H/E) of the same hypothesis given some item of relevant evidence E is given by equation (4)

$$P(H/E) = \frac{P(E/H)\ .\ P(H)}{P(E)} \tag{4}$$

If the probability P(E/not H) is known very accurately the probability P(E) can be calculated from equation (5), otherwise it can be calculated using equation (6) where H_i is every condition with which E might occur.

$$P(E) = P(E/H).P(H) + P(E/\text{not } H).P(\text{not } H) \tag{5}$$

$$P(E) = \sum_{i=1}^{n} P(E/H_i).(P_i) \tag{6}$$

If the system knowledge base holds details of each and every single cause with which a category of fault nature might occur and the current probabilities of these causes, then the probabilities P(E) and P(H/E) can be exactly estimated while they are constantly updated as $P(H_i)$ changes and being always accurate. Finally, it must be emphasised that the above mentioned methodology assumes independence of the variables being used which indicates that the developed knowledge base must consist of independent fault nature categories and failure causes.

Ideally, when the expert system requests a certain piece of information the user can give a straight answer. But it may be that the user is not sure of what answer to give. For this reason, uncertainty in users' response has been accommodated by allowing a reply R on an eleven point scale with +5 being Yes and -5 being No, 0 being Don't know and everything else being somewhere in between as it is shown in Figure 1. The calculations of the expert system proceed as before except that P(H/E) is replaced by the probability P(H/R) of H given the users' response R, which is calculated from equation (7) as a linear interpolation between P(H/E) and P(H/not E) subject to the constraint that if the response is nought then P(H/R) = P(H)

$$P(H/R) = P(H/E).P(E/R) + P(H/not\ E).P(not\ E/R) \qquad (7)$$

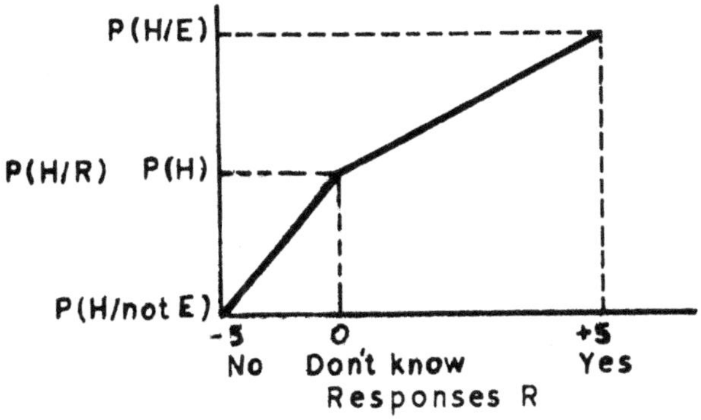

Figure 1. Uncertainty in users' responses

4.4. The Inference Engine

The inference engine of the expert system is the means the system is to use to drive around amongst the various inferences it might make. Both the forward and backward chaining methods [6] have been implemented and tested. The forward chaining approach proved to be very wasteful in computing time while the backword chaining method has serious disadvantages which stem from the problem of deciding in which order it will consider the various hypotheses open to it and then in which order it will try to obtain the relevant evidence for these hypotheses. For this reason the rule value approach [6] was adopted which for each item of evidence it assigns a value, the value of this rule in the process of inferencing, and asks that question with the highest value first. The value of each item of evidence RV is calculated from equation (8) as the total sum of the maximum probability shifts which it can induce in the whole of the hypotheses currently in the knowledge base. In this way an RV is developed for each item of evidence in the knowledge base. The procedure can then search for the maximum RV and request information on that item. Obviously, these rule values are not static since as the posterior probabilities $P(H_i/E)$ become continually updated they will continually cause the rule values to be altered. This method proved to be relatively easy to implement and not too slow.

$$RV = \sum_{i=1}^{n} \left| R(H_i/E) - P(H_i/\text{not } E) \right| \qquad (8)$$

For each hypothesis H there is an initial maximum P(H/E) and an initial minimum P(H/not E) that it could attain if all the remaining causes went in or against its favour respectively. The upper M1 and lower M2 threshold criteria for accepting a particular hypothesis are calculated as functions of these upper and lower bounds, P(max) and P(min), before any evidence has been accumulated at all. As evidence E is drawn in via responses R, P(H) will change and so will the maxima and minima possible P(H). Eventually, these three items will converge on a final P(H). Once P(max) drops below M2 then H may be rejected without any further evidence. Once P(min) rises above M1 then H may be accepted without any further evidence. As soon as P(min) is found for some H which exceeds P(max) for any other H then a most likely hypothesis has been found. As long as P(max) exceeds M2 and P(min) is less than M1 then there is uncertainty left in the situation which requires the input of further evidence.

4.5. Computational Algorithm

The computational algorithm of the developed expert system for fault diagnosis has the following main steps :

Step 1 : For each type of power system components to be considered :
- separate them in groups according to their nominal power rating or voltage rating
- define the nature of fault categories which may occur on the components and the associated failure initiating and contributing causes
- input the prior probabilities of each fault nature category and the corresponding conditional probabilities to the system knowledge base; this information for each group of components are extracted from the results obtained applying the computational algorithm of section 2 of the paper and presented in Tables I-VI.
- define the disconnection procedures which must be followed after the fault occurrence in order to isolate the component and for each of them determine the possible failure causes and categories of fault nature; input this information to the knowledge base.
- for each failure cause input the code number of the component disconnection procedure to the knowledge base.

Step 2 : Input the component group by indicating its power rating or voltage rating.

Step 3 : Determine the disconnection procedure which has been followed and the associated possible failure causes.

Step 4 : For each category of fault nature (hypothesis H) establish a prior probability P(H) by extracting the appropriate variable from the knowledge base.

Step 5 : For each failure cause (item of evidence H_i) relative to the determined disconnection procedure calculate a rule value (R.V.) by applying equation (8).

Step 6 : Of all RVs find the greatest RV to identify that cause which

can induce the greatest probability shift in all the categories of fault nature under consideration.
Step 7 : Interrogate the user on that cause using the question stored in the knowledge base; his response R being on a scale from -5 to +5.
Step 8 : Recalculate the probabilities of all fault nature categories which referenced that cause in their knowledge base and evaluate P(H/R) using equation (7).
Step 9 : Recalculate the rule values for all causes to allow for the change in probabilities that have taken place given the last response.
Step 10 : Calculate P(min) and P(max) of each category of fault nature and find the greatest P(min).
Step 11 : If there is a category with P(max) exceeding the greatest P(min) and all P(min) are less than M1 go back to step 6 and ask another question. If there is not, this is the most likely outcome and energise the summary routine to output the results and explain the details of the inferences the system has made.

It must be noticed that the developed knowledge base contains all the necessary information for all groups of components. In addition, information of all possible group combinations have been also inputted and therefore the users' responses to step 2 of the algorithm can either be the actual component group or a combination of appropriate groups.

4.6. Computer Program FANC1

The above described computational algorithm has been implemented very efficiently into the computer program FANC1 which has the following primary subprograms :

- User-program interface : the user interacts with the program by using a restricted set of words and figures as well as by answering menus which are very easily understood.
- Blackboard : it records intermediate results (decisions and data) created during the program utilisation. These data are useful both when trying to justify the program performance by using suitable explanations or when trying to check if a particular hypothesis is true or not.
- Scheduler : it determines the action that should be executed next by either estimating the effects of a rule application on the knowledge base according to steps 10 and 11 of the algorithm or by giving priority to a rule according to steps 5 and 6.
- Interpreter : it executes the action determined by the scheduler and interrogates the user according to step 7 of the algorithm, updates the blackboard elements according to steps 8 and 9 and calls the justifier.
- Justifier : it explains the actions of the program to the user. This explanation is based on standard linguistically expressed rules and on tracing backwards along blackboard solution elements from the questioned conclusion to the intermediate results or data that support it.

The computer program FANC1 was written in FORTRAN language and was tested on a low cost standard microcomputer.

5. FAULT DIAGNOSIS OF POWER SYSTEM TRANSFORMERS

The developed expert system FANC1 classifies the nature of faults on power system transformers into three categories, mechanical, electrical and magnetic failures. The transformer units of a power system have been divided into four groups according to their nominal power rating as shown in Table I together with their assigned code numbers. The causes which initiate and contribute to the failure of power transformers are shown in Tables III and IV together with their assigned code numbers. In addition, the causes which initiate and contribute to each category of fault nature are indicated in Table VI with a printed value of conditional probability of occurrence. The evaluated in section 3 of the paper reliability indices and probabilities of each transformer group and a selected number of combined groups have been inputted to the knowledge base of the expert system using the data formats described in section 4.2. The disconnection procedure of a faulted transformer is determined by the user by answering the following menu (step 3 of the algorithm) :

TRANSFORMER DISCONNECTION PROCEDURES
1. TRANSFORMER AUTOMATICALLY DISCONNECTED BY OVERCURRENT PROTECTION
2. TRANSFORMER AUTOMATICALLY DISCONNECTED BY GAS AND OIL ACTUATED RELAY (BUCHHOLZ RELAY)
3. TRANSFORMER AUTOMATICALLY DISCONNECTED BY DIFFERENTIAL PROTECTION
4. TRANSFORMER AUTOMATICALLY DISCONNECTED BY DIFFERENTIAL AND GAS RELAY OPERATION
5. TRANSFORMER MANUALLY DISCONNECTED
6. EXIT

SPECIFY AN OPTION NUMBER :

These disconnection procedures have been inputted to the developed knowledge base together with the corresponding failure causes and categories of fault nature. Table VII presents this information for the Buchholz relay disconnection procedures (option number 3 of the above menu).

6. CONCLUSIONS

The paper describes computational methods for estimating the reliability indices of power system components, their confidence limits and the associated probabilities of fault occurrence from commonly available data. Using the developed methods and the historic operating records of the Hellenic Transmission System, the paper presents the evaluated indices of system transformers. The major advantage of this work is that the required data for reliability studies of the Hellenic Transmission system are now available. Furthermore, these results can be used for developing the knowledge base of an expert system designed to aid in fault diagnosis of power system components. The paper describes the development of an expert system which can find the most likely nature characteristics (mechanical, electrical, magnetic, etc) of faults occurring on power system components. The fault diagnosis procedure is performed by computing the posterior probability of a category of a component fault occurrence given the prior probability of the same category and some items of relevant evidence about the failure cause. The procedure interrogates its users in order to acquire enough items of evidence to reach a conclusion by accepting or rejecting fault categories and causes. The main merits of the developed system are :

- its simplicity which makes it suitable for use by engineers without prior knowledge of expert systems
- its applicability to low cost standard microcomputers
- its efficiency to recognise real operational and constructional features of power system components.

Furthermore, the methodology can be of great help to the power system engineers in order to determine the expected outage time of the failure being occurred and select the most appropriate operational actions for performing the component repair and providing alternative routes of supply to the affected customers.

7. ACKNOWLEDGEMENTS

The authors wish to express their gratidute to the Hellenic Ministry of Research and Technology for providing the financial support and to the Hellenic Public Power Corporation for providing the necessary data of the work.

8. REFERENCES

[1] IEEE Committee Report, "Bibliography on the Application of Probability Methods in Power System Reliability Evaluation", IEEE Transactions, PAS-97, 1978, pp. 2235-2242.

[2] Allan, R.N., Billinton, R., Lee, S.H., "Bibliography on the Application of Probability Methods in Power System Reliability Evaluation", IEEE Transactions, PAS-103, 1984, pp. 275-282.

[3] Allan, R.N., Dialynas, E.N., Homer, I.R., "Modelling and Evaluating the Reliability of Distribution Systems", IEEE Transactions, PAS-98, 1979, pp. 2181-2189.

[4] Dialynas, E.N., Allan, R.N., "Local Generating Facilities in the Reliability Evaluation of Power Distribution Systems", IEEE/PES 1986 Winter Meeting, New York, 1986, paper No. 86WM058-2.

[5] Billinton, R., Allan, R.N., Reliability Evaluation of Power Systems, Longmans (formerly Pitman Books), London, 1984.

[6] Forsyth, R., Expert Systems, Principles and Case Studies, Chapman and Hall, London, 1984.

[7] Hayes-Roth, F., Waterman, D.A., Lenat, D.B., Building Expert Systems, Addison-Wesley Publishing Co., London, 1983.

[8] Weiss, S.M., Kulikowski, C.A., A Practical Guide to Designing Expert Systems, Chapman and Hall, London, 1984.

[9] Dialynas, E.N., Machias, A.V., Papadias, B.C., "Evaluating the Component Reliability Indices and the Cost-Benefit of Reinforcement Schemes in Power Transmission Networks", IASTED International Journal of Energy Systems, Vol. 6, 1986.

APPLICATION OF DYLAM METHODOLOGY TO THE RELIABILITY ANALYSIS OF A BATCH CHEMICAL PLANT

Z. Nivolianitou
National Technical University of Athens
Grant-holder of the Commission of the European Communities
Joint Research Centre - Ispra Establishment
21020 Ispra (Va) - Italy

N.A. Labath
CONICET (Consejo Nacional de Investigaciones Cientificas y Technologicas de la Republica Argentina)
Visiting scientist of the Commission of the European Communities, Joint Research Centre - Ispra Establishment
21020 Ispra (Va) - Italy

ABSTRACT. Safety analysis of chemical plants requires the application of techniques which are able to evaluate potentially dangerous transient conditions together with the reliability of mitigating or protective systems. The DYLAM methodology offers a solution to the problem and this is shown by its applications to a rather common type of batch chemical plant. For all components and equipment included in the plant, models according to the DYLAM requirements were developed. A dynamic simulator able to follow the variations along the time of all the process variables of interest was developed for normal plant operation and for the different combinations of equipment and component failures. The possibility of operator corrective actions was included in order to show how it works one of the best alternatives that the DYLAM technique provides. In order to avoid a large number of dynamic simulations, a strategy for the analysis and some time-saving criteria were applied. These are intended to eliminate most of the unnecessary sequences whose simulation does not produce significant information. Finally, probabilistic results are obtained, complemented with adequate cost data which permit the analyst to choose the most appropriate design criteria and parameters.

1. INTRODUCTION

The increasing complexity of chemical sites nowadays, combined with the diversity of dangers in various parts of them, calls for a more synthetic and representative way of plant and process description for risk and safety analysis as well as fault detection purposes. Towards this direction the DYLAM methodology /1/ offers a solution since it

S. Tzafestas et al. (eds.), System Fault Diagnostics, Reliability and Related Knowledge-Based Approaches, Vol. 1, 343–353.

takes account at the same time of process physics and system performance, two domains which are normally treated separately by current probabilistic techniques.

Since accidents proceed according to the values assumed by certain physical variables, such as temperature, pressure, concentrations, etc. as a function of time and to the occurrence of logical events such as lack of intervention of protective systems, control actions, human operator faults (which happen at random times), dynamic analyses are more appropriate for representing such phenomena because they depict more adequately the real accident sequence. This advantage is greatly reinforced when responses of control systems and/or of human operators are of concern because the real time period for their intervention can be quite precisely defined. In such a way, design parameters and operating procedures can be analysed in detail, helping the decision making process and the fault diagnosis for accident prevention.

2. DESCRIPTION OF THE DYLAM TECHNIQUE

The DYLAM technique and its related computer algorithm is a methodology which has been developed at the Joint Research Centre of the European Community, Ispra, Italy, and is still under amelioration. The basic characteristics of the method have been extensively described elsewhere /1/ and in this paper we are only going to shortly summarize and stress these aspects which become important for the elaboration of our example, as they are synthetically represented in Figure 1. As can be deduced from this figure, the technique has two main parts: the first one (steps 1 to 4) consists of generating all the possible system state configurations by the combination of the different component states. Every one of these configurations is finally represented by giving appropriate parameter values to a set of parametric equations, which contain both logical and physical information. The second part (steps 5 to 7) solves numerically the corresponding equation set in order to foresee the physical consequences of the event sequences generated, and it verifies if the TOP conditions are satisfied (i.e. if some physical variable has reached certain threshold value).

The process under analysis is described by a set of component models, obtained by a quantitative failure mode and effect analysis, resulting in logical analytical relationships characteristic of each component. Such models should be developed for each single component, but if this tends to be very CPU-consuming by increasing considerably the number of sequences to be analysed, the grouping of single components in macro-components is advisable.

DYLAM provides all the nominal sequences that produce one or more TOP conditions and the evolution of the physical variables of interest for those sequences. On the other hand, the probability distributions of the TOP conditions are given as a function of time, taking into consideration both statistical and functional failure dependencies. It must be stressed that the functional dependency consideration is made possible just because of the step-by-step procedure in the evaluation of the physical variables along the transient state and is

1) Event sequence generation (based on component models).
2) Sequences are generated until the maximum order specified is reached.
3) Analysis of passive states to cancel those sequences where all components are in passive state.
4) Verification that sequence probability is not lower than the cut-off value established by the user.
5) Numerical solution of equation set or dynamic simulation of the plant to predict the physical consequences of the respective sequence of events.
6) Verification of TOP event conditions and selection of minimal TOP TIME PATTERNS.
7) Results:
   i) Reliability parameters;
   ii) Physical evolution of every minimal sequence that reached at least one of the TOP event conditions.

Figure 1. Steps of DYLAM procedure.

being implemented in the new version of the code /2/.

In order to minimize the number of numerical solutions, which are generally most CPU consuming, steps 2, 3 and 4 of Figure 1 are performed as soon as an event sequence has been generated and before proceeding to the physical consequence evaluation. In that sense, the maximum order of simultaneous failures to be considered and the cut-off probabilistic limits are key factors in the total CPU time of the analysis.

The first and the fifth steps presented in Figure 1 are strongly interrelated because components should be modelled according to the way in which the numerical solution requires parameter values to calculate the system performance.

The numerical solution of the equation set, characteristic of each system, may be performed in several ways according to the complexity of the system, as applications in the nuclear and chemical field can show. There are the following cases:

a) The process equipment is represented by simplified formulas including some delay times to take account of transient situations /3/.
b) The system is represented by transient simulators /4/, the running times of which are considerably low as to be used directly, without simplification. The requirement of this alternative is that the simulation running time must be reduced to as little as possible for any combination of failures. The accuracy of this code should be compatible with the uncertainties which are present in the other steps of risk and safety analysis.

This paper will describe some applications of DYLAM to chemical plants, which generally fall in the second case just described. To that scope, some chemical process configurations are being analysed in order to check the code in several examples, to widen its limits of application and to improve its efficiency as much as possible according to the cases under study.

The examples try to comprise several typical cases of the chemical industry such as batch, semicontinuous and continuous processes, several process equipments (reactors, distillation towers, storage tanks, heat exchangers, pumps, control devices, etc.), which require the numerical solution of ordinary and/or partial differential equations, the possibility of operator actions (on site or from control rooms), to diagnose failures in order to correct or prevent dangerous conditions, etc.

3. PRESENTATION OF THE CASE STUDY

The chemical plant analysed is a batch plant for sulfolane production; it is divided into two parts: the addition unit and the hydrogenation unit. The process can be summarised in the following chemical reactions:

Addition reactor: $SO_2 + C_4H_6 \rightleftarrows$ Sulfolene (1)

SO_2: Sulfur dioxide; C_4H_6: butadiene

Hydrogenation reactor: Sulfolene + $H_2 \rightarrow$ Sulfolane (2)

Sulfolane: (tetrahydrotiofene-1-1 dioxide)

Reaction (1) is exothermic and reversible; the higher the temperature, the smaller the amount of products obtained.
Reaction (2) is irreversible, exothermic, and needs a catalyst to proceed at an acceptable rate.

The safety analysis described in this report regards only the addition reaction and the respective facilities. A schematic flow diagram of this plant section can be seen in Figure 2.

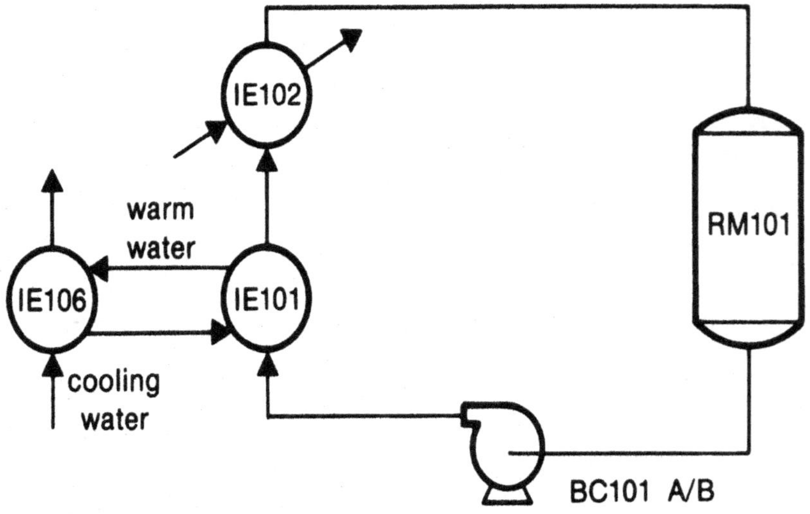

Figure 2. Simplified flow diagram (IE/ heat exchanger; BC: centrifugal pump; RM: reactor).

The reactants are charged at 25°C and the mixture is heated in IE 101 with hot water (70°C) to initiate the reaction. The reaction is self-accelerated by the heat produced and in order to avoid a runaway reaction, a second heat exchanger IE 102 is used when the temperature reaches the value of 93°C. For that scope, cooling water of 25°C is used, although after 70°C the hot water in IE 101 acts as "cooling" water as well. If no component fails, the reaction goes on normally and the final temperature of 70°C is reached, where the product concentration is nearly optimum.

There are several design parameters and safety measures to be defined, e.g. the cooling water flow to IE 102, the hot water flow to IE 101, the reactor design pressure and temperature, the temperature to activate IE 102, the necessity of a stand-by pump to circulate the reaction mixture, the number of safety valves and their pressure sets, etc. Many of these parameters can be chosen by the use of the dynamic simulator alone, although for certain decisions it is nearly always necessary to get the reliability parameters associated with the available technical options. For instance, the reactor design pressure may vary between 20 bars (normal operation) and 66 bars (worst-case failure).

The DYLAM technique, in contrast with the traditional fault-tree methods which are not able to treat adequately the complexity and the interrelation among the different variables, offers the possibility of working with these complex interrelations making rather simple changes in component models and in the simulation algorithm to account for different design configurations and operating procedures.

A dynamic simulator has been developed for this sulfolane batch plant /5,6/ which includes the following models:

- Reactor: lumped parameter model (all properties are uniform in every phase inside the reactor). Vapour and liquid phases are considered, including mass transfer rates between the two phases and latent heats of vaporisation in thermal balances. The reactor walls are included in the thermal balances, and the reactor is considered adiabatic.

- Double-tube heat exchangers IE 101 and IE 102: distributed parameter model (the temperature changes with time and distance). The exchangers are considered adiabatic and their walls included in the respective thermal balances.

- Pipe walls in the circuit are adiabatic and they were included in the thermal balances.

The equipment dynamic models were basically taken from /5/, with very few modifications, as well as all the necessary process data.

Several failures were included in the modelling of the different plant components and equipment, e.g. partial or complete lack of cooling water to IE 102 (caused by various reasons), partial or complete lack of warm water in IE 101 after the start of the reaction, changes in water temperature, recirculating pump failures, control devices failures and/or delays, operator delays, etc.

4. RESULTS OBTAINED

Two complete analyses were made, the difference between them lying in the consideration or not of the operator corrective actions for certain emergency situations.

The following parameters were considered:

- 13 component and event variables (without operator corrective actions);
- 18 component and event variables (with operator corrective actions);
- maximum TOP sequence order: 4;
- sequence probability threshold: $1x10^{-8}$.

The TOP event conditions were four: all imply the vessel rupture due to failure of the safety valve to open when the pressure reaches a certain value. Each TOP event is related with a certain design pressure: 27 bars, 36 bars, 46 bars and 56 bars. These pressure values were chosen in order to establish the vessel adequacy to resist the pressures developed by certain types of failures and they were identified with the help of the dynamic simulator.

The maximum pressure reached in normal conditions is around 20 bars. In cases of lack of cooling water in IE 102, the pressure rises to 55 bars, but when there is no recirculation due to failure in pumps BC 101 A/B, the pressure can reach even 66 bars. The pressure values of 36 bars and 46 bars are two intermediate ones selected for the scope of the study.

The operator corrective actions regard the opening of the by-pass hand valve in IE 102 and the by-pass hand valve IE 101. These interventions are necessary when there are failures in the control systems of the two devices.

Table 1 summarises the probability values calculated by DYLAM. It can be seen that operator corrective actions reduce the failure probabilities of TOP event 2 and even more of TOP event 3, as it can be expected. This reduction is not too drastic though, because of the very little time available to the operator in order to take corrective measures and reflects the high operator failure probability chosen for the evaluations.

Figure 3 represents a time pattern of the minimal TOP sequences for the first 0.810 hours of system operation without the operator intervention.

The number of TOP sequences for each order, which produce the TOP events at any time is marked over the diagram lines, for example, TOP event 1 is produced by 19 sequences of order 3 at time .50 hr, by 6 sequences of order 3 at .55 hr, by 6 sequences of order 2 at .48 hr, and so on.

The same type of figure can also be drawn for the case of operator intervention and there one can see that the sequences of order 2 simultaneously present are diminished whereas those of higher order are increased.

Another important parameter for a TOP time pattern is also the total TOP probabilities. Such an example is given in Table 2 for the TOP condition 1 and the TOP sequences of order 4.

TABLE 1 - Failure probability values

Sequence order	Failure probabilities (per year) TOP EVENT 1 Pressure > 27 ata	TOP EVENT 2 Pressure > 36 ata	TOP EVENT 3 Pressure > 46 ata	TOP EVENT 4 Pressure > 56 ata
Without operator intervention				
Order 2	5.4×10^{-4}	5.3×10^{-4}	5.3×10^{-4}	-
Order 3	5.3×10^{-4}	0.3×10^{-4}	0.2×10^{-4}	1.5×10^{-5}
Order 4	1.7×10^{-4}	0.07×10^{-4}	0.03×10^{-4}	0.006×10^{-5}
Total probabil.	1.3×10^{-3}	5.7×10^{-4}	5.6×10^{-4}	1.5×10^{-5}
With operator intervention				
Order 2	1.4×10^{-6}	1.4×10^{-6}	1.4×10^{-6}	-
Order 3	8.3×10^{-4}	2.5×10^{-4}	6.7×10^{-5}	1.5×10^{-5}
Order 4	3.4×10^{-4}	1.1×10^{-4}	1.5×10^{-5}	0.004×10^{-5}
Total probabil.	1.2×10^{-3}	3.6×10^{-4}	8.4×10^{-5}	1.5×10^{-5}

TABLE 2 - Example of a TOP-time pattern with the respective total TOP probabilities (Order 4 minimal TOP sequences)

Time (hr)	TOP seq.	Total TOP probabil.	Time (hr)	TOP seq.	Total TOP probabil.
3.000E-01	0	0.000E+00	5.300E-01	120	3.381E-04
3.300E-01	0	0.000E+00	5.400E-01	94	1.298E-04
3.600E-01	0	0.000E+00	5.500E-01	75	1.201E-04
3.900E-01	0	0.000E+00	5.600E-01	34	7.234E-06
4.200E-01	0	0.000E+00	5.700E-01	8	3.631E-07
4.300E-01	0	0.000E+00	5.800E-01	5	1.637E-07
3.300E-01	2	3.713E-08	5.900E-01	5	1.637E-07
4.500E-01	22	1.507E-05	6.000E-01	5	1.637E-07
4.600E-01	44	2.138E-05	6.300E-01	5	1.637E-07
4.700E-01	78	1.210E-04	6.600E-01	0	0.000E+00
4.800E-01	116	3.368E-04	6.900E-01	0	0.000E+00
4.900E-01	130	3.406E-04	7.200E-01	0	0.000E+00
5.000E-01	131	3.407E-04	7.500E-01	0	0.000E+00
5.100E-01	132	3.423E-04	7.800E-01	0	0.000E+00
5.200E-01	122	3.395E-04	8.100E-01	0	0.000E+00

The response of the system as obtained by the application of the DYLAM algorithm is depicted in Figure 4, where T-t curves are drawn for several plant conditions. Curves like these could be obtained for every sequence of events that has been simulated.

It is interesting, however, to point out in which sense the DYLAM algorithm helps the fault detection in a system. This can be done in the following way. Once the sequences that produce the dif-

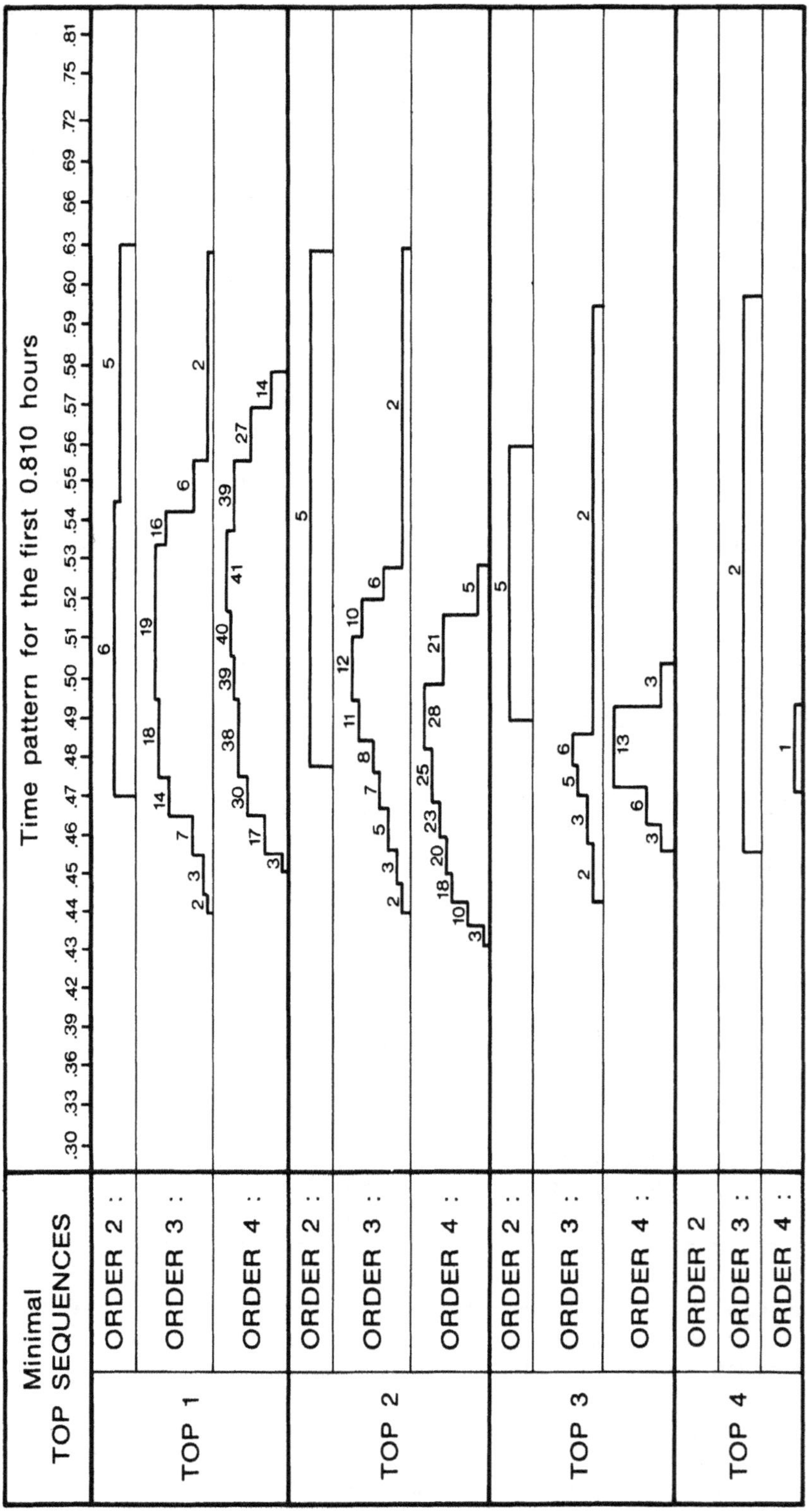

Fig. 3 : Time pattern for minimal TOP sequences

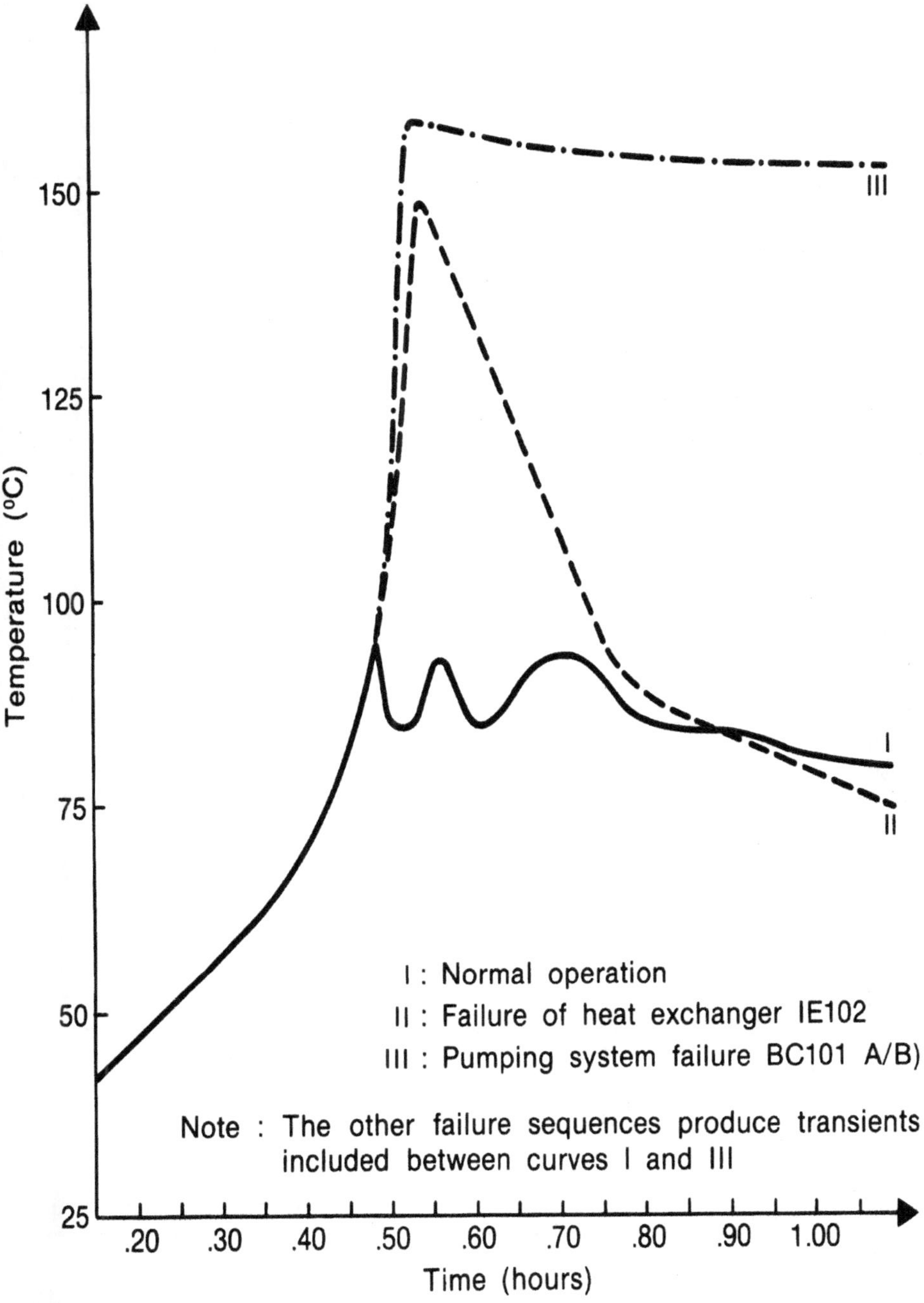

Fig. 4: Curves T-t obtained for normal operation and some equipment failures (no operator corrective actions are considered)

TABLE 3 - Number of component failures for the different sequences and TOP events*) (no operator corrective actions are considered).

Component failure	TOP 1	TOP 2	TOP 3	TOP 4	Component failure	TOP 1	TOP 2	TOP 3	TOP 4
	6 times	6 times	5 times	no time					
Order 2 sequences									
SST1	1	1	1	-	SCV2	1	-	-	-
SFT1	1	1	1	-	SCV3	1	1	1	-
SCT1	1	1	1	-	CWF1	1	1	1	-
	19 times	12 times	6 times	2 times					
Order 3 sequences									
SST2	5	2	1	-	SHT1	3	1	1	-
SCV1	4	1	-	-	SHT2	3	1	1	-
SCV2	10	8	4	-	SSPL1	1	1	1	1
SPM1 1	3	2	2	2	SPM2 1	1	1	1	1
SPM1 2	1	1	-	-	CWF2	1	1	-	-
SHW1	3	3	1	-	CWS1	1	1	-	-
SHW2	1	1	-	-					
	41 times	28 times	13 times	1 time					
Order 4 sequences									
SST2	26	18	7	1	SHW2	7	2	2	-
SCV1	12	8	1	-	SHT1	7	4	1	-
SCV2	12	12	12	1	SHT2	19	12	6	-
SPM1 1	10	5	3	-	CWF2	6	3	1	-
SPM1 2	7	7	2	-	CWS1	8	5	1	-
SHW1	9	8	3	1					

*) Note: As the failure of the safety valve is included in all these sequences, its failure numbers are not reported in this table.

ferent TOP events are identified, the contribution of each component failure to these sequences is calculated, and then classified according to their importance for the system operating mode. Table 3 above reports the various component failures present in the different orders of sequence and TOP event conditions with their total number of occurrence. The component failures are quoted with the codified names of the event variables in the component models.

As noticed above, the failure of the safety valve is included in all sequences, so that its failure numbers are equal to the times that any TOP event is present.

With all this information, the analyst can decide which are the most acceptable design parameters and/or where design changes should be made to improve the plant reliability.

5. CONCLUSIONS

The DYLAM methodology was applied to a chemical plant in order to analyse the risks involved in its operation and detect the possible failure sequences in a process where a runaway reaction may be triggered.

The complete probabilistic and physical related information for all the possible and significant combinations of component states are obtained, giving thus the parameters which help the reliability engineer to judge about the adequacy of the different design alternatives from the point of view of plant safety.

The simulation of a large number of possible accidental sequences which is made possible by the automated technique implemented in DYLAM, offers the designer a very strong body of knowledge to understand malfunctions and their effects. Therefore, it may give a great aid in applications directed towards operator support systems and fault diagnosis.

6. REFERENCES

1. Amendola A., Reina G., "DYLAM-1: A software package for event sequence and consequence spectrum methodology". Eur 9224 EN, 1984.
2. Amendola A., Reina G., "DYLAM-2: Description and how to use". EUR report, to be published.
3. Nivolianitou Z., Amendola A., Reina G., "Reliability analysis of chemical processes by the DYLAM approach". Reliability Engineering 14 (1980), 163-182.
4. Cacciabue P., Amendola A., Mancini G., "Accident simulator development for probabilistic safety analysis". ANS/ENS Top. Meet. on Probabilistic Safety Methods and Applications, San Francisco, February 25 - March 1, 1985.
5. Schluik A., "Dynamic simulation of a sulfolene reactor". PSO/MB-IN-001, 1983. INGAR, Santa Fe, Republica Argentina. Internal report.
6. Labath N., Nivolianitou Z., "Application areas of the DYLAM approach to the safety analysis of chemical processes". EuReDatA Conf. 1986, Springer Verlag, Heidelberg.

STEADY STATE RELIABILITY OF HYPERCUBE MULTICOMPUTERS TOLERATING SINGLE OR DOUBLE FAULTS

W. G. Schneeweiss
Fernuniversität
P.O. Box 940
D-5800 Hagen

ABSTRACT. This is a report on the derivation of basic reliability characteristics of maintained fault tolerant hypercube multicomputer systems. It is shown how fault tolerance can be traded for connectedness of the net working system. For practical reasons only the cases of single and of double fault tolerance are treated at full detail giving values of availability, MTTF, and hardware overhead. Key Words : Reliability, hypercube computer, fault tolerance, fault tree, availability, mean time to failure.

INTRODUCTION. This is a report on a very interesting type of highly symmetrical computer networks, which may be useful for a variety of purposes [1],[2]. Since fault tolerance can be rather easily incorporated into such structures it seems adequate to discuss the cases of tolerating single faults (1-FT) and double faults (2-FT), i.e. two faulty units (components), at some detail. Higher fault tolerance is of little interest in regularly maintained systems.

As is well known, fault tolerance makes little sense without adequate system self-diagnosability [3],[4]. Hence, at first sight, the idea of building powerful fault tolerant multicomputer networks, starting with a triangle, and adding more and more nodes, each of which is connected to the hitherto existing ones, looks promising. However, this is, for any number of nodes, the complete graph, which is, usually, not needed. Practical demands tend to lie inbetween the complete graph and the simple ring (cycle) structure. The most strongly interconnected non-complete graph structures of high topological regularity (or high degree of symmetry, if taken verbally) are structures based, not on triangles, but on squares. These are known [1],[2] as hypercube networks.

In an n-dimensional hypercube computing system every single computer is connected with n others. In this report, in order to ease the correspondence between the technical (physical) system and the graph model, only (duplex or half-duplex) connections by single links are discussed. For other connections see [1].

For ease of discussion (and to limit the length of this report) the following assumptions are made throughout :

S. Tzafestas et al. (eds.), System Fault Diagnostics, Reliability and Related Knowledge-Based Approaches, Vol. 1, 355–374.

1) Nodes of the hypercubes consist of a communications microprocessor and a host computer. The latter is the by far most unreliable unit. All computers (hosts) have the same unavailability U_h and the same MTTF (MTBF), viz. $(MTTF)_h$.

2) All system components are "good" or "bad" stochastically (s-) independent of each other. (The validity of this assumption must be checked carefully in cases of high-density packing of IC's and a tendency to limit the number of repairs and/or that of independent service personnel.)

3) Switching of spares is impossible with probability U_c being the unavailability of a communications microprocessor including half the links to all its neighbours.

4) Spare computers work as hot stand-by modules.

5) Indirect replacement, i.e. replacing a host, which is not a neighbour, is not practicable.

6) Each host is either a main unit or a spare. The system fails if at least one main unit fails which cannot be replaced by a proper spare. In other words : Any further cannibalization for further degradation is ruled out.

To shorten the presentation, only in sections 1 and 2 the cases of ideal and non-ideal links and communications processors are presented.

NOTATION

h,c are the indices for host computer & communications processor, respectively

$S_{i,j,\alpha}$ is a hypercube system of dimension i with j-FT of type α (if any).

ν_S is mean system failure frequency [5].

λ_i is the failure rate of unit i (of a hypercube).

μ_i is the repair rate of unit i.

X_i is the fault tree input variable of host computer i :

$$X_i = \begin{cases} 0 \text{ , if host i is "good" (up or intact)} \\ 1 \text{ , if host i is "bad" (down or defective)} \end{cases}$$

X_i' is the fault tree input variable of communications processor i.
$\vee$ is the conjunction (logical OR) symbol.
See appendix I for some more reliability notation.
The terms unit i, (system) component i, C_i are used as synonyms here. The term node, if used in a non-strictly graph theoretic meaning comprises the communications processor and its associated host. The numbering of the nodes is found by interpreting the binary number equivalents of the coordinate n-tuples, i.e.

$$(y_n, y_{n-1}, \ldots, y_3, y_2, y_1) \iff y_n\, y_{n-1} \cdots y_3 y_2 y_1$$

as binary numbers; see also fig. 1 for $n \leq 3$.

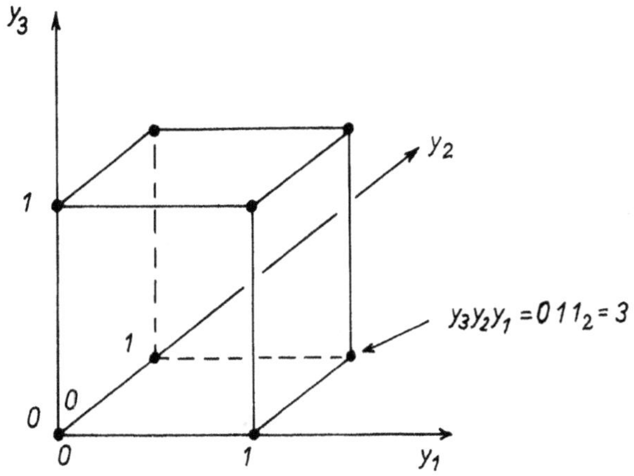

Figure 1. Notation for the vertices of the binary unit cube.

1. THE FIRST DIMENSION

1-FT :
Tolerating single faults in a one-dimensional hypercube can be trivially modeled by the single computing unit with one spare; see fig. 2.

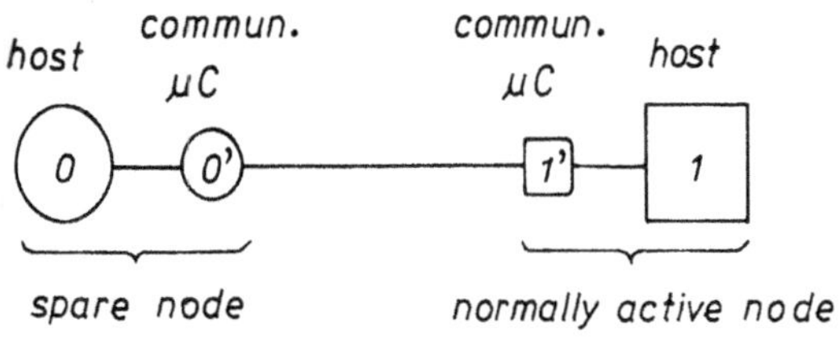

Figure 2. System $S_{1,1}$: 1-dimensional hypercube with 1 spare for 1-FT.

The relevant reliability theory for different stand-by and/or maintenance policies can be found e.g. in [6],[7],[8]. Let it suffice to give the redundancy structure function, i.e. the function for the top event of the fault tree : For ideal (infallible) communications micros and links

$$X_{S_{1,1}} = X_0 X_1 . \qquad (1\text{-}1)$$

Hence, with U for the unreliability or unavailability, for s-independent components, on taking the expected values of the X's of (1-1) :

$$U_{S_{1,1}} = U_h^2 \, . \tag{1-2}$$

Further, with the results of appendix I system-MTTF (MTBF) is

$$(\mathrm{MTTF})_{S_{1,1}} = \frac{1-U_h^2}{2U_h^2\mu_h} \approx \frac{1}{2U_h} \cdot (\mathrm{MTTF})_h \quad \text{for} \quad U_h \ll 1, \tag{1-3}$$

where $(\mathrm{MTTF})_h$ is the MTTF of any host. In words : Due to redundancy, system-MTTF grows by the factor $1/(2U_h) \gg 1$, if $U_h \ll 1$.

For non-ideal communications micros and links

$$X_{S_{1,1}} = X_1(X_0 \vee X_0' \vee X_1')$$

$$= X_1(X_0 + X_0' + X_1' - X_0X_0' - X_0X_1' - X_0'X_1' + X_0X_0'X_1'), \tag{1-1a}$$

whence

$$U_{S_{1,1}} \approx U_h^2 + 2U_hU_c \approx \text{for } U_h, U_c \ll 1$$

$$\approx U_h^2 \quad \text{for} \quad U_c \ll U_h \, . \tag{1-2a}$$

Note that a 2-units system is not self-diagnosable [3]. Clearly, there is also no 2-FT in the first dimension.

2. THE SECOND DIMENSION

1-FT :

Fig. 3a shows the case of the 2-dimensional hypercube, where the spares are not connected with each other, but only with the nodes to whom they belong as dedicated spares. Here the Boolean fault tree output function is (for ideal links and communications micros) :

$$X_{S_{2,1,a}} = X_0 X_1 \vee X_2 X_3$$

$$= X_0 X_1 + X_2 X_3 - X_0 X_1 X_2 X_3 \, . \tag{2-1}$$

As in the 1-dimensional case, for s-independent units, on taking the expected value of the last form of $X_{S_{2,1,a}}$ and using the obvious fact, that for Boolean X_i the expected value of X_i equals its probability of being 1, viz.

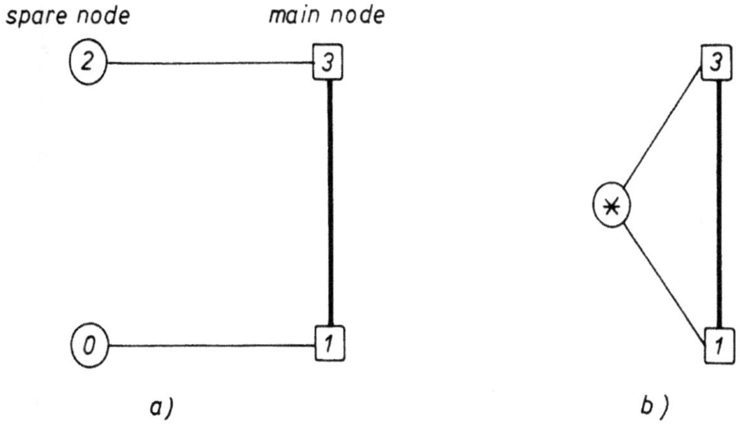

Figure 3. a) 2-dimensional hypercube with 2 spares for 1-FT. The main nodes 0 and 3 or their stand-by modules are supposed to work.
b) 2 main nodes with a single spare.

$$E\{X_i\} = P\{X_i=1\} = U_i \ : \tag{2-2}$$

$$U_{S_{2,1,a}} = 2U_h^2 - U_h^4 \approx 2U_h^2 \quad \text{for} \quad U_h \ll 1 \ . \tag{2-3}$$

Of course, also this result is plausible; see fig. 3. Further, from appendix I the mean frequency of system failures or restarts is

$$\nu_{S_{2,1}} \approx 4U_h^2\mu_h \quad \text{for} \quad U_h \ll 1 \ , \tag{2-4}$$

and the mean time to system failure is

$$(\text{MTTF})_{S_{2,1,a}} \approx \frac{1-2U_h^2}{4U_h^2\mu_h} \approx \frac{1}{4U_h\lambda_h} = \frac{1}{4U_h} (\text{MTTF})_h \quad \text{for} \quad U_h \ll 1 \ . \tag{2-5}$$

To achieve at least 1-diagnosability [3] there should be a link between nodes 0 and 2. 1-diagnosability is no problem in case of a common spare as indicated in fig. 3b, where one can speak of mergeing the two spares of fig. 3a. Obviously, the triangular structure is impossible in the framework of the links of a rectangle. However, the mergeing of nodes makes sense from an engineering point of view. As to addresses, the address of any of the two merged nodes will do.

Fig. 3b shows a typical 2-out-of-3 system, which is extremely well-known. Hence, without a lengthy derivation, we have (for ideal links and communications micros) :

$$U_{S_{2,1,b}} = 3U_h^2 - 2U_h^3 \approx 3U_h^2 \quad \textbf{for} \quad U_h \ll 1 , \tag{2-6}$$

$$\nu_{S_{2,1,b}} \approx 6U_h^2\mu_h \approx 6U_h\lambda_h \quad \text{for} \quad U_h \ll 1 , \tag{2-7}$$

and

$$(\text{MTTF})_{S_{2,1,b}} \approx \frac{1}{6U_h} (\text{MTTF})_h \quad \text{for} \quad U_h \ll 1 . \tag{2-8}$$

For non-ideal links and communications processors, by (1-1a), for nodes 1 and 3 to be operative

$$X_{S_{2,1,a}} = X_1 (X_0 \vee X_0' \vee X_1') \vee X_3 \; (X_2 \vee X_2' \vee X_3') , \tag{2-1a}$$

whence

$$U_{S_{2,1,a}} \approx 2U_h^2 + 4U_hU_c \quad \text{for} \quad U_h \ll 1 , \tag{2-3a}$$

and

$$\nu_{S_{2,1,a}} \approx 4U_h^2\mu_h + 4U_hU_c(\mu_h + \mu_c) \approx 1/(\text{MTTF})_{S_{2,1,a}} . \tag{2-4a}$$

If also the link between nodes 1 and 3 has to be 1-FT, then this link and the attached comm. processor must be duplicated. The approximation (2-3a) would still hold in this case.

Furthermore, as concerns fig. 3b,

$$X_{S_{2,1,b}} = (X_1 \vee X_1')(X_3 \vee X_3') \vee X_1 \; (X_* \vee X_1' \vee X_*') \vee X_3(X_* \vee X_3' \vee X_*') ,$$

whence

$$U_{S_{2,1,b}} \approx 3U_h^2 \quad \text{for} \quad U_c \ll U_h \ll 1 , \tag{2-6a}$$

which equals the approximation of (2-6). For $S_{2,1,b}$ a duplication of the link between nodes 1 and 3 is not necessary to keep these nodes connected, since there exists a 2nd path between these nodes via node *.

2 FT :

As to 2-FT, there are two basic possibilities, viz. using a) 3 or b) all 4 nodes of the square (2-dimensional hypercube); see fig. 4 a and b.

Case a) is simply that of the single node with 2 spares, yielding (for

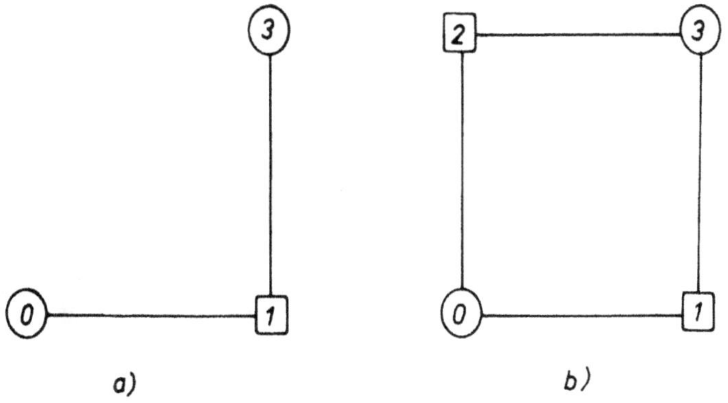

Figure 4. 2-FT 2-dimensional hypercubes.

ideal links and comm. processors) system unavailability

$$U_{S_{2,2,a}} = U_h^3 \, . \tag{2-9}$$

Mean system failure frequency is in this case

$$\nu_{S_{2,2,a}} = 3U_h^3\mu_h \approx 3U_h^2\lambda_h \, , \tag{2-10}$$

and mean time to system failure is in this case

$$(MTTF)_{S_{2,2,a}} \approx \frac{1}{3U_h^2} (MTTF)_h \, , \quad \text{for} \quad U_h \ll 1 \, . \tag{2-11}$$

To achieve system 1-diagnosability nodes 0 and 3 should be connected by another edge [3].

Case b) of fig. 4 can be modeled by a 2-out-of-4:G system.[1)] By elementary probability calculus (for ideal links and comm. processors):

$$U_{S_{2,2,b}} = 4U_h^3(1-U_h) + U_h^4 = 4U_h^3 - 3U_h^4$$

$$\approx 4U_h^3 \quad \text{for} \quad U_h \ll 1 \, . \tag{2-12}$$

1) This system fails if 3 or all 4 components fail.

Again, by appendix I

$$\nu_{S_{2,2,b}} \approx 12U_h^2\lambda_h \ , \tag{2-13}$$

and

$$(\mathrm{MTTF})_{S_{2,2,b}} \approx \frac{1}{12U_h^2} (\mathrm{MTTF})_h \ . \tag{2-14}$$

Note that the above model does not allow for any combined working of nodes 1 and 2 in fig. 4 b; however, this may pose no problems in dedicated applications of nodes 1 and 2.

In the framework of the PMC model [3] $S_{2,2,b}$ is only 1-diagnosable, whereas here 2-diagnosability would be more appropriate. On the other hand, the assumption of s-independent component states does not allow for simultaneous failures in two units with any appreciable probability. Hence with a very high probability (the higher, the quicker system self-diagnosis is possible [3]) the first faulty unit will be detected before a second one fails.

For non-ideal links and communications processors, instead of (2-9)

$$X_{S_{2,2,a}} = X_1(X_1' \vee X_3' \vee X_3)(X_1' \vee X_0' \vee X_0) \ ,$$

whence

$$U_{S_{2,2,a}} \approx U_h^3 + 4U_hU_c + 4U_c^2 \quad \text{for} \quad U_h \ll 1 \ , \tag{2-9a}$$

and

$$\nu_{S_{2,2,a}} \approx 3U_h^2\mu_h + 4U_hU_c(\mu_h + \mu_c) + 8U_c^2\mu_c \approx 1/(\mathrm{MTTF})_{S_{2,2,a}} \ . \tag{2-10a}$$

Furthermore,

$$\begin{aligned} X_{S_{2,2,b}} = {} & X_1(X_1' \vee X_3' \vee X_3)(X_1' \vee X_0' \vee X_0) \\ & \vee X_2(X_2' \vee X_3' \vee X_3)(X_2' \vee X_0' \vee X_0) \\ & \vee X_1X_2(X_2' \vee X_3' \vee X_3) \vee X_2X_1(X_1' \vee X_0' \vee X_0) \ ^{1)} \end{aligned}$$

---

1) In this case it is assumed (without loss of generality) that, with highest priority, nodes 0 and 3 try to replace nodes 1 and 2, respectively.

yielding, instead of (2-12),

$$U_{S_{2,2,b}} \approx 4U_h^3 + 12U_h^2U_c + 8U_hU_c^2 \quad \text{for} \quad U_h \ll 1 \; , \qquad (2\text{-}12a)$$

and

$$\nu_{S_{2,2,b}} \approx 12U_h^2\mu_h + 12U_h^2U_c(2\mu_h + \mu_c) + 8U_hU_c^2(\mu_h + 2\mu_c) \; . \qquad (2\text{-}13a)$$

3. THE THIRD DIMENSION

The strict sense cube is possibly a highly attractive building block for mass production. It is easily layed out as a graph that is planar, i.e. without intersections of edges; see fig. 5.

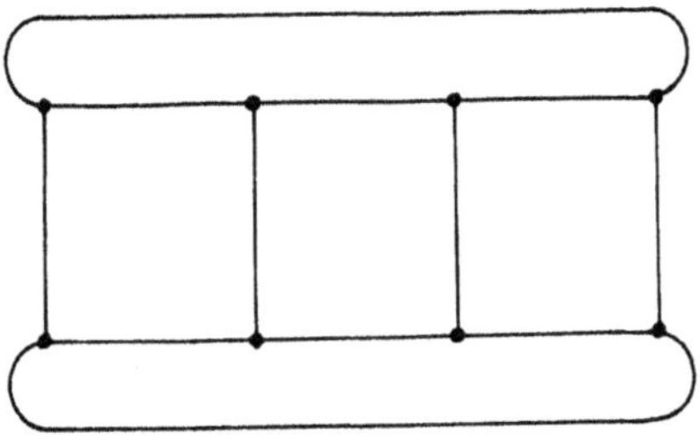

Figure 5. Planar figure of a cube as a cyclic "ladder".

1-FT :
Fig. 6 shows the cube (3-dimensional hypercube), where nodes 0 and 7

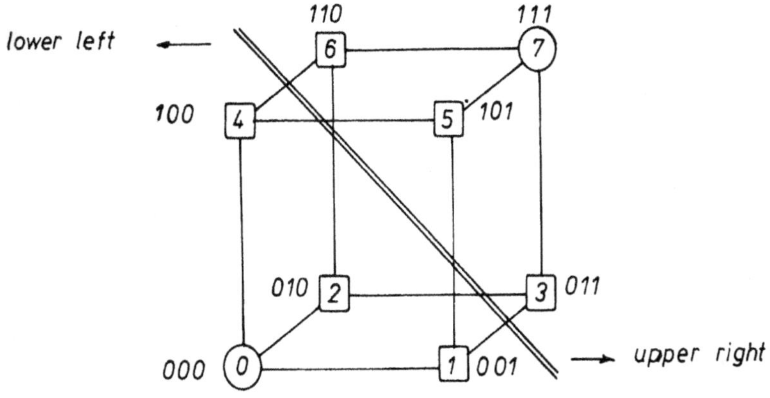

Figure 6. Cube with 2 spares (round nodes) for 1-FT.

have been chosen as spares. It is important that both spares serve non-intersecting sets of neighbours. All the 8-2=6 main units (nodes) are connected by a single cycle of links; see also fig. 7, which is a planar version of fig. 6 and an alternative of fig. 5.

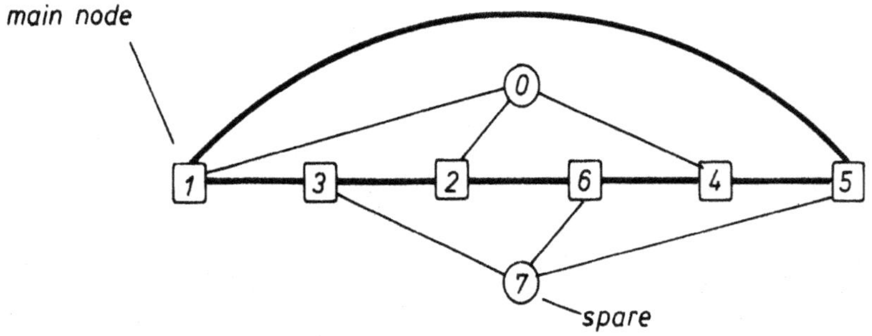

Figure 7. Plane version (plane graph) of fig. 6. Fat links connect main units.

Obviously, the system is in a satisfactory (useful) state as long as in the subsystems in the lower left and the upper right half, respectively, of fig. 6 at least 3 of the 4 nodes are faultless. Since for the mentioned subsystems (for ideal communications processors and links) by appendix II

$$U_{\ell\ell} = U_{ur} \approx 6U_h^2 \quad \text{for} \quad U_h \ll 1 , \tag{3-1}$$

here system unavailability in case of s-independent component states is

$$U_{S_{3,1}} = U_{\ell\ell} + U_{ur} - U_{\ell\ell}U_{ur} \approx 12U_h^2 \quad \text{for} \quad U_h \ll 1 . \tag{3-2}$$

Using the results of appendix I

$$(\text{MTTF})_{S_{3,1}} \approx \frac{1}{24U_h} (\text{MTTF})_h \quad \text{for} \quad U_h \ll 1 . \tag{3-3}$$

Note that, but for the 12 untolerable, there are 16 tolerable double faults, since $\binom{8}{2}=28$. Clearly, for the fat ring of fig. 7 to be 1-FT, all its links must be duplicated.

2-FT :

As in the 2-dimensional case, there are several options for 2-FT. A closed ring of main units is not possible, since the degree of any node is 3, and each node is to be linked to (at least) two spares. Hence, at most connected pairs of main units can be achieved; see fig. 8 a and b, which corresponds to fig. 4 a and b.

In fig. 8a two edges should be added, linking any pair of spares to achieve at least 1-diagnosability [3]. For 2-diagnosability such edges are desirable, by which one obtains cycles of minimal length 5 [3]; e.g. edges connecting vertices 0 and 4, and 3 and 7, respectively.

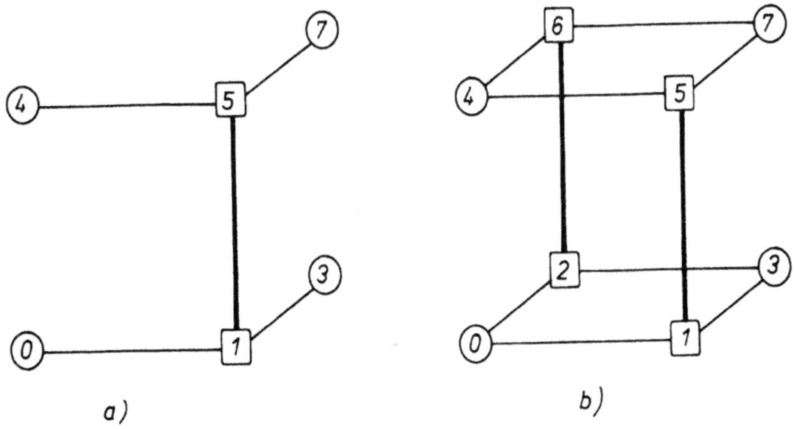

Figure 8. Use of a cube to achiev 2-FT for one (a), and two (b) pairs of linked computers.

As concerns fig. 8, one might think of reducing the number of spares from 4 to 2 by proper merging of the left and the right hand pairs of spares, respectively. Fig. 9 shows planar versions of the result. The system of fig. 9a is 2-FT but only 1/2 of the 8 units of a cube are used, and the system of fig. 9b is not 2-FT, since simultaneous faults, of e.g., nodes 0 and 1 are not tolerable, i.e. in that case there would be no substitute for C_1. Finally, as a 1-FT system this system is no interesting alternative to that of fig. 8b, since with standard mass produced (8-node) cubes the latter is a better choice. Disregarding the last argument, fig. 9c would be a 2-FT compromise between the systems of figs. 9b and 8b.

The system modeled by fig. 8b is 1-diagnosable. To achieve 2-diagnosability, i.e. cycles of minimum length 5 [3], only one further link needs to be added.

If the connectedness of main units is of no importance, better use of a given degree of redundancy can be made in the extremely symmetrical

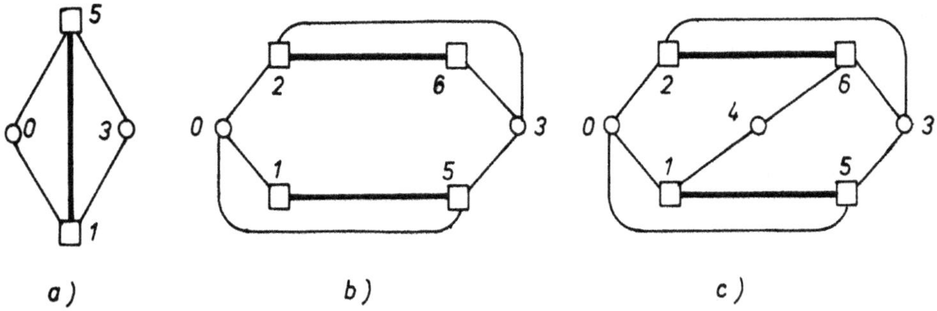

Figure 9. Fig. 8 after reducing the number of spares.

topology of fig. 10, where every spare can replace one of 3 main units or other spares, which were switched in before and have failed in the mean time. After a second look, it becomes obvious that the topology of fig. 10 is even 3-FT. Yet it is not 4-FT, even though all 4 main units are allowed to be faulty at the same time. In the subsequent discussion

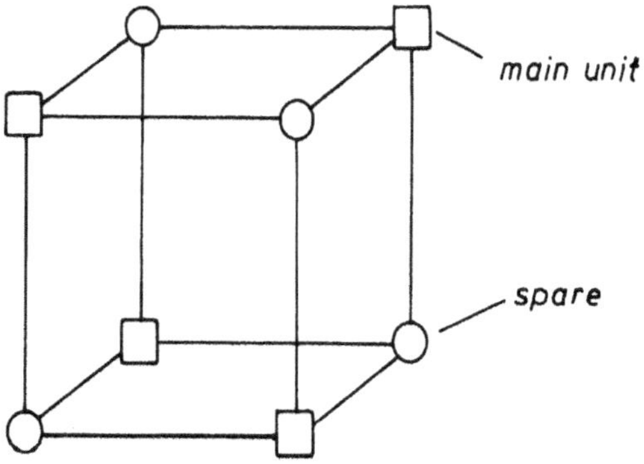

Figure 10. Extremely symmetrical spatial distribution of main units and spares.

of system reliability the 3-FT case will be omitted and the systems modeled by fig. 8 a and b, respectively, will be $S_{3,2,a}$ and $S_{3,2,b}$, respectively.

Since in $S_{3,2,a}$ the upper or the lower 1-out-of-3:G systems of fig. 8a must fail for system failure (in case of ideal links and comm. processors)

$$U_{S_{3,2,a}} = 2U_h^3 - U_h^6 \approx 2U_h^3 \quad \text{for} \quad U_h << 1 . \tag{3-4}$$

Hence, by appendix I

$$(MTTF)_{S_{3,2,a}} \approx \frac{1}{6U_h^2} (MTTF)_h \quad \text{for} \quad U_h << 1 . \tag{3-5}$$

Since in $S_{3,2,b}$ the upper or the lower 2-out-of-4:G system of fig. 8b must fail for ideal links and communications processors) from (2-12):

$$U_{S_{3,2,b}} = 2(4U_h^3 - 3U_h^4) - (4U_h^3 - 3U_h^4)^2 \approx 8U_h^3 \quad \text{for} \quad U_h << 1 . \tag{3-6}$$

Hence, by appendix I

$$(MTTF)_{S_{3,2,b}} \approx \frac{1}{24U_h^2} (MTTF)_h \quad \text{for} \quad U_h << 1 . \tag{3-7}$$

4. THE FOURTH DIMENSION

A perspective pictures of the 4-dimensional hypercube is given in fig. 11, whereas fig. 12 is a plane representation of the same graph

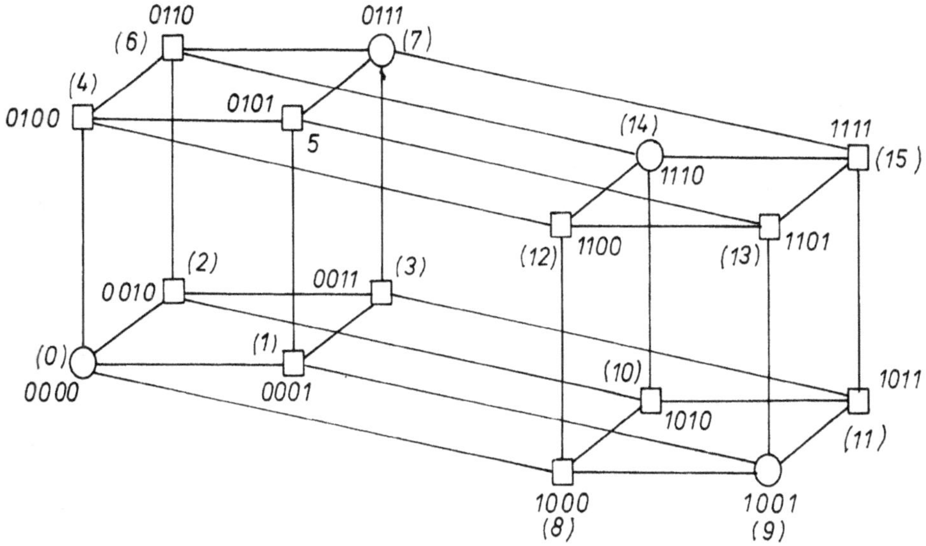

Figure 11. 4-dimensional hypercube.

emphasizing the links between main units by fat lines.

1-FT :

Figure 11 shows a plausible positioning of the spares for 1-FT. In the numbering (addresses) of the nodes, with every new dimension another bit is appended at the left, as was already practiced in sections 2 and 3.

The very first problem to be solved is the economical definition of spare nodes. For the upper cube in fig. 11 this is clear from fig. 6. For the lower one not the "same" corners, i.e. not nodes 8 and 15 should be chosen, since the old stand-by nodes (of the upper cube) can serve as stand-by nodes also for them. Hence, one is free to choose nodes 9 and 14 or 11 and 12 for stand-by nodes for the lower cube. In fig. 11 the first choice is visualized.

Fig. 12 shows, better than fig. 11 does, the special cyclic "ladder" type of the primarily working components. This certainly

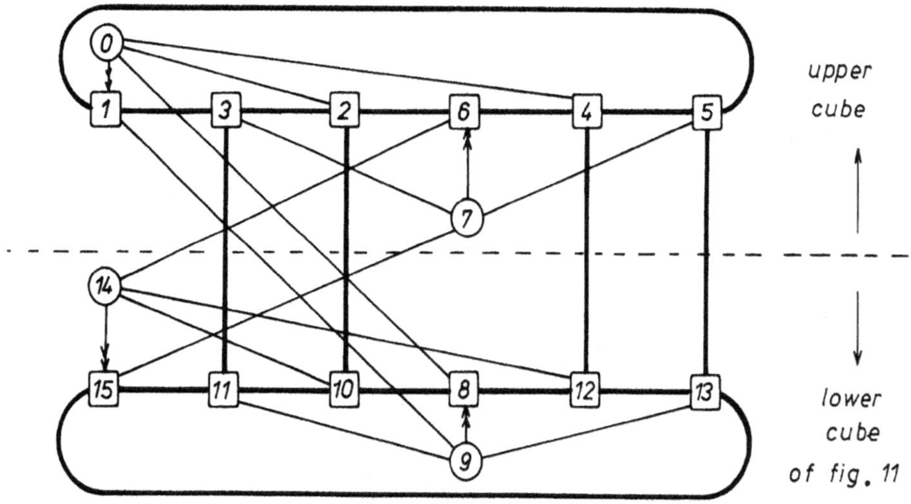

Figure 12. Plane (not planar) version of fig. 11.

reveals, that not too much of the original degree of interconnectedness had to be sacrificed to 1-FT. For ideal links and comm. processors

$$U_{S_{4,1}} \approx 12U_h^2 \quad \text{for} \quad U_h << 1 \, . \qquad (4\text{-}1)$$

This can be verified easily, using fig. 12, as follows :

1) No system failures occur if (simultaneouly) only single faults in the upper and the lower half of fig. 12 exist.
2) In each (upper or lower) half of fig. 12 there are 2 pairs of untolerable faults in primary components (main units) and 4 pairs of untolerable faults of a primary module and a spare module.

All in all this amounts to 12 untolerable double faults, yielding (4-1). Hence, from appendix I for $U_h \ll 1$

$$(MTTF)_{S_{4,1}} \approx \frac{1}{24U_h} (MTTF)_h \ . \qquad (4\text{-}2)$$

Notice that, as to unavailability and MTTF, there is little difference between $S_{4,1}$ and $S_{3,1}$.

Reconfiguration is not always trivial for multiple faults (which are supposed to occur sequentially). It may be necessary to readjust former replacements. For instance, in case C_1 fails first and is replaced by C_0, and C_2 fails soon after, then, if C_9 is still free and good, C_9 will replace C_1, and C_0 will replace C_2, since, otherwise, the system would fail because C_2 could not be replaced. Hence, the replacement of C_i, which has only one possible spare, has highest priority.

2-FT :
Fig. 8 a can be (trivially) extended to yield fig. 13. Note that the main system is a cycle, which may be important for certain applications.

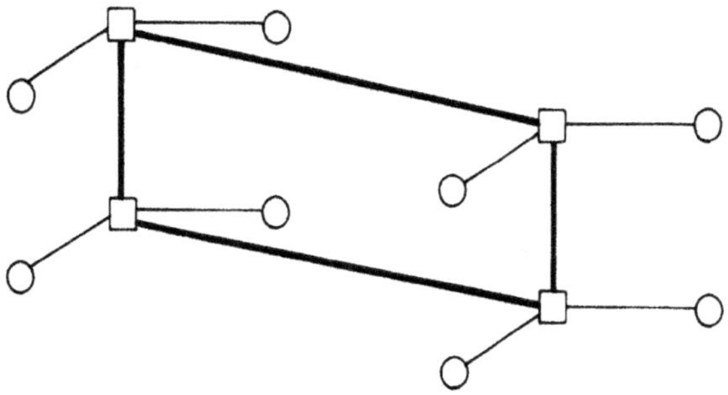

Figure 13. Extension of fig. 8 a to the fourth dimension.

The standard reliability analysis of this report (for ideal links and comm. processors) is extremely simple here, since all the spares remain dedicated spares. So, from (3-4)

$$U_{S_{4,2,a}} = 2U_{S_{3,2,a}} - U^2_{S_{3,2,a}} = 2(2U_h^3 - U_h^6) - (2U_h^3 - U_h^6)^2$$
$$\approx 4U_h^3 \ \text{for} \ U_h \ll 1 \ . \qquad (4\text{-}3)$$

Hence,

$$(MTTF)_{S_{4,2,a}} \approx \frac{1}{12U_h^2} (MTTF)_h \ . \qquad (4\text{-}4)$$

Fig. 8 b is readily extended to yield fig. 14, where the work (main) system is modeled by a disconnected graph with two rings.

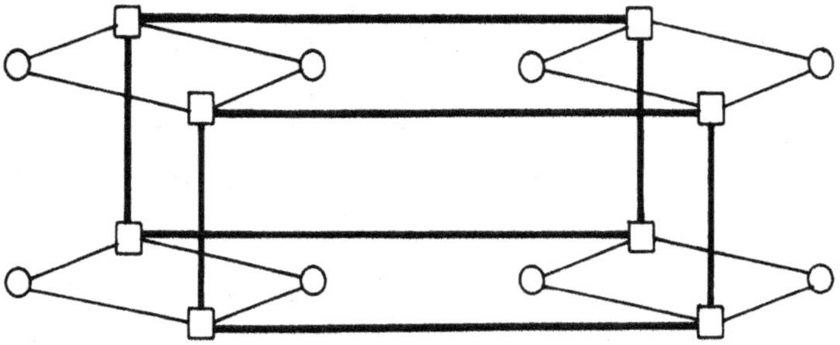

Figure 14. Extension of fig. 8 b to the 4-th dimension.

Since, again, all the spares are only linked to nodes in the same cube, by (3-6)

$$U_{S_{4,2,b}} = 2U_{S_{3,2,b}} - U_{S_{3,2,b}}^2 \approx 16U_h^3 \quad \text{for} \quad U_h << 1 \ . \qquad (4\text{-}5)$$

Hence,

$$(MTTF)_{S_{4,2,b}} \approx \frac{1}{48U_h^2} (MTTF)_h \ . \qquad (4\text{-}6)$$

Once again one might think of reducing the number of spares in figs. 13, 14, e.g. by pairwise merging the inner and the outer pairs of spares. The results would be - in planar version - figs. 15 a and b. Fig. 15a shows a 2-FT "rest" of a 4-dimensional hypercube. However, like the system of fig. 9b, the system of fig. 15b is not 2-FT, and as a 1-FT system it is of little interest in case of low-price standard VLSI hypercubes. Again disregarding the last argument (as in the 3-dimensional case) fig. 15 c shows a 2-FT-compromise with 4 spares instead of the 8 spares of fig. 14.

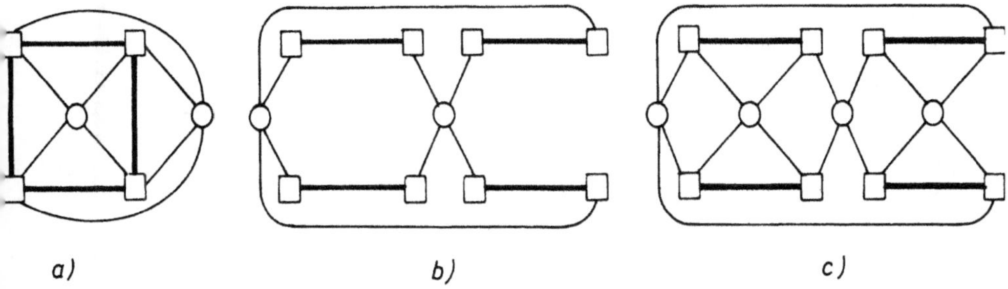

Figure 15 . Simplified versions of figs. 13 (a), and 14 (b).

As in the case of the configurations of fig. 9, those of fig. 15 are not discussed here as to reliability parameters.

5. HIGHER DIMENSIONS

From fig. 11 it is fairly obvious how n-dimensional hypercubes (n>4) are constructed. At every incrementation of the value of n from k to k+1 the k-dimensional hypercube is first duplicated and then all the old binary node-numbers (addresses) get a leading 0, the new ones get a leading 1, and, finally, every old node is linked to that new one whose address differs only in the most significant bit.

As to (a possible future) VLSI implementation a very good topology for n=3 is given in fig. 5, and for n≧4 one could use cyclic systolic arrays,as the one of fig. 16 for n=4,as building blocks [1].

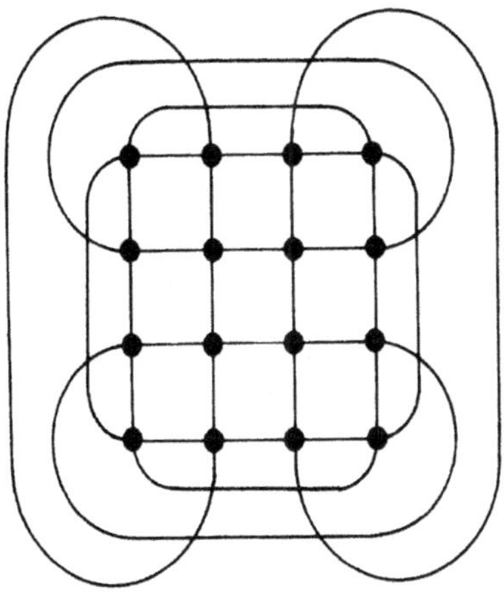

Figure 16. Plane, though not planar, version of fig. 11.

Fault tolerance can always be achieved by first designing the "work" system, consisting of the main nodes, say as a k-dimensional hypercube and then adding further dimensions. Clearly, for 1-FT and 2-FT incrementing k by one would suffice. Moreover, in principle, for j-FT j spares would do, each of which would have to be linked with all the main units. However, in the (k+1)-dimensional hypercube any node is linked with at most k+1 others. Hence, the minimum number of spares for j-FT is (for 2^k main units)

$$N_{sp,min} = 2^k \lceil j/(k+1) \rceil. \tag{5-1}$$

Proof : In case of j-FT there are j links per main unit ending in spares. By assumption, there are 2^k main units. In the (k+1)-dimensional hypercube every node has degree k+1. Finally observe, that the number $j2^k$ of links going to spares must equal the number coming from N_{sp} spares, at most k+1 from any of them. Hence

$$(k+1)N_{sp} \geq j\,2^k , \tag{5-2}$$

from which (5-1) is a trivial consequence, q.e.d.. Since, by assumption 5, only the above incrementation of the dimension from k to k+1 makes sense, the maximum number of spares for the k-dimensional work system is (for j-FT)

$$N_{sp,max} = 2^k . \tag{5-3}$$

Note that j-FT means that from every main unit at least j links are adjacent to spares, so that at most k+1-j links can be adjacent to other main units' nodes. Hence the maximum j of j-FT is

$$j_{max} = k+1 , \tag{5-4}$$

in which case the main units are isolated.

Rough unavailability approximations are usually not very difficult to find. However, for more accurate analyses the construction of the relevant fault trees may not be trivial.

As to the number of n-dimensional hypercube systems the following is obvious : Since each of the 2^n nodes can be either a main unit, a spare unit or "empty" [1], there are 3^{2^n} different n-dimensional hypercube configurations. This number also includes the cases where nodes are merged since also in such cases the degree of all nodes is at most n.

---

1) synonymous of non-existent

APPENDIX

I) RELATIONS BETWEEN BASIC TERMS OF REPARABLE SYSTEMS RELIABILITY THEORY

With ν_S for mean system failure frequency (or restart frequency)

$$\nu_S = \frac{1}{(MTTF)_S + (MTTR)_S}, \tag{A1-1}$$

$$U_S = \frac{(MTTR)_S}{(MTTF)_S + (MTTR)_S} \approx \frac{\lambda_S}{\mu_S}, \tag{A1-2}$$

where

MTTF = {mean time to failure}={mean up-time} = $1/\lambda$,
MTTR = {mean time to repair}={mean down-time} = $1/\mu$,
λ = {failure rate},
μ = {repair rate},

one gets system MTTF (sometimes called MTBF) as

$$(MTTF)_S = (1-U_S)/\nu_S \approx 1/\nu_s \quad \text{for} \quad U_S \ll 1 . \tag{A1-3}$$

As proved in [5], if U_S is a polynomial in the unavailabilities U_1, U_2, ... of its components, ν_S follows from U_S by multiplying every term U_iU_j ... by $(\mu_i+\mu_j+\ldots)$, and U_i^j by $j\mu_i$. (A1-4)

II) UNAVAILABILITY AND MTBF OF THE 3-OUT-OF-4:G SYSTEM[1)]

By elementary probability calculus (and combinatorics) for 4 s-independent equal components (with 3of4 to abbreviate 3-out-of-4)

$$\begin{aligned} U_{3of4} &= 6U^2(1-U)^2 + 4U^3(1-U) + U^4 \\ &= 6U^2 - 12U^3 + 6U^4 + 4U^3 - 4U^4 + U^4 \\ &= 6U^2 - 8U^3 + 3U^4 \\ &\approx 6U^2 \quad \text{for} \quad U \ll 1 . \end{aligned} \tag{A2-1}$$

Clearly, for very small U this approximate value is too big, and for bigger U, viz. for U>8/3 it would be too small. However, as a probability U≦1 always. Hence

1) This system is down if at least two components are down, and there are 6 pairs of component, and 4 triples of components in this system.

$$U_{3of4} < 6U^2. \tag{A2-2}$$

From (A1-4)

$$\nu_{3of4} = (12U^2 - 24U^3 + 12U^4)\mu$$

$$\approx 12U^2\mu \approx 12U\lambda \ . \tag{A2-3}$$

By the same type of argument as for (A2-2), here

$$\nu_{3of4} < 12U^2\mu \ . \tag{A2-4}$$

Finally, by (A1-3)

$$(MTTF)_{3of4} \approx \frac{1}{12U} (MTTF) \quad \text{for} \quad U \ll 1 \ . \tag{A2-5}$$

REFERENCES

[1] Rennels, A.: On implementing fault-tolerance in binary hypercubes. Proc. FTCS (fault tolerant computing systems) 16, July 1986, New York : IEEE-Press, pp. 344-349.

[2] Wallich P., Zorpette G.: Minis and mainframes. IEEE Spectrum, Jan. 1986, 36-39.

[3] Preparata F., Metze G., Chien R.: On the connection assignment problem of diagnosable systems. Trans. IEEE, vol. EC-16 (1967), 848-854.

[4] Kuhl J., Reddy S.: Fault diagnosis in fully distributed systems. IEEE : Proc. FTCS-11 (1981), 100-105.

[5] Schneeweiss, W.: Addendum to computing failure frequency via mixed products of availabilities and unavailabilities. Trans. IEEE-vol. R-32 (1983), 461-462.

[6] Barlow R., Proschan F.: Statistical theory of reliability and life testing (probability models). New York : Holt, Rinehart & Winston 1975.

[7] Singh C., Billinton R.: System reliability modelling and evaluation. London : Hutchinson 1977.

[8] Schneeweiss W.: Zuverlässigkeits-Systemtheorie (Reliability systems theory). Köln : Datakontext, 1980. (in German).

NETWORK OPTIMIZATION SUBJECT TO RELIABILITY CONSTRAINTS

A.N. Venetsanopoulos and I. Singh
Department of Electrical Engineering
University of Toronto
Toronto,Ontario,Canada, M5S 1A4

One factor which influences the reliability and cost of a computer network is its topology. In this paper the problem of optimizing the topology of a network, subject to reliability constraints is examined. A reliability measure is defined and a new heuristic optimization procedure is developed. The method is demonstrated with one example.

1. STATEMENT OF THE PROBLEM

In this paper, we shall be concerned with the following problem:

Given	-	The node locations of a computer network
Minimize	-	The total communications cost
Over the design variables	-	Topology of the network
Subject to	-	Reliability constraints

Our objective is to design minimum cost computer networks satisfying specified reliability constraints. In doing so, a new reliability measure is proposed. In addition, a new heuristic procedure using optimization techniques is established.

2. ASSUMPTIONS MADE

The following assumptions, similar to those used by Leggett et al [1], were made about these networks:

(1) The nodes of the network are perfect: i.e., each node exists with probability one.

(2) The branches are uncorrelated: i.e., the state of one branch does not affect the state of any other branch.

(3) Branch i is entirely ON with probability p_i and OFF with probability $q_i = 1 - p_i$.

(4) Branch capacities are not considered. They are assumed to be always adequate.

(5) The system is stationary: i.e., it has reached a steady state.

Practical networks satisfying these assumptions are conceivable. For example, node availability approaching one is easily achieved by duplicating the processing unit. Although this is an expensive

S. Tzafestas et al. (eds.), System Fault Diagnostics, Reliability and Related Knowledge-Based Approaches, Vol. 1, 375–386.

practice it is frequently done, where high security is required.

3. TERMINAL RELIABILITY OF A NETWORK

Two very efficient methods for determining the technique reliability of a network, whose components fail independently, are the minimal cut-set and tie-set methods [2-7]. These techniques were developed from the properties of the reliability graph.

The term cut-set is defined as a set of branches, which interrupts all connections between the source node and the destination node when it is removed from the graph. The minimal cut-set is a group of distinct cut-sets containing a minimum number of terms. All failures can be represented by the removal of at least one minimal cut-set from the graph. The probability of system failure is, therefore, given by the probability that at least one minimal cut-set fails. If C_1, C_2, ..., C_n represent the n minimal cut-sets and $\bar{C}_n$, the failure of the nth cut-set, then the probability, P_f, of system failure is given by:

$$P_f = P(\bar{C}_1 \cup \bar{C}_2 \cup \cdots \cup \bar{C}_n) \qquad (1)$$

Consequently, the probability of system success, P_s, or its reliability R is given by:

$$R = P_s = 1 - P_f = 1 - P(\bar{C}_1 \cup \bar{C}_2 \cup \cdots \cup \bar{C}_n) \qquad (2)$$

In the tie-set method, the system fails when all minimal tie-sets or paths are broken. A minimal tie-set is defined as a tie-set having a minimum number of branches and is obtained by traversing a node only once in tracing out the path from the source node to the destination node. If a system has n minimal tie-sets denoted by $T_1, T_2, \ldots, T_n$, then the system has a connection between the source and destination nodes, if there is at least one minimal path intact. The probability of having such a connection or the probability of success P_s, is given by:

$$P_s = P(T_1 \cup T_2 \cup \cdots \cup T_n) \qquad (3)$$

if the minimal paths are mutually exclusive, then the system reliability R, is given by:

$$R = P_s = P(T_1) + P(T_2) + \cdots + P(T_n) \qquad (4)$$

A simple and efficient method for finding the terminal reliability of a network, using the tie-set approach, is given in [3]. Basically, the method consists of the following steps:

(1) Determine all minimal tie-sets (paths).
(2) Generate the mutually disjoint paths.
(3) Calculate the reliability by using the equation for R, above.

4. OVERALL NETWORK RELIABILITY

The terminal reliability discussed in the previous section is widely used as a measure of the overall network reliability. This is a good measure, if the pair of nodes selected for study is the one most likely to be disconnected. The difficulty with this method lies in the selection of the pair of nodes in the worst condition. For analysis this may not cause trouble but for synthesis no tractable analytical expression exists, since the pair of nodes in the worst condition may change by slightly changing the network's topology or the unavailability of the network components. To get around this problem other measures have been developed. For example, the "connectivity" or "probability of connectedness" of the network, defined as the probability that the network is simply connected [8]. However, computation of this probability is very complex. Kelmans [8] investigated networks using connectedness as the reliability criterion and concluded that the topology of maximally reliable networks depends on the magnitude of the link failure probability. This complicates the problem of constructing optimum networks and makes it difficult, if not impossible, to derive an analytical solution.

5. PROPOSED RELIABILITY MEASURE

In this work, a new reliability measure is proposed. This measure is a simple extension of the terminal reliability measure and is defined as "the average probability of any two nodes in the network being connected". Often, in a practical network, the resources and computational load of each computer vary, thus having certain computer centres connected may be more desirable than having others. In view of this, the overall distributed network reliability is given by:

$$R(N) = \frac{\sum_{ij} W_{ij} P_{ij}}{\sum_{ij} W_{ij}} \quad (i = 1, N-1, \; j = i + 1, N) \qquad (5)$$

where N is the number of computer centers or nodes in the network, W_{ij} is the weight associated with the connection between the ith and j^{th} nodes and P_{ij} is the probability of having a connection between nodes i and j.

6. COMMUNICATION LINK RELIABILITY

The factor which affects the overall network reliability is the reliability of the individual communication links. In general, a link consists of a series arrangement of terminal equipment, such as transmitters, receivers, equalizers, etc., and a communication channel, wire or radio (microwave). Associated with each component is a failure probability. which contributes to the overall link reliability.

For this study a simplified model is used. In this model, the terminal equipment, local loops, interface equipment, etc., are lumped together with an equivalent failure probability assigned to them.

Assume that terminals A_i and B_i, of link i, have unavailabilities q_{A_i} and q_{B_i} respectively and the channel C_i has an unavailability that is a function of its length, ℓ, $q_{C_i}(\ell)$, then the availability of link i, P_i, is given by:

$$P_i = (1 - q_{A_i})(1 - q_{B_i})(1 - q_{C_i}(\ell)) = P_{T_i}(1 - q_{C_i}(\ell)) \quad (6)$$

where

$$P_{T_i} = (1 - q_{A_i})(1 - q_{B_i}) \quad (7)$$

P_{T_i} is the availability of all equipment associated with link i.

In general, the unavailability of the equipment associated with each link is assumed to be constant. This is a reasonable assumption if it is assumed that the equipment has already passed through its "burn-in" period and has not yet reached its "wear-out" stage. That is, the equipment availability is affected only by random failures.

The channel availability, however, is a function of its length. Some models of the channel were presented in [2]. In one of these models the channel was assumed made of a series of elementary hops (microwave connection) of equal length λ, where the failure of each hop was statistically independent. Assuming the experimental function relating the hop length to its unavailability is $h(\lambda)$, then the unavailability of channel i of length L_i is given by:

$$q_{C_i} = 1 - [1 - h(\lambda_i)]^{L_i/\lambda_i} \simeq (L_i/\lambda_i)\,[h(\lambda_i)] \quad (8)$$

Experimental results given in [2] show for hop lengths of 7.5 miles and 15 miles, unavailabilities of 5.4×10^{-5}/mile and 2.0×10^{-4}/mile were found respectively.

7. NETWORK COST

The overall network cost is primarily dependent on the associated hardware cost (storage facilities, concentrators, etc.), the software cost (control algorithms) and the communication cost. Since the communication cost constitutes the major part of the network cost, it is reasonable to consider this cost only in the optimization process. Thus, the network cost C(N) is given by:

$$C(N) = \sum_{i=1}^{n-1} \sum_{j=i+1}^{n} s_{ij} c_{ij} \quad (9)$$

where: s_{ij} represents the i,j entry in the connection matrix $S(N_k)$

$s_{ij} = 1$ indicates a link exists between nodes i and j

$s_{ij} = 0$ indicates no link exists between nodes i and j

c_{ij} denotes the i,j entry in the cost matrix [C], and represents the cost of establishing a link i,j between nodes i and j. The cost matrix is symmetric, signifying the cost of establishing link i,j and j,i is the same. Physically, for full or half duplex communication i,j and j,i designate the same link. In practical situations the link cost would be determined by the applicable tariffs and would depend on the line capacity, type of connection and line leased, distance between nodes, etc.

8. THE OPTIMIZATION ALGORITHM

The network optimization problem considered in this paper requires the minimization of the communication cost, subject to reliability constraints, with the network topology being the design variable. The constraint imposed is the network reliability, R(N), be greater than or equal to some value ε, less than 1, ($R(N) \geqq \varepsilon, 0 < \varepsilon < 1$). In solving this problem there are many practical considerations, which make it extremely difficult to derive computationally feasible algorithmic solutions. Consequently, heuristic approaches are often used. In this section, a simple heuristic algorithm has been constructed to solve the optimization problem.

The algorithm proposed is based on two optimization techniques, namely: Marginal analysis and branch-and-bound. Basically, given a starting network configuration, the marginal analysis technique is used to arrive at a feasible solution, then the branch-and-bound technique takes over in an attempt to find the local optimum solution. Repeating this process for different starting network configurations and selecting the best local solution, gives an approximation of the true or global optimum solution. A block diagram of the procedure is shown in Figure 1. A detailed description of the procedure is available from the authors.

9. EVALUATION OF THE ALGORITHM

To get an appreciation for the saving in computer time, the algorithm is compared with a "restricted" form of the exhaustive search method. In the full exhaustive search method, all the topologies generated by the addition of one or more links to the starting network are examined. However, in the "restricted form, only the topologies generated by the addition of a specified number of links to the starting network are investigated. To some extent, this is similar to the branch exchange method, where given an arbitrary starting configuration, it searches through all the local solutions by eliminating one or more old links and inserting one or more new links.

Table 1 shows the computer time required for solving some small networks, using both methods, the algorithm proposed in this work and the "restricted" form of the exhaustive search method. The starting configuration, the reliability constraint and the node-pair information were the same for both methods. It is worthwhile to point out the solutions obtained were identical for both cases. It is clearly seen that the algorithm proposed in this paper is more efficient, with the saving in computer time increasing considerably with the size of the network.

10. EXAMPLE

Consider a network with nodes located on cities across Canada, as shown in Figure 2. The node-pair information is given in Table 2. Note, the weights were chosen on the assumption that it is very important for Toronto to communicate with all other cities. Also, the cost of the links were calculated on the basis of Telepak rates for a channel capacity of 19.2 kbits/sec (1 x 19.2 kbits/sec - termination cost $850/month and mileage cost $2.50/month/mile). The total link availability is based on a terminal reliability of 0.9 and channel unavailability of 5.4×10^{-5}/mile (based on 7 mile hops). For the starting configuration shown in Figure 2, the cost-reliability relationships for varying degrees of redundancy are given in Table 3 and the optimum solutions for different reliability constraints were given in Table 4. The corresponding cost-reliability plots are shown in Figure 3. From which it is possible for the designer to evaluate the network cost for a required reliability.

11. CONCLUSIONS

The topology of computer networks influences both their cost and reliability. In this paper the optimization of the network's topology, subject to specified reliability constraints was studied. The reliability measure proposed here is a simple extension of the terminal reliability method and takes into consideration the effects of all the terminal pair reliabilities in the network. In addition, this measure is very simple to compute. Like other methods used to compute network reliability, the disadvantage with this method is for large distributed networks, where the computer time could be long, since for a network with n nodes, ½ (n) (n-1) terminal reliabilities must be evaluated. However, by considering only the important (higher weighted) node-pairs, approximate solutions could be obtained with considerable reduction in computer time.

The results confirm that the topology of maximally reliable networks depends on the availability of the individual network links. As this makes it difficult to derive computationally feasible algorithms, heuristic techniques are used. In general, the heuristic generated optimal solution is only an approximation of the true or global optimum. The heuristic procedure presented here employs the marginal analysis and the branch-and bound optimization techniques. This procedure reduces the search process and has proven efficient and practical. Additional results are given in [9].

12. REFERENCES

[1] L.D. Leggett and S D. Bedrosian, "Synthesis of Reliable Networks", IEEE Trans. on Circuit Theory, August 1969.

[2] L. Fratta and U.G. Montanari, "Synthesis of Available Networks", IEEE Trans. on Reliability, Vol. R-25, No.2, June 1976.

[3] S. Rai and K.K. Aggarwal, "An Efficient Method for Reliability Evaluation of a General Network", IEEE Trans. on Reliability, Vol. R-27, No.3, August 1978.

[4] K.K, Aggarwal and K.B. Misra, "A Fast Algorithm for Reliability Evaluation", IEEE Trans. on Reliability, Vol. R-24, No.1, April 1975.

[5] Y.H. Kim, K.E. Case and P.M. Ghare, "A Method for Computing Complex System Reliability", IEEE Trans. on Reliability, Vol. R-21, No.4, November 1972.

[6] J. DeMercado, N. Spyratos and B.A. Bowen, "A Method for Calculation of Network Reliability", IEEE Trans. on Reliability, Vol. R.25, No.2, June 1976.

[7] A.C. Nelson, J.R. Batts and R.L. Beadles, "A Computer Program for Approximating System Reliability", IEEE Trans. on Reliability, Vol. R-19 No.2, May 1970.

[8] A.K. Kelmans, "Connectivity of Probabilistic Network", Automat. Remote Contr. No. 3, 1967.

[9] A.N. Venetsanopoulos and I. Singh, "Topological Optimization of Communication Networks Subject to Reliability Constraints", Problems of Control and Information Theory, in press.

TABLE I
COMPUTING TIME FOR DIFFERENT SIZED SMALL NETWORKS

Network Size (No. of Nodes)	Computer Time (Sec.)	
	Proposed Algorithm	Exhaustive Search (Restricted Form)
5	2	2
6	8	28
7	28	283
8	130	>1200

TABLE 2
NODE PAIR INFORMATION FOR EIGHT-NODE DISTRIBUTED NETWORK-CANADA

Links Source-Dest. Node Pairs	Approximate Distance (Miles)	Weights	Cost ($)	Link Availability
1-2	247	100.0	1467	0.8880
1-3	334	100.0	1685	0.8838
1-4	1095	100.0	3587	0.8468
1-5	1233	100.0	3932	0.8401
1-6	1590	100.0	4825	0.8227
1-7	2030	100.0	5925	0.8018
1-8	2707	100.0	7617	0.7684
2-3	113	30.0	1130	0.8946
2-4	893	30.0	3082	0.8566
2-5	1300	30.0	4100	0.8363
2-6	1658	30.0	4995	0.8194
2-7	2098	30.0	6095	0.7980
2-8	2775	30.0	7787	0.7651
3-4	761	30.0	2752	0.8630
3-5	1357	30.0	4242	0.8340
3-6	1713	30.0	5132	0.8167
3-7	2151	30.0	6227	0.7955
3-8	2886	30.0	8065	0.7597
4-5	1995	30.0	5837	0.8030
4-6	2351	30.0	6727	0.7857
4-7	2844	30.0	7960	0.7618
4-8	3460	30.0	9500	0.7318
5-6	356	30.0	1740	0.8827
5-7	797	30.0	2842	0.8613
5-8	1474	30.0	4535	0.8284
6-7	487	30.0	2067	0.8763
6-8	1118	30.0	3645	0.8457
7-8	711	30.0	2777	0.8625

TABLE 3
COST-RELIABILITY RELATIONSHIP FOR EIGHT NODE LOOP NETWORK-CANADA

Number of Redundant Links	Network Reliability	Network Cost ($)	Normalized Cost
0	0.8348	25658	1.000
1	0.8690	29245	1.140
2	0.9026	33177	1.293
3	0.9188	34307	1.337
4	0.9389	36047	1.405
5	0.9620	38799	1.512
6	0.9725	41576	1.620
7	0.9864	45221	1.763
8	0.9946	51316	2.000

TABLE 4
LOCAL OPTIMUM SOLUTIONS FOR THE STARTING CONFIGURATION SHOWN IN FIGURE 2 (CANADA)

Reliability Constraint (R_o)	Links Added		Network Reliability	Network Cost ($)	Normalized Cost
0.800	-		0.8348	25658	1.000
0.900	1,2		0.9100	27125	1.057
0.930	1,2	6,7	0.9405	29192	1.138
0.950	1,2	3,6	0.9543	32257	1.257
0.970	1,2	6,7	0.9723	34324	1.338
	3,6				
0.980	1,2	6,7	0.9807	40377	1.574
	4,8	1,3			
0.990	1,2	6,7	0.9919	44381	1.730
	1,3	1,6			
	5,7	4,5			

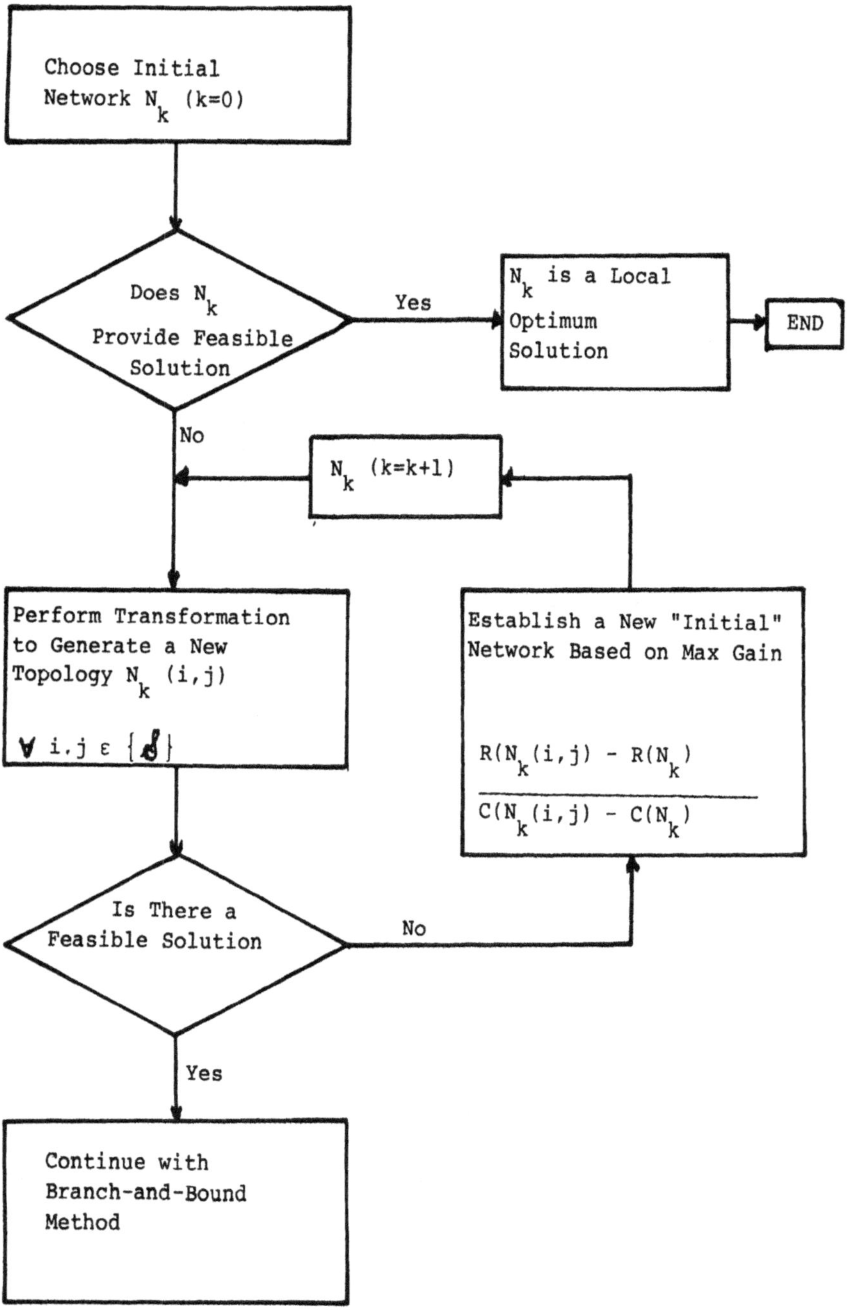

FIGURE 1
BLOCK DIAGRAM OF MARGINAL ANALYSIS PROCESS

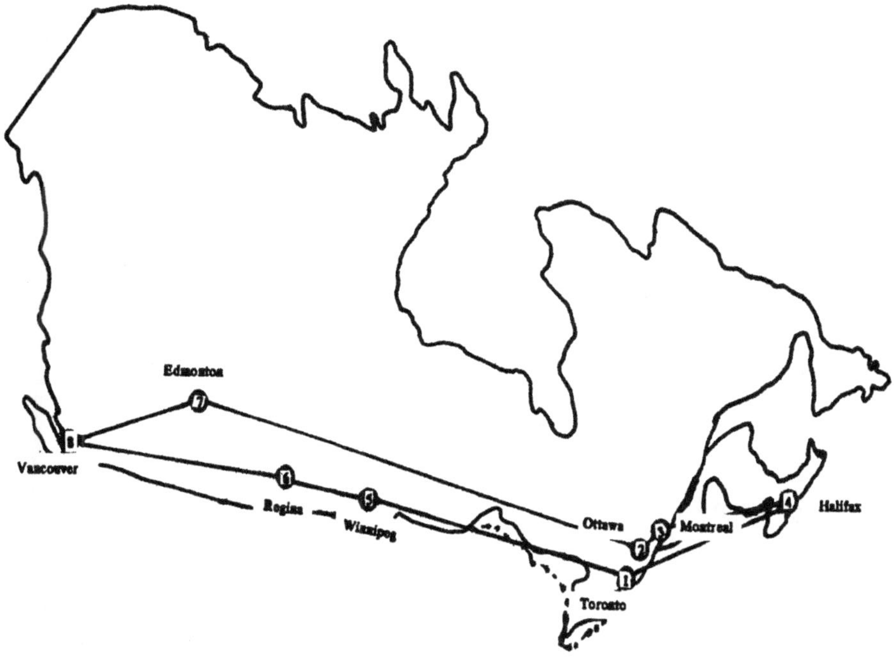

FIGURE 2
MAP OF CANADA SHOWING LOCATION OF NODES

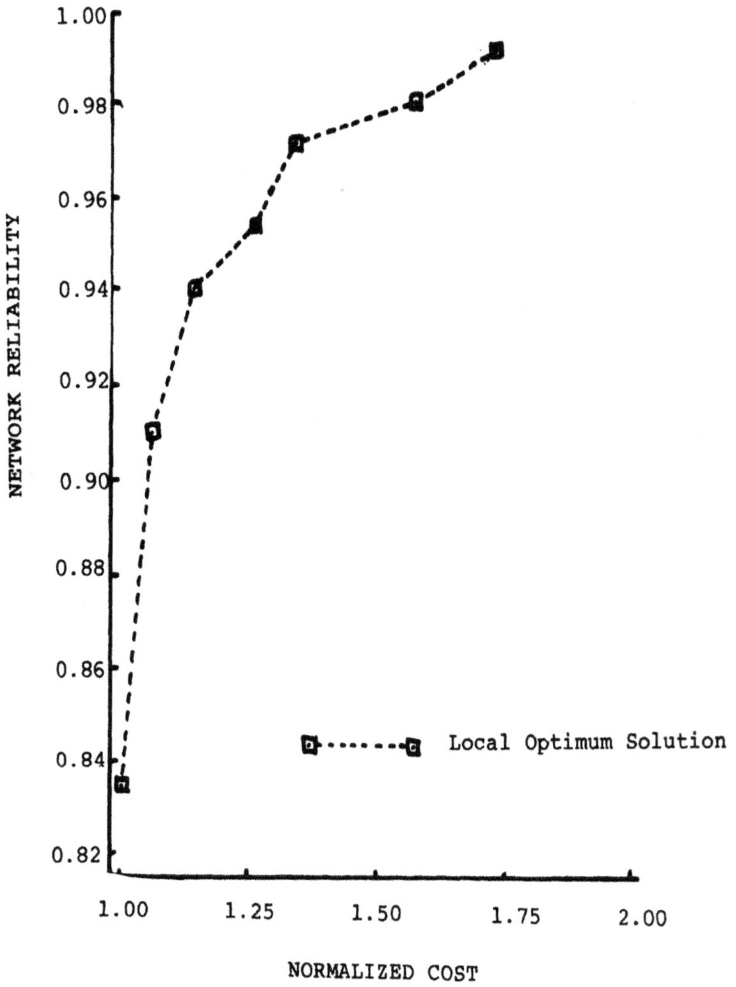

FIGURE 3
COST RELIABILITY RELATIONSHIP FOR 8-NODE PRACTICAL NETWORK (CANADA)

ARQ PROTOCOLS FOR RELIABILITY ENHANCEMENT IN COMPUTER NETWORKS

M. E. Anagnostou, E. N. Protonotarios
Division of Computer Science
National Technical University
Zografou, Athens 157 73
Greece

ABSTRACT. Reliable data communication over a noisy channel is achieved by means of two techniques, namely Forward Error Correction (FEC) and/or usage of Automatic Repeat Request (ARQ) protocols. In the first part of this paper a survey of the most well known ARQ protocols is presented.In the second part performance criteria and corresponding analysis methods are described.

1. ARQ PROTOCOLS

1.1. Introduction

The ideal error-free channel has not been devised so far. Approximations to the ideal channel by using high quality cables, improved modulation-demodulation techniques etc. are very expensive. In fact, this kind of improvement is much more expensive than the usage of FEC and/or ARQ schemes (for the same amount of improvement).Moreover,already established data networks use low quality (including telephone quality) lines, and this is a good reason to try to live with noise.

Before Shannon's work it was thought that channel errors were reduced by reducing data rate. Shannon proved that every channel has a fixed capacity C, and that for any any rate R less than C, there exist codes of rate R, which have an arbitrarily small decoding error probability. He did not give any methods of achieving this small error rate. FEC codes and retransmission (ARQ) schemes offer substantial improvement towards the direction shown by Shannon, although their performance is clearly below his theoretic limit. In systems using forward error correcting codes the information rate is steady because the number of redundant symbols used by the code is a constant percentage of the number of information symbols. Therefore, a fixed amount of bandwidth

S. Tzafestas et al. (eds.), System Fault Diagnostics, Reliability and Related Knowledge-Based Approaches, Vol. 1, 387–397.

is wasted when FEC is used. On the contrary, the bandwidth wasted when ARQ is used depends on the frequency of the transmission errors. Indeed, in ARQ systems data packets, which are detected in error by the receiver, are retransmitted. Retransmissions slow the information rate. During error free periods retransmissions cease to occur, and the system adapts to the situation better than pure FEC systems. Clearly, FEC is an open loop procedure, whereas ARQ schemes take advantage of feedback. The use of feedback in ARQ protocols is obvious, since the receiver should be able to notify the transmitter of the status of each arriving packet (for details see below). However, the delay suffered by packets in systems using ARQ is not constant. In satellite channels retransmissions may cause long delays, and this may lead to unacceptable response times in interactive applications. Stringent delay requirements are also imposed on integrated services networks, which carry digitized voice and video signals. Delays might produce gaps in voice communication and synchronization problems in the reproduction of tv images. A combination of ARQ and FEC is often most efficient.

Moreover, not all applications need the same grade of protection against errors. Banks are not satisfied with the reliability, which would be enough for ordinary data. On the contrary, digitized voice and video signals may not need any protection at all, because of their inherent redundancy.

There are no general rules to decide on the best error control scheme, given the requirements of a system. Instead, there is a great variety of retransmission protocols, and an even greater variety of codes. Neither the performance of all codes, nor the performance of all retransmission schemes has been evaluated. A choice based on the global cost of the system should take into account both the cost added by the quality of the service offered by the error control scheme, and the cost induced by the complexity of the soft- and hardware components needed for its implementation. This is impossible to achieve in a world, where it is much easier to propose a new code or a new retransmission scheme, or a combination of them, than to evaluate even simple performance measures for it. The usual procedure in order to choose an error control scheme usually relies on very incomplete information about the existing possibilities.

1.2. The role played by ARQ protocols in computer networks

In principle, ARQ schemes can be used in any application, where information (in digital form), must be transferred from one device to another. For example, a trivial version of ARQ is used when data are readed from a magnetic disc. When there are errors in the data, they are re-read. On the contrary in compact disc players, where variable delays cannot be tolerated, the cross-interleaved Reed-Solomon codes (CIRC) are used.In somewhat more serious applications, ARQ has been used in voice-grade modems, which married data terminals to telephone lines. But these applications are now considered trivial, and require the usage of extremely simple retransmission protocols. Indeed, a key parameter, which greatly influences the complexity of the ARQ scheme, is the round trip delay, that is, the time interval which begins with the end of the transmission of a block of data, and ends with the reception at the transmitter of a message by which the receiver either confirms the error free reception of the block, or requires retransmission of the same block. The length of the data block may be the length of just one character, or several hundreds or thousands of bits, corresponding to a packet of a computer network. If the round trip delay, measured in block transmission times, is almost zero, the situation is simple. If, however, it is greater than one, the time of the round trip delay might be used for the transmission of other blocks, instead of being wasted. A long round trip delay may significantly slow down the performance of a system, if the proper ARQ scheme is not used. This problem happens, for example, in satellite links, where the naive Stop-and-Wait protocol, which makes no use of the round trip delay time,is usually not the best choice. The reason of long round trip delays on satellite links is obviously due to the no more negligible transmission delay. The increase of transmission rates may cause the same problem on terrestrial links too. This is more likely to happen in future integrated services networks, which will use very high transmission rates. Long round trip delays may also be caused by processing.

Nowadays data communication services are organized into computer networks. Computer networks make extensive use of ARQ protocols, and incorporate satellite links, and high speed channels. Processing delays are often important. The functions performed by computer networks are extremely complicated. In order to deal with this complexity a structuring technique, which is called layering [1],has been widely accepted. The communication functions are partitioned into a hierarchical set of layers. According to the Open Systems Interconnection (OSI) Reference Model, which is a set of Recommendations under compilation by the International Standards Organization for the architecture of computer networks, there are seven layers. The first (lowest) layer is concerned with the transmission of a bit stream over a physical link. It involves signal voltages, bit durations, etc. The second layer, which is called the Data Link layer, is used to give the impression to the higher layers that the (unreliable) transmission channel is reliable. This is achieved by means of ARQ protocols (a set of rules is a protocol). Retransmissions are used in conjunction with sliding window protocols. In continuous ARQ schemes several packets may be transmitted before the transmitter gets any information about possible transmission errors. The number of such outstanding packets is usually limited by a number, which is called the window width. This is due to the fact, that a specific control field for the enumeration of the packets exists in each packet. If this field has places for α bits, the window width equals 2^{α}. When a confirmation for correct reception of a previously transmitted packet arrives at the transmitter, the corresponding number can be reused for the enumeration of a new packet. If the round trip delay is always less than the window width, there are always available numbers (window open). Otherwise all numbers may be given to packets before confirmation arrives for any of those packets (window closed), and packets arriving at the transmitter will have to wait until the window opens again [1]. Usually the window width is designed to be sufficient, so that the window rearly becomes closed.

The data link layer protocols of a device perform handshaking functions too, in order to make sure that the other device is ready to communicate. After the handshaking it enters the information transfer phase. During this phase the retransmission of corrupted packets is the main job of the data link layer.

1.3. Basic ARQ protocols

Consider two communicating nodes of a computer network. There may be two way communication between them, but we are interested in packets transmitted from one of the nodes, which will be called transmitter, to the other node, which will be called receiver. For each packet sent by the transmitter to the receiver, the latter checks it for possible errors. If the packet is error free the receiver sends to the transmitter an acknowledgement (ACK) message. Otherwise a negative acknowedgement message (NAK) is created. Upon reception of a NAK the transmitter, which retains a copy of every packet, for which no ACK has been received, retransmits the corresonding packet. This is the general form of an ARQ scheme.

Several things concerning the above scheme must be made clear. How does the receiver detect errors in the packet? Both nodes employ an error detecting code. Several additional bits are required for the implementation of the code, and this is a fact, which should be taken into account when estimating the performance of the system. Occasionally a code with both detecting and correcting capabilities is employed.Codes can detect more errors than they can correct. Then the receiver corrects as many errors as it can, and, if it detects uncorrectable errors, it requires retransmission of the packet.

How does the receiver notify the transmitter of the integrity of the packet? Early data transmission systems used a single channel for both information packets and ACK/NAKs. Implicit in this discipline was that the line had to be turned around twice for each block. The channel invertion took about 300 milliseconds. Later a separate low speed channel was used for the ACK/NAKs, and the turnaround was no more necessary. Today, in bidirectional communication taking place in computer networks, ACK/NAK messages corresponding to packets of one direction are incorporated into packets of the other direction. If no information packets are available to carry the ACK/NAKs, special control packets are created.

What does the transmitter do if the ACK/NAK is lost? After the expiration of a fixed time interval the packet is retransmitted.

What does the transmitter do in the interval between the transmission of a packet, and the reception of its ACK/NAK? In fact, this is the most important question, because the behaviour of the transmitter during this interval makes all the difference between different ARQ protocols. Although there are numerous ARQ protocols, they usually are variations and/or combinations of the following three schemes:(1) The Stop-and-Wait (SW), (2) the Go-Back-N (GBN), and (3) the Selective-Repeat (SR) ARQ protocol.

In a system using the SW protocol the transmitter remains idle after the transmission of a packet, until it gets an ACK/NAK. In case of an ACK, it transmits the next packet, otherwise it retransmits the same packet. If the round trip delay is long, the transmitter wastes the greatest part of its time.

The transmitter does not remain idle in systems using the GBN protocol. Instead, it transmits packets continuously without waiting for an ACK/NAK. When the receiver detects (uncorrectable) errors in a packet, this packet and all packets following it are rejected. Packet rejection stops when a correct version of the packet found in error is received. Upon reception of a NAK message, the transmitter goes back to the corresponding packet, and retransmits this packet and all the packets, which were transmitted after the distorted packet. Observe, that these packets are retransmitted, although they may have arrived at the receiver without errors. The time for their first transmission is obviously wasted.

According to the SR protocol, packets are continuously transmitted, but only packets, for which a NAK arrives, are retransmitted. Obviously no time is wasted in idling or in unnecessary retransmissions. However, the order of the packets arriving at the receiver is not preserved. The receiver must be able to buffer and reorder the incoming packets.

Obviously, if the round trip delay is negligible, then there is no difference between the three protocols. If the channel were error free the GBN and the SR would work in just the same way, whereas the SW protocol cannot eliminate its idle periods.

The implementation of the GBN is more complex than the implementation of the SW protocol, not only because of the go-back function, but also because of increased buffering requirements. The SR is also more complex than the GBN, mainly because of the reordering process at the receiver.

1.4. Modifications and combinations of the basic protocols

Protocols, which differ from the three basic ARQ protocols, fall in one of the following categories, or in both of them: (a) Modified versions and combinations of the basic protocols; (b) protocols, which are tied with specific error correcting codes.

ARQ protocols, which employ only error detecting codes, and rely solely on retransmissions for the correction of errors, are called pure ARQ protocols. Protocols, which make use of error correcting codes, and use retransmission only for uncorrectable errors, are called hybrid ARQ protocols. Any pure ARQ protocol may be converted into a hybrid one by replacing the error detecting code by an error correcting code. However, not every hybrid protocol is convertible into a pure one. The protocols of category (a) are either pure or hybrid. Their description is independent of the code used. (b) type protocols are always hybrid and cannot be converted into pure ones.

The protocols, which will now be briefly described belong to the (a) category. The lever, which the designer has at his disposal for the improvement of the protocol is only the retransmission strategy.

A.R.K.Sastry proposed in 1975 a modification of the basic protocols, which improves their performance under certain high error rate conditions [2]. SW systems can be improved by transmitting each packet more than once when a NAK is received. In GBN systems the packet is retransmitted continuously till it is received correctly. Improvement is observed for packet error probabilities larger than 0.5. Of course the SR performs better than both protocols, but it again requires more complex circuitry for the rearrangement of packets. In 1978 J.M.Morris proposed an improvement [3] on Sastry's scheme. The new scheme avoided the rejection by the receiver of correct packets following distorted ones.

In 1979 D.Towsley suggested [4], that the performance of GBN systems can be bettered by transmitting a block more than once whenever there are no other packets waiting for transmission or retransmission. Clearly, this strategy makes use of the idle periods of the receiver in the basic GBN.

K.Mase et al. have modified the GBN ARQ so as to be applicable to point-to-multipoint communication [5]. They also presented a 'tandem error control scheme' for satellite communication. In this scheme the uplinks and the downlinks employ separate ARQ mechanisms.

E.J.Weldon proposed a modification [6] of the SR ARQ. Distorted packets are repeated multiple times. The number of times increases as the receiver buffer approaches overflow. Y.Chang and C.Leung suggested [7], that under certain circumstances it is more beneficial to use multiple attempts even for new packets, whereas in [6] new packets are transmitted once.

H.Bruneel, and M.Moeneclay propose a family of extensions of the GBN protocol [8]. According to their analysis the optimum strategy is either to send each packet until an ACK is received for it, or to send it a constant number of times.

Obviosly, there are two main ideas underlying these protocols, namely (a)to use the idle time of the receiver, and (b)to transmit multiple copies of the each packet in order to anticipate delayed retransmissions.

1.5. Hybrid schemes using specific FEC codes

The error control schemes described in the previous section are independent of the code, which they use. They could be either pure ARQ schemes by using any error detecting code, or hybrid schemes by using any block error correcting code. Any such code could cooperate with those protocols, and the encoding-decoding process would be independent of the ARQ strategy. Moreover, there is no dependence between different packets. There no dependence even between the decoding of copies of the same packet. Therefore, it seems quite natural to ask, whether specific ARQ schemes could be designed in order to have a more close cooperation with certain codes. The dual question is, whether specific codes could be selected or designed, in order to be more suitable for use in hybrid ARQ schemes. Not all hybrid protocols go this far.

Before describing specific protocols, it is worthwhile paying some attention to the difference between block and convolutional codes. A block code is suitable for the protection of a block of information symbols. The encoder transforms each input block of k symbols into a longer sequence of n symbols. k and n are fixed parameters of the code. Each output block depends only on the corresponding input block. On the contrary, a convolutional encoder processes a continuous information sequence and produces a code sequence. At any time unit the encoder accepts a small block of information symbols and produces a greater block, but the output block depends also on previous blocks.

Most hybrid schemes employ block codes, and each packet

is decoded independently of other packets. Convolutional codes cannot be easily used in SR schemes, because of the rearrangement of packets.

A.R.K.Sastry, and L.N.Kanal proposed in 1976 a hybrid protocol using convolutional codes [9]. The ARQ protocol selected was the GBN. As for the coding, the generalized burst trapping codes (GBT) were employed. These are convolutional codes capable of correcting both random and burst errors.

B.Arazi suggested, that instead of retransmitting the whole message, only those bits whose parity check failed should be retransmitted.In order to do this, the transmitter is provided with a piece of information much more complex than a NAK, namely the detected syndrome.

C.S.Leung, and A.Lam have examined in Ref. 10 the performance of the BCH error correcting codes in conjunction with the SW protocol. A.Drukarev, and D.J.Costello proposed the usage of sequential decoding in hybrid SW, GBN, and SR protocols [11].

P.S.Sindhu in Ref.12, and J.J.Metzner, and D.Chang in Ref.13 have suggested that the information of the first transmission of a packet be combined with the information in a retransmission in order to be achieved correct decoding. The resulting scheme is called memory ARQ.

The ARQ schemes mentioned so far do not constitute a complete catalog of hybrid ARQ schemes. However it is hoped that all the main ideas employed for the construction of hybrid protocols are adequately represented in these works.

2. ANALYSIS OF ARQ PROTOCOLS

2.1. Introduction

Usually what one wants to know about a protocol are (i)whether it is capable of handling all possible situations and (ii) whether it works efficiently. The analysis aiming at an answer to the first question is called verification of the protocol. The kind of analysis corresponding to the second question is called performance evaluation.

The idea underlying verification techniques is to explore all the possible states of the system, and look for undesirable states. Finite state automata, Petri nets, and other discrete mathematics tools are used for this purpose [1].

The performance analysis answers questions about the effective utilisation of resources, the response times in systems using the protocol etc. Because of the statistical nature of packet arrivals at a network station, of errors in a channel, and so on, the mathematical tool used for this kind of analysis is the probability theory, and especially queueing theory [14].

2.2. Performance criteria

There is an obvious flaw in the description of protocols in the previous sections. When an improvement was claimed, it was not explained, what exactly was improved. What is really needed, is to refer to well defined performance measures, when one speaks about improvement. The most common measures are the following: (i) Throughput, which equals α/β, where α is the mean transmission time of a packet, and β is the average time dedicated by the transmitter to that packet. Obviously the throughput is less than 1. (ii) Delay, which is equal to the length of a time interval beginning at the arrival of a packet at the transmitter, and ending when a correct copy of the packet arrives at the receiver. Delay is a random variable. An important part of the delay is the waiting time a packet spends before transmission at the transmitter buffer. The number of errors, which remain undetected in packets accepted by the receiver is important too, but this is mainly a performance measure of the code.

2.3. Performance evaluation of ARQ protocols

The throughput analysis of ARQ protocols is an easy task for the basic protocols. It is usually simple for modified versions of these schemes, and for hybrid strategies. On the contrary, the computation of the delay statistics is a fairly complex mathematical adventure, even for the basic ARQ protocols. That is why most protocol designers prefer to make comparisons based on the throughput performance, although several works for the analytical evaluation of delays have appeared. It should be noticed, however, that a throughput improvement does not always imply a delay decrease.

In Refs. 15-20 the throughput and delay of the SW and GBN protocols have been evaluated. A careful study of these works reveals the fact, that the same queueing model model can be used for both protocols, the only difference being in the service time desription of each model. If geometric arrivals are assumed, closed form expressions [18-19]for the generating function of the delay, and hence for its mean value, can be obtained. For more general distributions for the description of the arrival process simple computational algorithms exist [20].

Although the throughput evaluation of the basic SR schemes is trivial, the delay analysis is quite the contrary [15,19,21-23]. The complexity of exact models grows exponentially with the round trip delay, therefore approximate models have also been devised.

For non basic ARQ protocols only throughput evaluations are available, apart from a few exceptions.

REFERENCES

[1] A.S.Tanenbaum, Computer Networks, New York: Prentice Hall, 1981.

[2] A.R.K.Sastry, 'Improving ARQ performance on satellite channels under high error rate conditions', IEEE Trans. on Communications, pp.436-439, April 1975.

[3] J.M.Morris,'On another Go-Back-N ARQ technique for high error rate conditions', IEEE Trans. Comm., Vol.COM-26, No.1, pp. 187-189, Jan.1978.

[4] D.Towsley, 'The stutter Go-Back-N ARQ protocol', IEEE Trans. Comm., Vol. COM-27, No.6, pp.869-875, June 1979.

[5] K.Mase,T.Takenaka,H.Yamamoto,and M.Shinohara,'Go-Back-N ARQ schemes for point-to-multipoint satellite communications', IEEE Trans. Comm., Vol.COM-31, No. 4, pp. 583-589, April 1983.

[6] E.J.Weldon,'An improved Selective-Repeat ARQ strategy', IEEE Trans.Comm.,Vol.COM-30,No.3,pp.480-486,March 1982.

[7] Y.Chang, C.Leung,'On Weldon's ARQ strategy',IEEE Trans. Comm., Vol.COM-32, NO.3, pp.297-300, March 1984.

[8] H.Bruneel, M.Moeneclaey,'On the throughput performance of some continuous ARQ strategies with repeated transmissions', Vol.COM-34, No.3,pp.245-249,March 1986.

[9] A.R.K.Sastry, L.N.Kanal, 'Hybrid error control using retransmission and generalised burst trapping codes', IEEE Trans.Comm.,Vol.COM-24,No.4,pp.385-393,April 1976.

[10] C.S.Leung, A.Lam,'Forward error correction for an ARQ scheme',IEEE Trans. Comm., Vol.COM-29, No.10, pp. 1514-1519, October 1981.

[11] A.Drukarev,D.J.Costello,'A comparison of block and convolutional codes in ARQ error control schemes', IEEE Trans. Comm., Vol.COM-30,No.11,pp.2449-2455, Nov. 1982.

[12] P.S.Sindhu, 'Retransmission error control with memory', IEEE Trans.Comm.,Vol.COM-25,No.11, pp.473-479,May 1977.

[13] J.J.Metzner, D.Chang, 'Efficient Selective Repeat ARQ strategies for very noisy and fluctuating channels', IEEE Trans. Comm., Vol.COM-33,No.5,pp.409-416,May 1985.

[14] L.Kleinrock, Queueing Systems, Vol.1, New York: Wiley, 1976.

[15] R.J.Benice,A.H.Frey,'An analysis of retransmission systems', IEEE Trans. Comm. Syst., Vol. CS-12, pp.135-144, Dec. 1964.

[16].R.J.Benice,A.H.Frey, Comparisons of error control techniques',IEEE Trans. Comm. Syst., Vol.CS-12, pp.146-154, Dec. 1964.

[17] G.Fayolle, E.Gelenbe, G.Pyjolle,'An analytic evaluation of the performance of the Send and Wait protocol', IEEE Trans. Comm., Vol.COM-26, No.3, pp.313-319, March 1978.

[18] D.Towsley, J.K.Wolf, 'On the statistical analysis of queue lengths and waiting times for statistical multiplexers with ARQ retransmission schemes', IEEE

Trans. Comm., Vol.COM-27, pp.693-702, Apr. 1979.
[19] A.G.Konheim,'A queueing analysis of two ARQ protocols', IEEE Trans.Comm., Vol. COM-28, pp.1004-1014, July 1980.
[20] M.E.Anagnostou, E.N.Protonotarios, 'Steady state and transient delay analysis of ARQ protocols', Computer Communications, Vol. 7, pp.23-30, Febr. 1984.
[21] M.J.Miller,S.Lin,'The analysis of some selective repeat ARQ schemes with finite receiver buffer', IEEE Trans. Comm., Vol. COM-29, pp.1307-1315, Sept. 1981.
[22] B.H.Saeki, I.Rubin,'An analysis of a TDMA channel using stop-and-wait, block, and select-and-repeat ARQ error control', IEEE Trans. Comm., Vol.COM-30, pp. 1162-1173, May 1982.
[23] M.E.Anagnostou, E.N.Protonotarios,'Performance Analysis of the Selective Repeat ARQ protocol',IEEE Trans. Comm. Vol. COM-34, No.2, pp.127-135, Febr. 1986.

PREVENTIVE MAINTENANCE POLICIES FOR REFINERIES

Dr. D.E. Ventzas
Instrumentation Eng.
Hellenic Aspropyrgos Refinery
Kazantzaki 23 - Ano Pefki
Athens 15 121 GREECE

ABSTRACT. This paper examines and compares the maintenance techniques met in refineries; it focus in the preventive maintenance methods and their statistics and technology. The comparison is extended to electronic / mechanical components - systems met in chemical and general production plants. Interaction of maintenance to other production phases is discussed. Maintenance has a strategic role inside a company due to safety, productivity and economic demands of their increasingly antagonistic environment. Cost analysis and efficient operation of the installed machinery is a must. The rapid recognition of developing problems and health monitoring during the operation should be a common practice of a modern factory.

1. INTRODUCTION

Maintenance is highly correlated with cost, technological advancement and complexity, safety, productivity and efficiency. In a continuous cycle production plant we can choose among the following maintenance policies:

(a) On failure or breakdown
(b) Preventive
  (b1) based on statistics methods
    (b11) at constant working age
    (b12) at constant data
    (b13) opportunistic
  (b2) on condition (after inspection)
(c) Corrective.

1.1 Failure or breakdown maintenance

A machinery is left to work until it is broken and then it is repaired; the method is expensive in terms of production losses, risk to destroy the machinery and safety.

S. Tzafestas et al. (eds.), System Fault Diagnostics, Reliability and Related Knowledge-Based Approaches, Vol. 1, 399–413.

1.2. Preventive maintenance

With this general method we prevent the failure before it occurs. The "best" method to intervene is calculated in a statistic way based on historical data or following on-line data (including inspection) showing the real wear state, see fig. 1. In case (bl1) the component is substituted each time it reaches a tetermined working age (real working time). In case (bl2) we substitute a component at constant period not depending on the real working time. In case (bl3) we substitute a component after shutting down the system because it has reached the allowed / maximum real working time or as a preventive maintenance before this time will be reached and during a shut-down caused by other components.

1.3. On condition (inspection) maintenance

If we measure, some parameters corresponding to wear, while the machinery is working, we can determine the components safe-work limits, see fig. 2. Vibrations is usually a major parameter in plant analysis maintenace and protection. We usually calculate a short period of time, close to the break-down point, during which we should repair on first opportunity the system. This interval is where the machinery moves from a good work condition to an unsafe condition during which a significant change occurs to its parameters.

Usually the on condition maintenance will not substitute but integrate the preventive maintenace at a certain period. With the corrective maintenance we increase the dynamic potential of repair, controlling in an active way the machinery lifecycle; all methods together reduce manpower and materials costs, increase quality of maintenance work and the whole production in total.

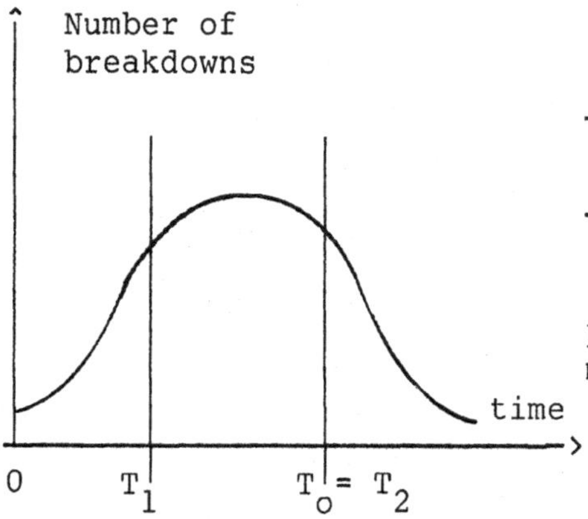

(a) Failures distribution and time interval calculation

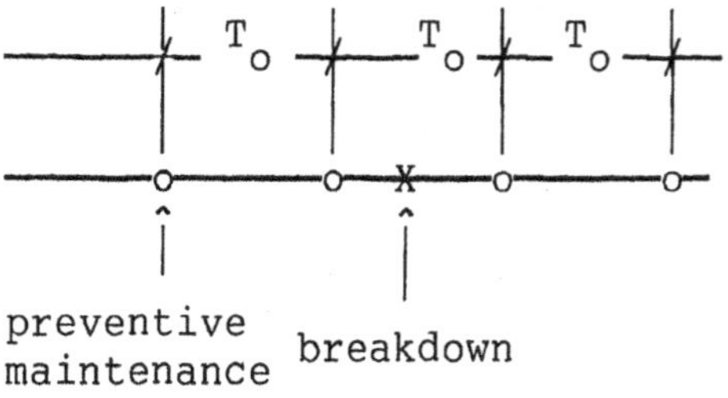

(b) Preventive maintenance at constant period

Fig. 1. Estimation of maintenance parameters

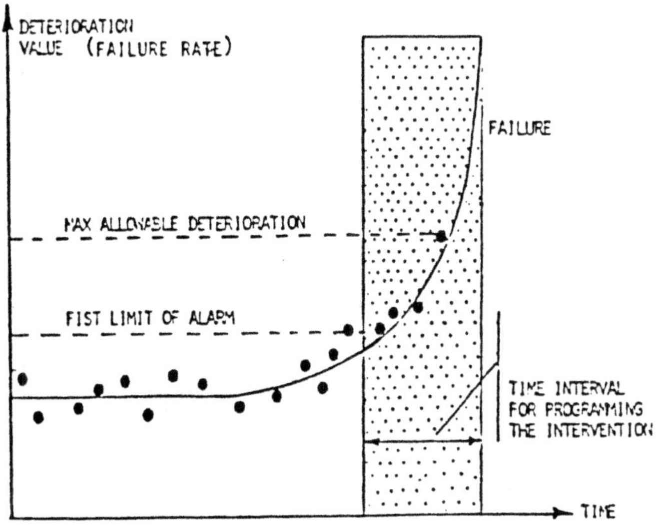

Fig. 2. On condition maintenance parameters definition

2. POLICIES SELECTING CRITERIA

To choose the correct maintenance policy we maust under consideration the following parameters:

<u>λ = failure rate</u>. Usually the number of failures grows proportionally to the working period. The failure rate follows the "bath-tube" curve shown in fig. 3. In this figure we can distinguish the infantile diseases, the normal operating (mature) period and the aging infirmities due to corrosion, relaxation and wear phenomena.

<u>Cp = total cost for preventive maintenance</u>

<u>Cb = total cost for break-down maintenance</u>

<u>Ci = cost of inspection</u>.

The policies followed for maintenance depend on failure rate:

(a) For λ decreasing with the "life" of the machinery the break down policy is prefered.

(b) For λ increasing with the machinery´s life the $Cp \gtreqless Cb$ cost is examined to decide between the type of the preventive maintenance.

(c) For λ statistically constant with time, behaviour that is typical for complex machinery, on condition maintenance is followed for $Ci < Cp < Cb$. This kind of policy is applied to the most critical machinery on the plant (needs for high reliability); "out of use" state of these machinery could lead to very high costs of production losses.

Table 1, shows the criteria for maintenance policies selection. Of course in a plant we always find a combination of

COSTS COMPARISON / λ TREND	$C_b \leq C_p$	$C_p < C_b$, $C_i \geq C_b - C_p$	$C_p < C_b$, $C_i < C_b - C_p$
DECREASING	BREAKDOWN	BREAKDOWN	PREDICTIVE
CONSTANT	BREAKDOWN	BREAKDOWN	PREDICTIVE
INCREASING	BREAKDOWN	PREVENTIVE STATISTIC	PREVENTIVE STATISTIC OR PREDICTIVE

Table 1. Maintenance policies selection summary

the above mentioned policies in order to reach the best results from both the economic and technical sides. In a refinery plant, where the mechanical differences among the machineries installed, the continuity of the production cycle and the severe running conditions create increased problems. Other parameters affecting maintenance selection policies are production cycles and their seasonality, stock levels, budgets, etc. First we must specify those equipments whose malfunction creates high risk situations. For fixed equipments and rotating equipment withound stand-by (compressors, turbogenerators, etc) it will be a compromise of "on condition", "preventive" and"opportunistic" maintenance. Critical equipments are systematically inspected by off-line or more expensive on-line systems. Putting intelligence into instruments, machinery, etc and wiring this intelligence by means of software or hardware to appropriate interlock systems is the modern trend (Ventzas, 1986 - Fell, 1985). For all the others machineries that have stand-by or may be by-passed, the on condition maintenance is followed. The regular equipment works for the 80-90 % of the time versus 20-10 % of the stand-by equipment. Small pumps are maintained by total substitution from its (not installed) stand by which is stored as spare part. Mechanical seals inpumps are usually maintained onabreak down basis. For control and regulation instruments, that are impossible to by-pass, are maintained during turnaround stops; when it is possible to be by-passed a break down or preventive policy can be followed. For electronic instruments with self-diagnosis an "on-condition" service is followed. For volumetric displacement meters a preventive maintenance policy at constant intervals is followed, since the efficiency of such equipment must remain within certain limits. For machineries not directly involved with the production processes (vehicles, lift trucks, fire-fighting equipment, workshop machines, auxiliary equipments, etc) preventive maintenance at constant working age is used. The tanks are maintained on an "on condition" basis. All these are resumed in Table 2. Integrated electronic and computerized (concentrared or distributed) systems combine all kinds of maintainance policies but since they include self-diagnostics functions an "on-condition" policy is followed.

EQUIPMENT TYPE		WHEN	MAINTENANCE POLICY				
			BREAKDOWN	PREVENTIVE			
				CONSTANT DATE	CONSTANT CYCLE	OPPORTUNITY	ON CONDITION (OR INSPECTION)
SINGLE AND NOT BY-PASSABLE EQUIPMENTS		TURNAROUND				X	X X X X
EQUIPMENTS WITH STAND-BY OR BY-PASSABLE		UNITS RUNNING			X X		X X X
MECHANICAL SEALS		UNITS RUNNING	X X			X X X	
INSTRUMENTS	INSIDE EQUIPMENTS NOT BY-PASSABLE	TURNAROUND				X X	X X X
	BY-PASSABLE	TURNAROUND OR UNITS RUNNING	X X X				X X
	BOARD INSTR.	TURNAROUND OR UNITS RUNNING	X X (1)	X X X (1)(2)			X X X X X (3)
	VOLUMETRIC METERS	UNITS RUNNING		X X X	X X		
TANKAGE		UNITS RUNNING					X X X X X
VEHICLES (4)		-			X X X X X		
LIGHTING		-	X		X X X	X	
AIR CONDITIONING		-		X X X X X			

PREFERENCE AMONG DIFFERENT MAINTENANCE POLICIES IS INDICATED WITH A FIVE X TOTAL

(1) - FOR CONVENTIONAL INSTRUMENTS
(2) - IN OCCASION OF PERIODICAL SETTING
(3) - FOR MICROPROCESSOR ELECTRONIC INSTRUMENTS (SELF-DIAGNOSTIC)
(4) - AND OTHER MOTOR-DRIVEN EQUIPMENTS

Table 2. Refinery maintenance policies

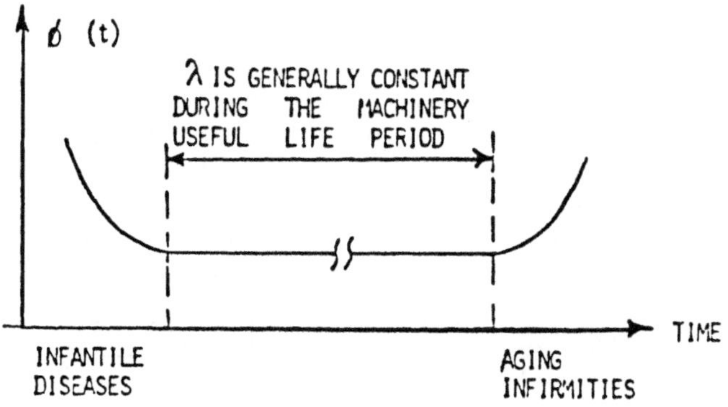

Fig. 3. The failure rate curve

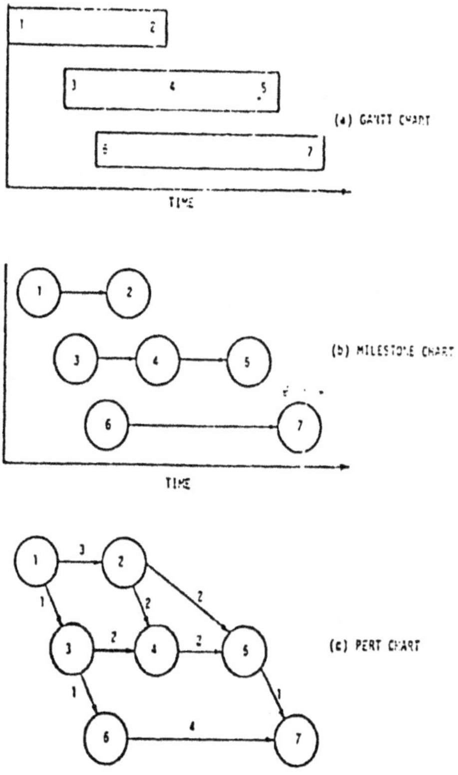

Fig. 5. GANT, MILESTONE and PERT charts

3. FUNCTIONS AND INTERACTIONS

The most frequent method of maintainance policies is the on condition one, then the preventive at constant working age and last the preventive at constant period. Modern refineries running on highly productive schedules are characterized by longer and longer operational cycles (2-3 years); after such a long period a very well planned shut-down should involve almost the whole refinery: operation, technical, administrative and personnel departments. The core of the planning system is the Maintenance Planning Office which collect coordinate and direct the information flow. The interactions of the M.P.O. and the rest of the refinery are very complex and are shown in fig. 4. Nevertheless the daily maintenance requires a remarkable managerial ability to be bestly administrated. The M.O.P. stimulates collaboration between the appropriate persons and departments. To reach this objective

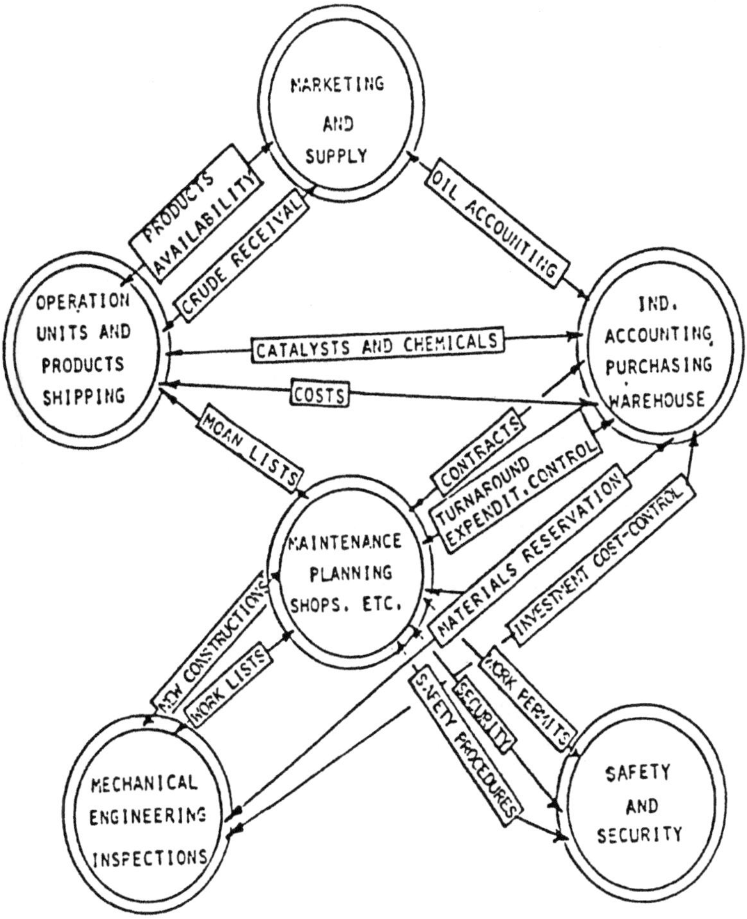

Fig. 4. Turnaround maintenance main functions and interactions

the M.O.P. will "use" staff advisors (inspectors, safety people, project offices, etc) and logistic support (warehouse, purchasing, etc.). The mean to communicate is the "Work Request" and the "Work Order". Planningn means better work in minimum time by less people.

3.1. Organization

The organization, hierarchy and structure can vary from refinery to refinery according to local needs. Main emphasis should be placed in the difference between the Inspection of fixed and rotating equipments; they should remain separated and under different superintendents. The inspection for fixed equipments should include checking, collect data, analyze and select materials depending on working conditions, set the safety valves, test repaired equipments, etc. The report should be towards the Technical Service and Production and Project Engineers. The "on condition" maintenance is a choice of the Maintenance Manager, who is responsible for technical and economic results. The inspector should not be a corrosion expert but a workshop specialist. He is not a "consultant" in staff position but a maintenance person in line with the Workshop Superintendent.

4. POLICIES

4.1. Definition - Policies and Procedures

By the word policy we mean the physiognomy given to a function and a correct guideline for the results we want to reach. We have to define the type of maintenance to be followed, the persons and functions involved, the actions to be done. Procedures are defined the precise rules for each function. The policies followed are:

(a) Constant Cycle Preventive Maintenance
(b) On Condition Maintenance
(c) Turnaround Works Preparation
(d) Daily Maintenance Planning
(e) Material Reservation
(f) Lubrication Planning
(g) Trucks, Cranes and other Vehicles Maintenance
(h) Historical break-down data collection and technical indexes elaboration.

A common structure for comparison reasons is set up by:

(i) Object
(ii) Field of application
(iii) Eventual remarks
(iv) Involved functions
(v) Recommended actions and procedures
(vi) Data collecting and results control.

The actions belonging to policies (c), (d), (e), (f), (g), (h) can be processed in real time.

5. OTHER METHODS

Better control techniques are introduced in order to cope with the complexities, massive data and tight deadlines of modern industries with their highly competitive environment. Modern project planning uses since World War II four integrated project management techniques. These are the program evaluation and review technique (PERT), the critical path method (CPM) and variations of them; much more modern methods are the line of balance (LOB) for production lines activities and the Gantt chart that has been used used for long time in industry. Fig. 5 shows these methods schematically. Finally pilot plants give invaluable experience in maintenance. The real challenge of a scientist is how much these mathematical methods can work in a realistic environment and how much is the quantitative change of significant parameters after their use.

6. RELIABILITY DATA BANK

Reliability data banks are set-up by big companies (such as EXON, ENI, etc.) for safety, availability (operational) and maintainance reasons. A petrochemical plant cannot be tested in the laboratory, with accelerated life cycles. The establishment and handling of such data banks is a difficult and cosly exercise which may give good results in the medium and long term; any hope for miraculous reliability improvements is a vain. A quantified knowledge of the reliability of the components could introduce in later stages further sophisticated in the system analysis techniques (fault-tree, Markovian chains, Colmogorov and X^2 tests, etc.). Table 3 shows a preliminary list of industrial items to be investigated and Table 4 gives the prior distributions for the failure rates of such an item with their parameters as a function of σ, μ, M, U, k and c, where:

σ is the standard deviation of the failures
μ mean value of theirs
M the 50th percentile (median)
k $\lambda_{recommended} / \lambda_{pessimistic}$
U 95th percentile
c σ / μ

7. NEW MAINTENANCE TECHNIQUES

Maintenance intents to achieve reliability and extended life, and to avoid the direct and indirect consequences of failure. Direct costs include downtime, lost production, machine replacement costs, etc. Indirect costs include delay-

Distribution and Parameters	c as a Function of the Parameters	Range of $k = U/M$; and Approximated Ranges of $c = \sigma/\mu$ and Parameters	Notes
Uniform (A, B)	$\frac{B/A-1}{\sqrt{3}(B/A+1)}$	$1.5 < k < 1.9$ implies: $0.33 < c < 0.57$ $0 < A < \infty$ $B/A > 3.5$	$B/A = (10k-1)/(19-10k)$ $A > 0$ implies $k < 1.9$
Loguniform (A, B)	$\left[\frac{\ln(B/A)\ln(B/A+1)}{2(B/A-1)} - 1\right]^{1/2}$	$1.5 \le k \le 100$ implies: $.26 < c < 2.0$ $0 < A < \infty$ $2.5 < B/A < 27.8 \cdot 10^3$	$B/A = k^{20/9}$
Lognormal (m, s)	$\left[\exp(s^2)-1\right]^{1/2}$	$1.5 \le k \le 100$ implies: $.26 < c < 50$ $-\infty < m < \infty$ $.25 < s < 2.7$	m and s are the parameters of the normal distribution of $y = \ln x$. $s = \ln(k/1.645)$
Gamma (a, b)	$1/\sqrt{b}$	$1.5 \le k \le 100$ implies: $.27 < c < 2.48$ $0 < a < \infty$ $.17 < b < 14.6$	$\Gamma(b) = \int_0^\infty e^{-z} z^{b-1} dz$ If $b=1$, then $k = k^* \approx 4.3$ and the gamma distribution reduces to the exponential distribution. If $b < 1$, then $(k > k^*)$, and the distribution is L shaped. If $b > 1$, then $(k < k^*)$, and the distribution is unimodal.

Distribution and Parameters	Range	Density Function	Parameters as a Function of μ and σ	Parameters as a Function of M and U
Uniform (A, B)	$A < x < B$	$\frac{1}{B-A}$	$A = \mu - \sqrt{3}\sigma$ $B = \mu + \sqrt{3}\sigma$	$A = \frac{19M-10U}{9}$ $B = \frac{10U-M}{9}$
Loguniform (A, B)	$A < x < B$	$\frac{1}{(\ln B - \ln A)\,x}$	not defined analytically	$A = \exp\left(\frac{19\ln M - 10\ln U}{9}\right)$ $B = \exp\left(\frac{10\ln U - \ln M}{9}\right)$
Lognormal (m, s)	$0 < x < \infty$	$\frac{1}{\sqrt{2\pi}\,s\,x}\exp - \left[\frac{(\ln x - m)^2}{2s^2}\right]$	$m = \ln\left[\frac{\mu^2}{(\mu^2+\sigma^2)^{1/2}}\right]$ $s = \left[\ln\frac{\mu^2+\sigma^2}{\mu^2}\right]^{1/2}$	$m = \ln M$ $s = \frac{\ln U - \ln M}{1.645}$
Gamma (a, b)	$0 < x < \infty$	$\frac{a^b}{\Gamma(b)} x^{b-1} e^{-ax}$	$a = \frac{\mu}{\sigma^2}$ $b = \frac{\mu^2}{\sigma^2}$	not defined analytically

Table 3. Prior Distributions for the Failure Rate of an item as a function of μ and σ, M and U, k & c.

VALVES	Safety valves Solenoids Control valves Venting valves
MACHINERY	Centrifugal pumps Centrifugal compressors Reciprocating compressors Electric motors Pump and Motor Units Generating networks Steam turbines Power transformers
SWITCHGEARS	Low voltage switchgears High voltage switchgears Relay panels Automatic change-over panels
INSTRUMENTS	Pressure switches Thermocouples Controllers Level switches Flame failure detectors Pressure / flow transmitters Flow switches Flow switches Limit switches Alarm annunciators Process analysers Non rotating power supplies
MISCELLANEOUS	Power cables Instrument / signal cables Safety heads

Table 4. Preliminary list of items to be investigated

ed production, late deliveries, unhappy customers, potential accidents, stained reputation, etc. Traditional methods of maintenance are largely inadequate. Detecting techniques, monitoring machine operation and condition, new instrumentation new technologies and know-how on new applications were introduced recently. The availability of computerized maintenance management system software and hardware decreased the amount of paperwork, service time and increased reliability. Now an even more advanced approach allows for monitoring the components in real time, and to note how well they operate within acceptable limits. Any significant deviation from the acceptable range is noted, tracked and monitored, analyzed if the deviation continues, diagnosis and prognosis performed, and then predict a probable time of failure and condition of failure. Channeling this information through the proper sequences will then result in the issuance of maintenance work orders for remedial action before failure occurs

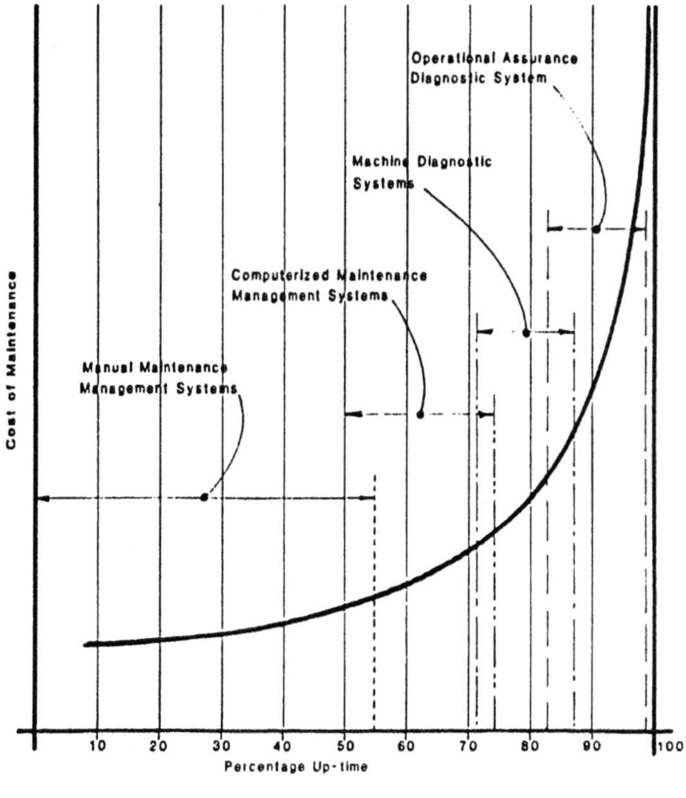

Fig. 6. Correlation between the cost of maintenance and the percentage uptime achieved by various maintenances

Fig. 6 illustrates acorrelation between the cost of maintenance and the percentage up-time for various types of maintenance and maintenance management techniques. Computerized systems manage the work effectively and limit the paperwork; therefore they provide more time for planning, managing and supervising. Combining expert knowledge, computers, instrumentation and diagnostics can bring the up-time availability close to 100 % by reducing or eliminating down-time due to unexpected failures. Detection systems can be of fixed or portable type. The analysis and diagnosis is based on human knowledge and experience. This research and decision making is a tedious and time consuming process. Combining instruments and computers is a further important step for real time condition monitoring and critical components diagnosis. By building a data base of experience, knowledge and interpretive rules a still further step can be taken: a real time expert system for maintenace needs.

7.1. Some Diagnostic Techniques

With current technology and instrumentation, over 20 different simple operating characteristics, and several dozen complex characteristics, of machinery and equipment can be measured and evaluated. Among these are: temperature, humidity, vibration, acceleration, cycling action, effluent discharge, mechanical function, electrical sequence, magnetic field characteristics, flow, stress and strain, noise/sound, decceleration, time, chemical condition, mechanical sequence, electro-mechanical function and sequence, pressure, mechanical motion / position, rate of change, accidity / pH, electrical function / characteristics (usage/consumption, operating characteristics, starting characteristics, temperature rise, leakage), oil condition (usage, lubricity, dilution, additives, viscosity, contamination, particulates, oxidation). No single measurement can provide adequate information for decision making; but it could be an indicator-guide. So any system must be built on acquiring multiple data measurements and characteristics. The most important parameters are (1) temperature, (2) vibration (3) oil condition.

7.2. Typical Diagnostic Instrumentation

Diagnostic instrumentation coincides with process control instrumentation. Selective use of it is necessary in order to make such a system cost effective. Data aquisition and real time processing require substantial computing power (computer hardware and peripherals). Modern instrumentation incorporates a microcomputer directly in the level of a transducer, leading to a data reduction in main computer. Some typical system arrangements include: instrumentation, reading, data aquisition and conditioning devices, computer aquisition and display, review and examination of the data, analysis of the data, decision making regarding the data, aquiring additional data, remedial/corrective action, informing the maintenance department, verifying availability of the machine, manpower, parts/materials, tools, estimating the cost, scheduling the work, issuing the work order, performing the work, verifying completion and reporting, feedback information to the data base.

The instruments commonly used are for: temperature - pyrometers, thermocouples, thermistors, infrared detectors, pressure and strain - pressure gages, transducers, strain gages, load cells, vibration and noise - vibration transducers, vibration analyzers, electrical - voltmeters, ammeters, multimeters, meggers, continuity testers, ohmmeters, insulation thickness gauges, oil condition - laboratory testers, spectrographic analyzers, portable oil test units, in line oil condition monitors, mechanical conditions - proximity devices, magnetometers, magnetic flux devices, accelerometers.

8. CONCLUSIONS

The existing maintenance techniques and their selection criteria were presented in this paper. The basic ideas about organizing the structure of a refinery for reliable maintenance weregiven. Some comments about availability of maintenance data banks and other techniques used in orginizing the maintenance better are included. The modern approach to integrated computerized detection systemand its cost effectivenes are finally discussed.

Present work and interest tend towards studying reliability models of whole systems and isolated equipment/devices in a refinery, maintenance techniques for the above, NDT in intelligent transducers, fault tolerant architectures for operational and availability purposes and the special field of NDT in rotating equipments.

9. REFERENCES - BIBLIOGRAPHY

9.1. Fell, R., E.H. Higham, A.Ajaya, "Signal analysis and intelligent flowmeters", Measurement + Control, Flowmeasurement Special Issue, June 1986, **vol. 19**, No. 5. pp. 47 - 50.

9.2. D.E. Ventzas, "Fault Diagnostics and SFT Analysis in Intelligent Flowmeters", 1st European workshop on Fault diagnostics, reliability and related knowledge based approaches, 1986, Rhodes, Greece.

9.3. Bruel & Kjaer "Machine-Health Monitoring", December 1985.

9.4. E. Marshall, "New maintenance techniques cut downtime", Industrial World, June 1986, pp. 6 - 32.

9.5. A.G. Colombo, R.J. Jaarsma, "Statistical Methods for the European Reliability Data System", 3rd European Reliability Data bank Seminar, University of Bradford, April 1980, **vol. 3**, pp. 157 - 173.

9.6. G. Bello, A. Bobbio, "A Reliability Data Bank in the Petrochemical Sector", 3rd European Reliability Data Bank Seminar, University of Bradford, April 1980, **vol. 3**, pp. 211 - 224.

9.7. COMERINT (Gruppo ENI), "An Introductive lecture on Program Evaluation and Review Technique (PERT), 1986.

9.8. COMERINT (Gruppo ENI), "Preventive Maintenance Policies", Preliminary, 1986.

9.9. W. O'Keefe, T.C. Elliot, "Solving Plant Problems - - Design, Operation, Maintenance", McGraw-Hill, 1984.

9.10. G.C. Carrol, "Industrial Instrument Servicing Handbook", McGraw-Hill, 1960.

CONTROLLED PROCESSES WITH EXPLICIT OR IMPLICIT BREAKDOWNS AND REPEAT ACTIONS

Boyan N. Dimitrov, Peter G. Petrov
Mathematical Institute Bulg. Acad. Sci.
P.O. BOX 373
1090 Sofia
Bulgaria

ABSTRACT. Some mathematical models leading to minimization of the total process duration by processing with unreliable process are considered. By the help of proper schedule of control tests, copies etc. one can achieve considerably reduction in spite of unreliability.

1. INTRODUCTION

Let during the operating time X of a process (for example by computing, exploitation, etc.) in a random way some undesirable events (break-downs, catastrophes) arise, which lead to process interruption or to incorrect final results.

We consider two kinds of breakdowns as
(i) explicit ones which lead to evident process interruptions,
(ii) implicit breakdowns, leading to incorrect final result. In the case (ii) it is necessary to make a control test to detect the appearance of the implicit breakdown.

If one wants to have a successful finish, the process must be repeated from the origin (Fig. 1).

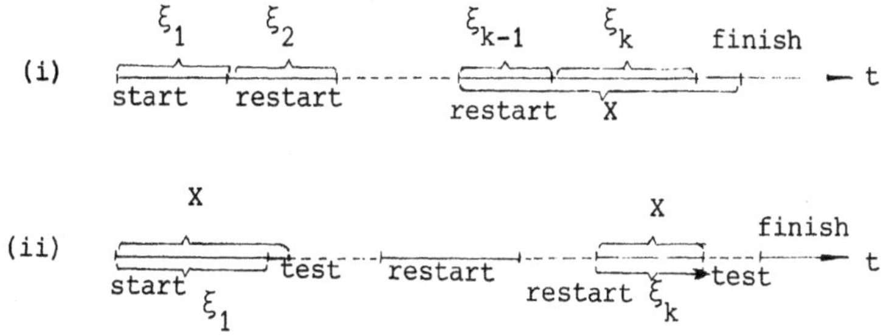

Figure 1. (i) explicit breakdowns lead to evident process interruption and the process must be repeated from the origin: it can be made immediately; (ii) implicit breakdowns need a control test to detect their appearance. To have a successful finish the process must be repeated from the origin until no breakdown be detected.

S. Tzafestas et al. (eds.), System Fault Diagnostics, Reliability and Related Knowledge-Based Approaches, Vol. 1, 415–428.

It is profitable to introduce a strategy for making intermediate copies (for the (i) case) or test control for correctness and copies (for the (ii) case) to remember the process states in some chosen moments. If a catastrophe appears or an incorrect event is discovered, the process continues from the last successful copied state. In the present paper some strategies for process control are proposed in the both cases (i) of explicit breakdowns and (ii) of implicit ones. These strategies minimize a functional of the total process operating time until its successful finish. Explicit results are obtained in the cases when the unreliability is exponential one, the pure required process duration X (if no breakdowns are supposed) is constant or exponentially distributed random variable, and the copy intervals and the control test times form two sequences of independent identically distributed random variables with given distributions.

The problem and its modeling arises in connection of a report of E. Gelenbe [7] and a paper of Kovalenko and Stoykova [9]. Certain results for the optimal computer control, the optimal blocking time by service with unreliable server in different particular cases were found with the work of Bulgarian mathematicians K. Yanev [3,4], N. Kolev [10], and the authors [5,6]. Similar results are found in papers G.L. Brodezki appearing after [2], for information service. There are some results of K. Barosov [1] concerning the applications of the considered models in priority queues as well as some generalizations. The present paper gives the main results for the main model and does not concern the possible applications and generalizations.

2. MATHEMATICAL DESCRIPTIONS. MODELING AND DENOTATIONS

Let us suppose that the process operation is absolutely reliable. Then if it requires time of durations X, its total duration (without any breakdowns) until the successful finish will be equal to X. We say X is the pure process duration and suppose X is a random variable (r.v.) with giveen distribution function

$$A(x) = P(X < x).$$

Let us suppose now that the process operation is unreliable, and its unreliability caused from a random flow of discrete events (breakdowns, catastrophes, sources of incorrectness, etc.). Let ξ_n be the time interval between the (n-1)-th and n-th consecutive events, n = 1,2,... and $P(\xi_n \geq x) = \exp(\gamma x)$, $\gamma > 0$, i.e., $\{\xi_n\}$ is a Poisson process with an intensity γ.

Remark: We assume that $\{\xi_n\}$ is process either of explicit breakdowns or of implicit ones, but never with mixed (explicit or implicit) breakdowns. We designate the first case as (i)-case and the second one- as (ii)-case.

Let the process starts at t = 0, its operations is unreliable, and the corresponding pure process duration is X. If the process restarts after avery breakdown (explicit or discovered implicit one) without any external interference, its total duration up to a successful finish $\tau(X)$ can be expressed in the form:

for the (i)-case

$$\tau_{(i)}(X) = \sum_{j=1}^{k-1} \xi_j + X; \tag{1}$$

for the (ii)-case

$$\tau_{(ii)}(X) = \sum_{j=1}^{k-1} (\delta_j + X) + \delta_k + X, \tag{2}$$

where k is the smallest number j for which $\xi_j \geqq X$ and δ_j is the time of the j-th test duration after the j-th attempt to restart the process and to receive correct final results in the (ii)-case. The corresponding diagram of the real time processing is given on Fig. 1. We suppose $\{\delta_j\}$ is a renewal process with distribution function

$$D(x) = P(\delta_j < x).$$

No breakdowns can occur during some kinds of test times.

Evidently in the both cases holds $P(\tau(X) \geqq X) = 1$. It is clear that introducing suitable schedule of execution copies (or tests and copies, if it necessary) one can economize some spenidng time for repetitions the processing because of unreliability.

Let us introduce a sequence $\{a_k\}$ of time intervals between copies (test and copies for the (ii)-case) and the sequence $\{\theta_k\}$ of copying time durations. After every time interval a_j one makes a copy of the process state in a time θ_j - for the (i)-case and a test and a copy (if there are no detected breakdown) in a time $(\delta_j+\theta_j)$ - for the (ii)-case. The implicit breakdown are supposed impossible during tests and copies. The process restart from the last copying state (Fig. 2). In that case we denote the total process duration until its successful finish by

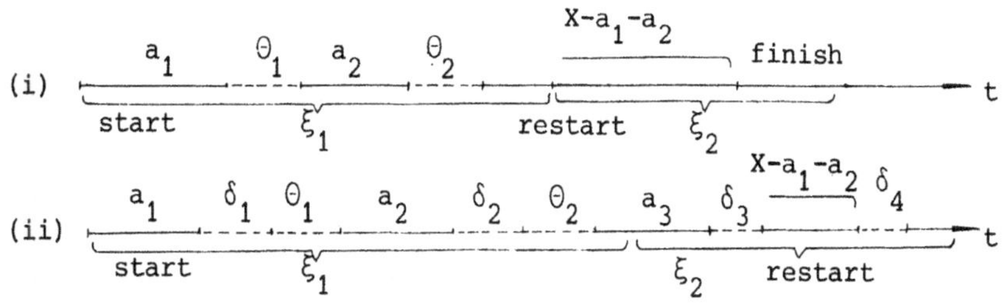

Figure 2. (i) - the copies of the process states help to economize some time spending for repetitions the process because of unreliability; (ii) - the tests and copy schedule help to economize the total process duration, because the implicit break happens in time interval a_3, and the restart is from the process state at the instant (a_1+a_2).

$\tau_{(i)}(X,\{a_k\},\{\theta_k\})$ in the (i)-case, and by $\tau_{(ii)}(X,\{a_k\},\{\delta_k\},\{\theta_k\})$ in the (ii)-case. Sometimes the short notations $\tau_{(i)}$ and $\tau_{(ii)}$ will be used. The both values $\tau_{(i)}$ and $\tau_{(ii)}$ are controlled processing time durations and obviously are r.v.

Our final assumptions is that $\{a_k\}$ is a deterministic process; $\{\theta_k\}$ forms a renewal process with distribution function $\Theta(x) = P(\Theta_k < x)$, and all the values X, $\{a_k\}$, $\{\delta_k\}$, $\{\theta_k\}$, and $\{\xi_k\}$ are mutually independent r.v. and have finite mathematical expectation.

The optimal control problem can be formulated in the next way: under the given assumptions how to choose the schedule $\{a_k^*\}$ so that

$$\inf_{\{a_k\}} E\tau_{(i)}(X,\{a_k\},\{\theta_k\}) = E\tau_{(i)}(X,\{a_k^*\},\{\theta_k\}); \tag{3}$$

$$\inf_{\{a_k\}} E\tau_{(ii)}(X,\{a_k\},\{\delta_k\},\{\theta_k\}) = E\tau_{(ii)}(X,\{a_k^*\},\{\delta_k\},\{\theta_k\}). \tag{4}$$

The sequence $\{a_k^*\}$ will be called as an optimal control schedule.

Further we shall use capital Latin letters for notation of distribution function; the corresponding Greece letters- for notation of its Laplace-Stilties transformations (L.S.T). For example

$$\alpha(s) = e^{-sX} = \int_0^\infty e^{-sx} dA(x); \quad \theta(s) = e^{-s\theta_k} = \int_0^\infty e^{-sx} d\Theta(x). \tag{5}$$

Specially we introduce the functions

$$\tau^*_{(1)}(s) = E\exp(-s\tau_{(1)}(X)); \quad \tau^{**}_{(1)}(s) = E\exp(-s\tau_{(1)}(X,\{a_k\},\dots)) \tag{6}$$

for (1) = (i) or (1) = (ii), which are the L.S.T of the probable distributions of r.v. $\tau_{(1)}(X)$ and $\tau_{(1)}(X,\{a_k\},\dots)$ accordingly.

3. SOME AUXILIARY RESULTS

Let us remind that the flow of events causing the breakdowns is the Poisson one with parameter $\gamma > 0$. It holds

<u>Lemma 1</u>. a) The L.S.T $\tau^*_{(1)}(s)$ of the total process duration (1), (2) are determined by the expressions

$$\tau^*_{(i)}(s) = \alpha(s+\gamma)/\{1 - \frac{\gamma}{s+\gamma}[1 - \alpha(s+\gamma)]\}; \tag{7}$$

$$\tau^*_{(ii)}(s) = \alpha(s+\gamma)\delta(s)/\{1 - [\alpha(s) - \alpha(s+\gamma)]\delta(s)\}. \tag{8}$$

b) The expected total durations (1), (2) are

$$E\tau_{(i)}(X) = [1/\alpha(\gamma) - 1]/\gamma; \quad (9)$$

$$E\tau_{(ii)}(X) = [EX+\delta]/\alpha(\gamma). \quad (10)$$

correspondingly where according to (5), $\delta(s) = Ee^{-s\delta_k}$ is the L.S.T. of the distributions function D(x) of a test duration, EX, δ, Θ, are the expected pure process duration, the control test duration and the copy time one correspondingly.

Proof: The proof is a simple consequence of the next recurrent relations:

$$\tau^*_{(i)}(s) = \alpha(s+\gamma) + \frac{\gamma}{s+\gamma}[1 - \alpha(s+\gamma)]\tau^*_{(i)}(s); \quad (11)$$

$$\tau^*_{(ii)}(s) = \alpha(s+\gamma)\delta(s) + [\alpha(s) - \alpha(s+\gamma)]\delta(s)\tau^*_{(ii)}(s) \quad (12)$$

One can derive (11) and (12), using the probabilistic interpretations of the L.S.T. for s > 0 8 . It is useful to observe that the breakdowns (in the (i)-case) and the ends of tests with detected breakdowns (in the (ii)-case) are points of regenerations for the processing times. The first summand in (11), (12) take into account the possibility to finish the process without any breakdowns, and the seconds one express the possibilities to finish after the first points of regenerations in the both cases correspondingly.

The expressions (7), (8) are the correspondingly solutions of the equations (11), (12); (9) and (10) follow from the proved expressions and from the rule

$$E\tau_{(1)}(X) = \frac{d}{ds}\tau^*_{(1)}(s)\Big|_{s=0}$$

doing the necessary calculations. The lemma is proved.

Corollary 1. If the pure process duration is a known constant, i.e., if P(X = x) = 1 for a given x > 0, then it is fulfilled

$$\tau^*_{(i)}(s) = (s+\gamma)/\{\gamma+s\exp[(s+\gamma)x]\};$$

$$\tau^*_{(ii)}(s) = \delta(s)/\{\exp[(s+\gamma)x] - [\exp(\gamma x) - 1]\delta(s)\};$$

$$E\tau_{(i)}(x) = [e^{\gamma x}-1]/\gamma; \quad E\tau_{(ii)}(x) = (x+\delta)e^{\gamma x}$$

Proof: It is a simple consequence from Lemma 1. and the fact that under our assumptions we have $\alpha(s) = \exp(-sx)$; EX = x.

It is important for remark that in the both cases the expected total proccess durations under certain unreliability is exponentially greater than the pure process durations, and it exponentially increases with x and γ.

Lemma 2. Under the assumptions of Corollary 1. and a given control schedule $\{a_k\}$ it holds:

$$\tau^{**}_{(i)}(s) = \frac{s+\gamma}{\gamma + s\exp[(s+\gamma)(x-bn_x)]} \prod_{k=1}^{n_x} \frac{(s+\gamma)\theta(s+\gamma)}{\gamma\theta(s+\gamma) + s\exp[a_k(s+\gamma)]};$$

$$\tau^{**}_{(ii)}(s) = \frac{\delta(s)}{\exp[(s+\gamma)(x-b_{n_x})] - [\exp(\gamma(x-b_{n_x}))-1]\delta(s)} \prod_{k=1}^{n_x} \frac{\exp[-(s+\gamma)a_k]\theta(s)\delta(s)}{1-[\exp(-sa_k) - \exp(-(s+\gamma)a_k]\delta(s)};$$

and

$$E\tau_{(i)}(x,\{a_k\},\ldots) =$$

$$1/_{\gamma} \sum_{k=1}^{n_x} [\exp(\gamma a_k)/\theta(\gamma) - 1] + [\exp(\gamma(x-b_{n_x})) - 1]/\gamma;$$

$$E\tau_{(ii)}(x,\{a_k\},\ldots) =$$

$$\sum_{k=1}^{n_x} \exp(\gamma a_k)(a_k+\delta) + n_x\theta + (x-b_{n_x})\exp(\gamma(x-b_{n_x})) + \delta.$$

Here the denotations

$$b_o = 0; \quad b_n = a_1+a_2+\ldots+a_n; \tag{13}$$

$$n_x = \max\{n : b_n \leqq x\}$$

are used.

Proof: First of all we mention that a schedule interval a_k and the next copy time θ_k (for the (i)-case) or the next test and copy times $\delta_k+\theta_k$ (for the (ii)-case) form an interval of length $\tau^k_{(1)} = \tau_{(1)}(a_k+\theta_k)$ in the total process durations $\tau_{(1)}(x,\{a_k\},\ldots)$. There are exactly n_x such of intervals, and the last one does not need a copy time; its pure duration is $(x-b_{n_x})$ correspondingly, but the corresponding partial total process durations is $\tau_{(1)}(x-b_{n_x})$. The "lack of memory" property of the exponential distribution shows, that the r.v. $\{\tau^k_{(1)}\}_{k=1}^{\infty}$ are independent and

$$\tau_{(1)}(x, a_k, \ldots) = \sum_{k=1}^{n_x} \tau_{(1)} + \tau_{(1)}(x-b_{n_x}). \tag{14}$$

Further we introduce $\tau^{(s)}_{(1)k} = Ee^{-s\tau^{k}_{(1)}}$. Using the equation

$$\tau^{**}_{(ii)}(s) = \delta(s)\theta(s)\exp[-(s+\gamma)a_k] +$$
$$+ [\exp(-sa_k) - \exp[-(s+\gamma)a_k)]\,\delta(s)\tau^{**}_{(ii)}(s) \tag{15}$$
$$\tau^{**}_{(1)}(s) = \prod_{k=1}^{n_x} \tau^{*}_{(1)k}(s)E\exp[-s\tau_{(1)}(x-b_{n_x})]$$

and the results of Lemma 1. we obtain the statements of Lemma 2. from (14) and (15).

Further we shall give only the results about the expected total process duration $E\tau_{(1)}(X,\{a_k\},\ldots)$ since the optimal control problem (3), (4) needs namely it. But one can derive (if it is necessary) explicit results and about the L.S transformations of the corresponding distribution function. It is true

Lemma 3. Under a given control schedule $\{a_k\}$ with $\sum_{k=1}^{\infty} a_k = \infty$ the expected total process duration by a given pure duration X and distribution function A(x) is correspondingly

$$E\tau_{(i)}(X,\{a_k\},\ldots) = 1/\gamma \sum_{k=1}^{\infty} [\exp(\gamma a_k)/\theta(\gamma)-1]\,(1-A(b_k)) +$$
$$+ 1/\gamma \sum_{k=0}^{\infty} \int_{b_k}^{b_{k+1}} [\exp(\gamma(x-b_k)) - 1]\,dA(x);$$

and

$$E\tau_{(ii)}(X,\{a_k\},\ldots) = \sum_{k=1}^{\infty} \exp(\gamma a_k)(a_k+\delta)(1+A(b_k)) +$$
$$+ \theta \sum_{k=0}^{\infty} k[A(b_{k+1}) - A(b_k)] +$$
$$+ \sum_{k=0}^{\infty} \int_{b_k}^{b_{k+1}} \exp(\gamma(x-b_k))(x-b_k+\theta)\,dA(x).$$

The notations (13) are used.

Proof: These results follow from Lemma 2., the Fubini's theorem and some simple transformation of the obtained eduations. It is sufficient to remark that $n_x = k$ where $x\ b_k,\ b_{k+1}$ and to use it in the expression

$$Eg(x) = \int_{o}^{\infty} g(x)\,dA(x) = \sum_{k} \int_{b_k}^{b_{k+1}} g(x)\,dA(x)$$

Definition: We say a schedule $\{a_k\}$ is an uniform one iff $a_k = a > 0$ for each $k = 1,2,\ldots$

In that case $b_k = ka, \quad k = 1,2,\ldots$

Corollary 2. If the pure process duration is exponentially distributed with parameter $\lambda > 0$ and if the control schedule $\{a_k\}$ is an iniform one, then the expected controlled total process durations are presented by the expressions

$$E\tau_{(i)}(X,\{a\},\{\theta_k\}) =$$

$$1/\gamma\{\lambda[\exp(a(\gamma-\lambda))-1]/(\gamma-\lambda) + \exp(a(\gamma-\lambda))/\theta(\gamma)-1\}/(1-\exp(-\lambda a))$$
$$\text{for } \lambda \neq \gamma;$$

$$1/\gamma[a\gamma + 1/\theta(\gamma)-1]/(1-\exp(-\gamma a)) \tag{16}$$
$$\text{for } \lambda = \gamma;$$

and

$$E\tau_{(ii)}(X,\{a_k\},\{\theta_k\},\{\delta_k\}) =$$

$$\{\lambda[(a+\delta)\exp(a(\gamma-\lambda)) - \delta + (1-\exp(a(\gamma-\lambda)))/(\gamma-\lambda)]/(\gamma-\lambda) +$$
$$+ (a+\delta)\exp(a(\gamma-\lambda)) + \theta\exp(-\lambda a)\}/(1-\exp(-\lambda a))$$
$$\text{for } \lambda \neq \gamma; \tag{17}$$

$$[\lambda a(\delta+a/2)+a+\delta+\theta\exp(-\gamma a)]/(1-\exp(-\gamma a))$$
$$\text{for } \lambda = \gamma$$

correspondingly.

Proof: In the considered case we have $A(x) = 1-e^{-\lambda x}$, $a_k = a$ and $b_k = ka$. By putting these values in the expressions of Lemma 3. and taking into account that $1 - A(b_k) = \exp(-k\lambda a)$; $A(b_{k+1}) - A(b_k) = \exp(-k\lambda a)(1-\exp(-\lambda a))$; $dA(x) = \lambda\exp(-\lambda x)dx$, after long, but not very complicated calculations one can convince oneself the truth of (16) and (17).

4. THE OPTIMAL CONTROL SCHEDULE

First of all let us note an optimal property of the uniform control schedule. It holds

Theorem 1. If $P(X=x) = 1$, then the optimal control schedule is the uniform one.

Proof: We shall use the notations of Lemma 2. The results of lemma 2. concerning the mathematical expectations terms of values b_k, using the equations $a_k = b_k-b_{k-1}$. The optimal control schedule $\{a_k^*\}$ gives the sequence $\{b_k^*\}$ which terms are among the solutions of the system of equations

stant, i.e., if $P\{\theta = \text{const}\} = 1$ then the thresholds $x^*_{(i)n}$ are depending only by the product $z = \gamma\theta$ and are the solution of the equation

$$e^{zx/(n+2)}(1+(n+1)e^{z}) - e^{zx/(n+1)}(1+ne^{z}) = 1.$$

For an arbitrary value of θ the corresponding thresholds are proportional to $x^*_{(i)n}$, i.e., $x_{(i)n}(\theta) = \theta x^*_{(i)n}$. Their numerical values as roots of equation (29) are given in Table 1., published in 5 . Unfortunately there are no similar simplification for the thresholds of the (ii)-case. It means that the solutions of the equation (30) determined a 3-parametrical table (the parameters are γ, δ and θ) for the numerical values of the thresholds $x^*_{(ii)n}$. Therefor for the practice it is not practical for use special tables for the shown thresholds and a general computer program, doing the necessary calculations for any particular case is preferable.

Now let us turn to the case of exponentially distributed pure process duration X with parameter $\lambda > 0$, and an external influence with uniform control schedule by a given constant a, i.e., let us consider the case determined by the conditions of the Corollary 2. It holds.

Theorem 3. Under the conditions of corollary 2. the optimal control schedule $\{a^*_{(1)}\}$ is determined as the unique solution of the equation:
a) for the (i)-case

$$\frac{e^{-\gamma a} - e^{-\lambda a}}{\gamma - \lambda} + \frac{1 - e^{-\lambda a}}{\lambda\theta(\gamma)} = \frac{\theta(\gamma)-1}{\gamma\theta(\gamma)}$$

if $\lambda \neq \gamma$

and $(\gamma a-1)\exp(-\gamma a) = \theta(\gamma)$ if $\lambda = \gamma$.
If $\lambda \neq \gamma$, the value $a^*_{(i)}$ exists iff the inequality

$$\lambda < \gamma/(1-\theta(\gamma))$$

is fulfilled; otherwise it is optimal to make no external control; for $\lambda = \gamma$ the control is preferable;

b) for the (ii)-case

$$e^{\gamma a}\{[1+\gamma(a+\delta)](1-e^{-\lambda a}) + \gamma^2/(\gamma-\lambda)^2 - \lambda\gamma(a+\delta)/(\gamma-\lambda)\} +$$
$$+ \lambda^2 e^{\lambda a}[\delta(\gamma-\lambda)-1]/(\gamma-\lambda)^2 = \lambda\theta(2-e^{-\lambda a})$$

when $\lambda \neq \gamma$ and

$$e^{\gamma a}[\gamma(\delta+a) + 1] = a\gamma(2+\delta\gamma+a\gamma/2) + \theta\gamma + 2\delta\gamma + 1$$

when $\lambda = \gamma$. The control is preferable, because the written equations always have a positive solutions with respect to the parameter a.

Proof: The optimal control schedule interval $a^*_{(1)}$ is determined as a solution of the equation

$$\frac{\partial}{\partial a}\tau_{(1)}(a) = 0$$

$$\frac{\partial}{\partial b_k} E\tau_{(1)}(x,\{a_k\},\ldots) = 0, \quad k = 1,2,\ldots \tag{18}$$

It is easy to see, that in the (i)-case the system (18) becomes the form

$$e^{\gamma(b_k - b_{k-1})} = e^{\gamma(b_{k+1} - b_k)}, \quad k = 1,2,\ldots \tag{19}$$

and in the (ii)-case respectively

$$[\gamma(b_k - b_{k-1}) + \gamma\delta + 1]\exp(\gamma(b_k - b_{k-1})) = [\gamma(b_{k+1} - b_k) + \gamma\delta + 1]\exp(\gamma(b_{k+1} - b_k)), \quad k = 1,2,\ldots \tag{20}$$

To be satisfied (19) and (20) for every $\gamma > 0$ it is necessary to have $b_k - b_{k-1} = b_{k+1} - b_k$ i.e., a_k is constant for all $k = 1,2,\ldots$; it proves the theorem.

For an abbreviation we introduce the notations

$$n_x = n(a,x) = [x/a]$$

for the integer part of the number x/a; $I_n(a) = 1$ iff $a \in (\frac{x}{n+1}, \frac{x}{n}]$, and $I_n(a) = 0$ otherwise. In addition we denote $T_{(1)n}(x,a) = E\tau_{(1)}(x,\{a_k\},\ldots)$ iff $a_k = a$ for all $k = 1,2,\ldots$ i.e.,

$$T_{(i)n}(x,a) = n[\exp(\gamma a)/\theta(\gamma) - 1]/\gamma + [\exp(\gamma(x - n_x a)) - 1]/\gamma; \tag{21}$$

and

$$T_{(ii)n}(x,a) = n_x \exp(\gamma a)(a+\delta) - n_x\theta + (x - n_x a + \delta)\exp(\gamma(x - n_x a)) \tag{22}$$

Now, the result of Lemma 2. in the case of an arbitrary uniform control schedule $\{a_k\}$ with $a_k = a$ for $a > 0$ can be rewritten as follows:

<u>Lemma 4</u>. If $P(X=x) = 1$ and the control schedule is uniform one with an arbitrary fixed $a > 0$, it holds

$$E\tau_{(1)}(x, a_k, \ldots) = T_{(1)}(x) = \sum_{n=0}^{\infty} I_n(a) T_{(1)n}(x,a).$$

Obviously, when $a > x$, no active control will be made and the expected total process duration is expressions of Lemma 1. When a is a small value, in comparison of the pure process duration x, then the control will be very frequent and the corresponding expected total process duration can arise of the expense of superfluous interference. The control make sense in the case $P(X=x) = 1$ iff for some $a < x$ is fulfilled $T_{(1)}(x) < T_{(1)0}(x)$.

Further let us consider the case $P(X=x) = 1$ more detailed. First of of all we remark, that the term $T_{(1)n}(x,a)$ represents the function $T_{(1)}(x)$ only over the interval $(x/(n+1),\ x/n]$. But if we fix the value of n, then the functions $T_{(1)n}(x,a)$ are defined for all $a > 0$. It holds.

Lemma 5. (i) The absolute minimum of $T_{(i)n}(x,a)$ is

$$t^*_{(i)n} = \exp\{\gamma[x+\ln\theta(\gamma)]/(n+1)\}[n/\theta(\gamma)+1]/\gamma - (n+1)/\gamma \tag{23}$$

and it gets in the point $\hat{a}_n = x/(n+1) + \ln\theta(\gamma)/(\gamma(n+1)) < x/(n+1)$, which is out (left) of the interval of the essentially of $T_{(i)n}(x,a)$;
(ii) The absolute minimum of $T_{(ii)n}(x,a)$ is

$$t_{(ii)n} = (x+(n+1)\delta)\exp(\gamma x/(n+1)) - n\theta \tag{24}$$

and it gets in the point $\tilde{a}_n = x/(x+1)$ which is the left final point of the interval of essentially of $T_{(ii)n}(x,a)$.

Proof: One can easy denive (23) and (24), using the ordinary analytical techniques with the first and second derivatives of the investigated functions. Some complications arise in the (ii)-case, where appears the equation

$$e^{\gamma[x-(n+1)a]} = \frac{\gamma(a+\delta)+1}{\gamma(x-na+\delta)+1}$$

for the optimal value of the parameter a, and when must be proved that $a = x/(k+1)$ is the unique root of this equation. We omit the details.

It is important to remark now that the ensential parts in the graphics of $T_{(1)n}(x,a)$, placed over the interval $(x/(n+1),x/n]$ have the form given of Fig. 3. and increase from left to right.

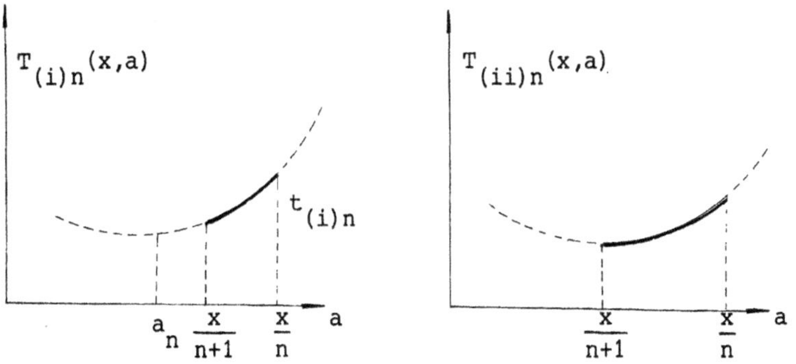

Figure 3.

Let us denote by

$$t_{(1)n-1} = \lim_{a\downarrow\frac{x}{n}} T_{(1)}(x); \quad \tilde{t}_{(1)n} = \lim_{a\uparrow\frac{x}{n}} T_{(1)}(x) \tag{25}$$

the left and right limit of the essential parts in the graphics of the functions $T_{(1)n-1}(x,a)$ and $T_{(1)n}(x,a)$ correspondingly; So we get

Lemma 6. The relations

$$t_{(i)n} = (n-1)[\exp(\gamma x/n)/\theta(\gamma)-1]/\gamma + [\exp(\gamma x/n)-1]/\gamma$$
$$\tilde{t}_{(i)n} = n[\exp(\gamma x/n)/\theta(\gamma)-1]/\gamma \tag{26}$$

and

$$\tilde{t}_{(ii)n} = (x+n\delta)\exp(\gamma x/n) + n\theta + \delta \tag{27}$$

are correct and always $t_{(1)n-1} < \tilde{t}_{(1)n}$, where $t_{(ii)n}$ are determined by (24).

Proof: Using (21) and (22) we find the shown limits. It is easy to verify the inequality by subtracting the both terms.

Lemma 7. For any fixed x and γ the minimal total process duration is determined by the relation

$$T^*_{(1)}(x) = \inf_{a>0} T_{(1)}(x) = \inf_{n>1} t_{(1)n} = t_{n_o}$$

where the integer n_o exists and is finite. The optimal copy schedule is determined by the value $a^* = x/(n_o+1)$.

Proof: According to the result of Lemma 6. one can remark that the lower points of the graphic of the function $T_{(1)}(x)$ form the sequence $\{t_{(1)n}\}_{n=0}^{\infty}$, i.e., $T_{(1)}(x)$ has the form shown on Fig. 4.

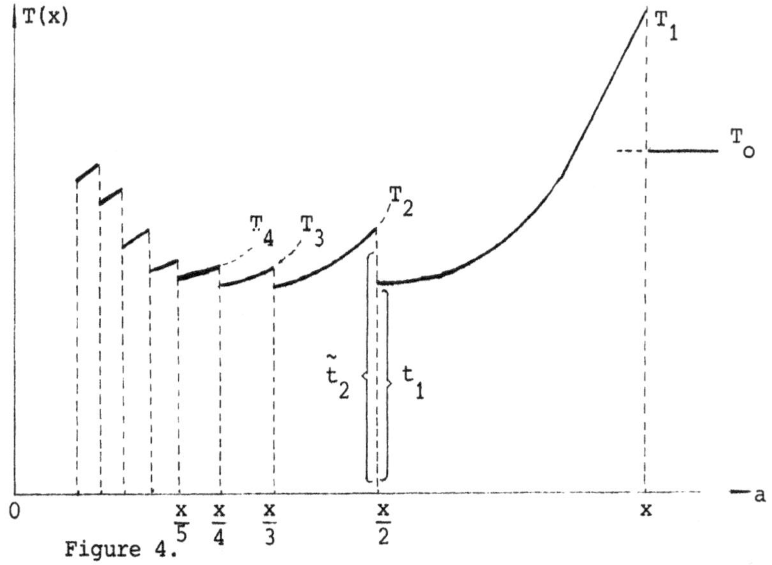

Figure 4.